PRINCIPLES OF MICROWAVE TECHNOLOGY

STEPHEN C. HARSANY
Mt. San Antonio College

Prentice Hall
Upper Saddle River, New Jersey   Columbus, Ohio

Library of Congress Cataloging-in-Publication Data

Harsany, Stephen C.
  Principles of microwave technology / Stephen C. Harsany.
     p. cm.
  Includes index.
  ISBN 0-13-205568-6 (casebound)
  1. Microwaves. 2. Microwave communication systems. 3. Microwave devices. I. Title
TK7876.H37 1997
621.381'3—dc20

Editor: Charles E. Stewart, Jr.
Production Editor: Rex Davidson
Production Supervision: Custom Editorial Productions, Inc.
Production Manager: Laura Messerly
Marketing Manager: Debbie Yarnell
Cover Designer: Scott Rattray
Cover Photo: Courtesy of NASA, Johnson Space Center.
  The cover picture shows the Defense Support Program (DSP) satellite being deployed from the NASA STS-44 Space Shuttle bay in November 1991. The Defense Support Program system is a survivable and reliable satellite-borne system that detects and reports on real-time missile launches, space launches, and nuclear detonations. Aerojet designed and built the infrared sensors for missile boost phase detection, and has been providing surveillance sensors for the DSP since the late 1960s. TRW Space & Technology Group builds the DSP satellites and integrates the sensor payload. The work is performed under contract to the U.S. Air Force Space and Missile Systems Center.

This book was set in Times Roman by Custom Editorial Productions, Inc., and was printed and bound by Quebecor Printing/Book Press. The cover was printed by Phoenix Color Corp.

©1997 by Prentice-Hall, Inc.
Simon & Schuster/A Viacom Company
Upper Saddle River, New Jersey 07458

Printed in the United States of America

10 9 8 7 6 5 4 3 2 1

ISBN 0-13-205568-6

Prentice-Hall International (UK) Limited, London
Prentice-Hall of Australia Pty. Limited, Sydney
Prentice-Hall Canada Inc., Toronto
Prentice-Hall Hispanoamericana, S. A., Mexico
Prentice-Hall of India Private Limited, New Delhi
Prentice-Hall of Japan, Inc., Tokyo
Simon & Schuster Asia Pte. Ltd., Singapore
Editora Prentice-Hall do Brasil, Ltda., Rio de Janeiro

This book is dedicated to my father, Joseph, and my stepfather, Gordon Grovier, whose stories about cigar box crystal radios and help in assembling radio kits were my inspiration for entering the field of electronics many years ago.

PREFACE

Principles of Microwave Technology is intended primarily for students enrolled in advanced communications course work in either vocational-technical schools, technical institutes, or community colleges. It is not limited to two-year programs, as I feel it will meet the needs of many four-year programs, especially electronics engineering technology programs. Because of the comprehensiveness of this text, it should serve as a good reference for others as well.

I believe that a good text should explain how things work, show appropriate applications, and provide complete topic coverage. It is my intent to provide descriptions and analyses that are as comprehensive and up-to-date as possible with specific applications in radar and communications, and, in particular, the new wireless technologies.

Principles of Microwave Technology begins with a general breakdown on microwave basics: frequency, bands, wavelengths, and the like. It includes the model of propagation and propagation paths and shows the link between transmission and reception via the transmission line. Next are chapters on passive microwave components, active microwave devices (both thermionic and solid state), antennas, and measurements. The final chapter covers communications applications.

All math is kept at the algebra or simple trigonometry level. The only electronics prerequisite is a background in alternating current (AC) theory. The use of complex number notation used in chapters 2 and 3 should be met by most AC courses.

Only chapters 2 and 3 need to be taught in order, since the transmission line formulas in chapter 2 can easily be shown as graphical solutions on the Smith charts of chapter 3. Otherwise, the chapters can be covered in any order that meets your needs. Each chapter begins with a set of objectives and ends with a summary, a list of key equations, and a problem set, questions, or both. The answers to all problems are found in the solutions section at the end of the book. The text includes a glossary and several appendices. Appendix C lists several no-cost magazines and some low-cost videos that greatly supplement the course.

A diskette provided with this text offers a tutorial on microwave thermionic devices. Many texts shortchange this subject, and many programs fail to adequately cover this important topic. For this reason, Litton Electron Devices, San Carlos, California, commissioned a software tutorial to aid new hires to their engineering staff. The tutorial is being made available to schools as well. My thanks to Tina Murphy, Advertising Manager, Litton Electron Devices, for her assistance in getting permission to provide the diskette with this text. We are sure that you will find it a useful and helpful addition.

Several educational vendors have microwave lab equipment that meets the needs of a laboratory component. I have used the Lab-Volt Systems microwave trainer (8090 Microwave Training System) in my own lab for more than seven years. It comes with its own lab manual that I would be hard-pressed to improve upon, thus, I chose not to write my own lab manual to accompany this text. If you can find the time and funds to include a lab component with your microwave course, I heartily endorse their product. You can get more information from them at:

Lab-Volt Systems
P.O. Box 686
Farmingdale, NJ 07727
(908) 938-2000
(800) 223-1057

ACKNOWLEDGMENTS

I would like to thank all the people at Prentice Hall Career and Technology and Custom Editorial Productions, Inc. who helped in the publication of this text. These include:

Charles E. Stewart, Jr., Senior Managing Editor
Mollie Pfeiffer, Assistant to the Executive Editor
Maureen Henry, Marketing Department
Rex Davidson, Production Editor
Laura Bofinger, Project Editor
Ann P. Royalty, Copy Editor

I also want to thank my reviewers for their insightful suggestions. Their input greatly enhanced the text. These include: Jeff Beasley, New Mexico State University; Michael Charek, SE Missouri State University; Anthony J. Iula, Wentworth Institute of Technology; Ronald F. Moody, Pima Community College; Leon J. Nicelly, ITT Technical School; Mark E. Oliver, Monroe Community College; Earl F. Owen, Brigham Young University; and H. Paul Shuch, Penn College of Technology.

Finally, thanks to my wife, Claudette, for her patience and support while I spent so much time in the research and writing of the text.

Stephen C. Harsany
Mt. San Antonio College

CONTENTS

1 ELECTROMAGNETIC FUNDAMENTALS

OBJECTIVES

1. To define "microwave" and other microwave terminology.
2. To differentiate how microwave technology differs from ordinary, high-frequency technology.
3. To describe the designations for various microwave bands.
4. To compare and contrast the differences between "lumped" and "distributed" parameters.
5. To calculate various wavelength, frequency, and velocity parameters associated with microwaves.
6. To describe propagation and propagation paths of electromagnetic (EM) waves.

1.1   INTRODUCTION

Microwave technology owes its origin to the development of radar, which started before World War II but mushroomed during World War II by necessity. Various investigators were trying to solve the problem of devices that could operate in the UHF/microwave bands with high power. At the heart of their investigation was the conventional vacuum tube, which at the time seemed to represent the best approach. The problem with the conventional vacuum tube is twofold. The first problem is interelectrode capacitance between elements within the vacuum tube. The second is a longer electron transit time. The interelectrode capacitance effectively shorting at higher frequencies and the longer transit time contribute to the necessity to use these devices only at a lower operating frequency.

A solution to the transit-time problem was proposed in 1920 by German scientists H. Barkhausen and K. Kurz. Their solution was the Barkhausen-Kurz oscillator (BKO), a special type of vacuum tube that generated higher frequency signals. Unfortunately, traditional vacuum tube construction limited available output power.

Another solution to these problems was proposed in 1921 by A.W. Hull, who used a magnetic field to influence the flow of electrons. His design was the original *magnetron* and modifications of his design are still in use today. Further development of the magnetron is credited to Randall and Boots in England in 1939.

The power-vs.-frequency dilemma seemed unsolvable for several years. In the mid-1930s, a solution to this problem finally arrived. Dr. W. W. Hansen and Drs. A. and O. Heil thought of turning electron transit time into an advantage with a mechanism called *velocity modulation*. Then in 1937, the Varian brothers extended Dr. Hansen's work into the development and production of the *klystron* vacuum tube. It could be used as either an oscillator or a power amplifier. With the production of these vacuum tube devices, radar was finally a commercial, albeit a military, success at microwave frequencies. The development of semiconductor devices saw similar, if not identical, problems. (More on this in chapters 6 and 7). Because of these developments, frequencies in the gigahertz (GHz = 10^9) range were opened up to communication engineers as well.

Military developments in the decades that followed continued to be in radar, while the commercial sector's use of microwaves was limited primarily to telephone companies. By the 1960s, microwave communications had replaced 40% of the telephone circuits between major cities. The consumer marketplace saw an explosion begin in the early 1980s with TVRO (television receive only) broadcast service to the home of satellite TV transmissions.

The 1990s have seen a continuous evolution of microwave developments, particularly in the consumer marketplace. Many institutions such as network television, hotel franchises, retail outlets, and newspapers have access to satellite transponders that provide effective customer services. Direct broadcast satellite services (DBS) to the home at even higher frequencies and power have occurred. One hundred fifty channels of programming to a receiver dish about 18 inches in diameter have become available (with even more channels of programming to come). Personal communicators, cellular phones, and the like, which are under the general umbrella of personal communication systems (PCS), continue to be heavy growth areas.

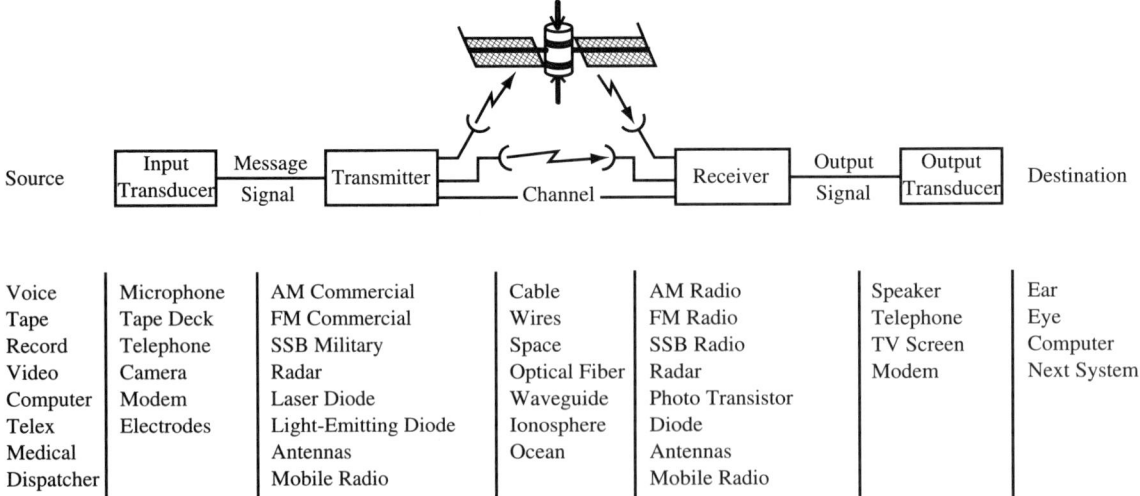

Figure 1.1   Model of a communications system.

Table 1.1   Metric Prefixes

Prefix	Power of Ten	Symbol
exa	10^{18}	E
peta	10^{15}	P
tera	10^{12}	T
giga	10^{9}	G
mega	10^{6}	M
kilo	10^{3}	k
milli	10^{-3}	m
micro	10^{-6}	μ
nano	10^{-9}	n
pico	10^{-12}	p
femto	10^{-15}	f
atto	10^{-18}	a

Microwaves have found applications in areas other than those in communications and in radar uses (surveillance, navigation, and meteorology). They are also used in medicine, in surveying land, in heating, in industrial quality control, in radio astronomy, in navigation via global positioning systems, and in power transmission.

Figure 1.1 shows a model of a typical communication system. Note that this model shows many types of frequency transmissions, not just microwave. This text emphasizes the "channel" in Figure 1.1 by covering transmission lines and space in chapters 2 through 4. The passive and active devices used in the transmitter and receiver stages, or in the lab, are covered in chapters 5 through 7. Antenna types and measurement techniques are covered in chapters 8 and 9. The primary focus on applications is in chapter 10.

1.2   UNITS AND PHYSICAL CONSTANTS

Following standard engineering practice, all units in this text will be in either the CGS (centimeter-gram-second) or MKS (meter-kilogram-second) systems unless otherwise noted. Because these systems depend on using multiplying prefixes on the basic units, a table is provided for the common metric prefixes (Table 1.1). Other tables are provided to give the standard physical constants of interest and some common conversion factors. These are Tables 1.2 and 1.3, respectively.

Table 1.2   Physical Constants

Constant	Value	Symbol
Boltzmann's constant	1.38×10^{-23} J/K	K
Electric charge (e-)	1.6×10^{-19} C	q
Electron (volt)	1.6×10^{-19} J	eV
Electron (mass)	9.12×10^{-31} kg	m
Permeability of free space	$4\pi \times 10^{-7}$ H/m	μ_o
Permittivity of free space	8.85×10^{-12} F/m	$\in_o$
Planck's constant	6.626×10^{-34} J $\cdot$ s	h
Velocity of electromagnetic waves	3.0×10^{8} m/s	c
Pi (π)	3.1416	π

Table 1.3 Conversion Factors

1 micron	=	10^{-6} m
1 inch	=	2.54 cm
1 inch	=	25.4 mm
1 foot	=	0.305 m
1 mile	=	1.61 km
1 nautical mile	=	6080 ft
1 statute mile	=	5280 ft
1 mil	=	2.54×10^{-5} m
1 kg	=	2.2 lb
1 neper	=	8.686 dB
1 gauss	=	10,000 teslas

Table 1.1 covers the range of numbers from 10^{18} (exa) down to 10^{-18} (atto). You will need to review this table if you are not already familiar with these common prefixes.

1.3   WHAT ARE MICROWAVES?

Microwaves are electromagnetic radiation of frequencies from several hundred MHz to several hundred GHz. Microwaves, by virtue of their high frequency, have extremely short wavelengths, therefore, the word "micro" in the name. Figure 1.2 shows the available frequency spectrum.

Note that microwaves start at 500 MHz and continue to 300 GHz. Various sources differ in the exact starting and ending frequencies. What can be standardized is that most microwave equipment falls in the range from 1 GHz to 100 GHz.

The infrared (heat) region is at the upper end of the spectrum with a range of 0.3 to 430 THz. Visible light is above infrared and falls between 430 THz and 1 PHz. This region also includes laser and fiber optics operation. The region 1 PHz and above includes x-rays, gamma rays, and cosmic rays.

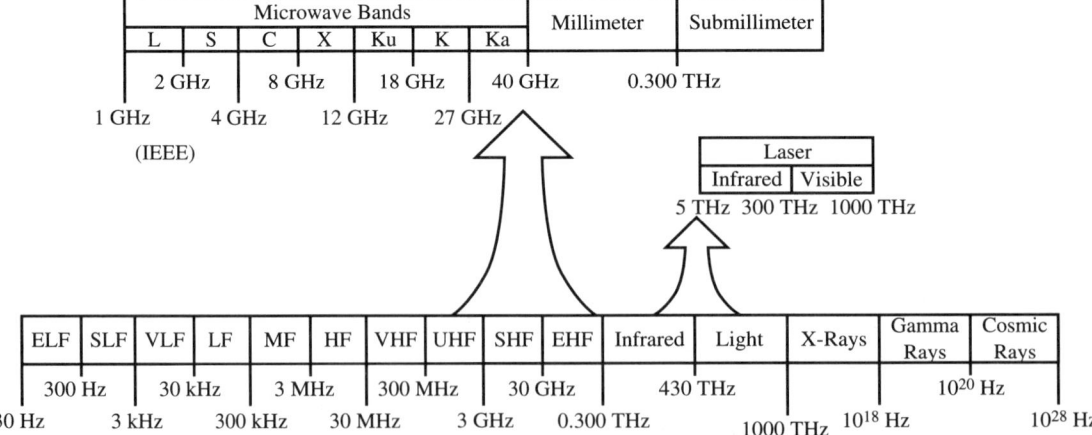

Figure 1.2 Available frequency spectrum.

Table 1.4 CCIR Band Designations

Band	Frequency Range	Band Designation
2	30–300 Hz	ELF (extremely low frequency)
3	0.3–3 kHz	SLF/VF (voice frequency)
4	3–30 kHz	VLF (very low frequency)
5	30–300 kHz	LF (low frequency)
6	0.3–3 MHz	MF (medium frequency)
7	3–30 MHz	HF (high frequency)
8	30–300 MHz	VHF (very high frequency)
9	0.3–3 GHz	UHF (ultra high frequency)
10	3–30 GHz	SHF (super high frequency)
11	30–300 GHz	EHF (extremely high frequency)
12	0.3–3 THz	Infrared light
13	3–30 THz	Infrared light
14	30–300 THz	Infrared light
15	0.3–3 PHz	Visible light
16	3–30 PHz	Ultraviolet light
17	30–300 PHz	X-rays
18	0.3–3 EHz	Gamma rays
19	3–30 EHz	Cosmic rays

Table 1.4 further defines the various frequency ranges in tabular form. These were designated by the International Radio Consultative Committee (CCIR).

Note that the beginning frequency of each band is a multiple of ten of the previous band or the next band. You are familiar with the MF band, as AM radio falls within this range, as well as the VHF band, as conventional TV and FM radio fall within this range. Some police radars are in the SHF band, as are most commercial cable and satellite services. Most growth is occurring in this band.

Table 1.5 shows the current microwave spectrum bands as designated by the Institute of Electrical and Electronic Engineers (IEEE); these are the industry standards. This includes bands from L to Ka (1–40 GHz) and the millimeter (40–300 GHz) and submillimeter bands (greater than 300 GHz).

Table 1.5 IEEE/Industry Standards Bands

Band	Frequency Range, (GHz)
HF	0.003–0.030
VHF	0.030–0.300
UHF	0.300–1.00
L	1.00–2.00
S	2.00–4.00
C	4.00–8.00
X	8.00–12.0
Ku	12.0–18.0
K	18.0–27.0
Ka	27.0–40.0
Millimeter	40.0–300.0
Submillimeter	greater than 300

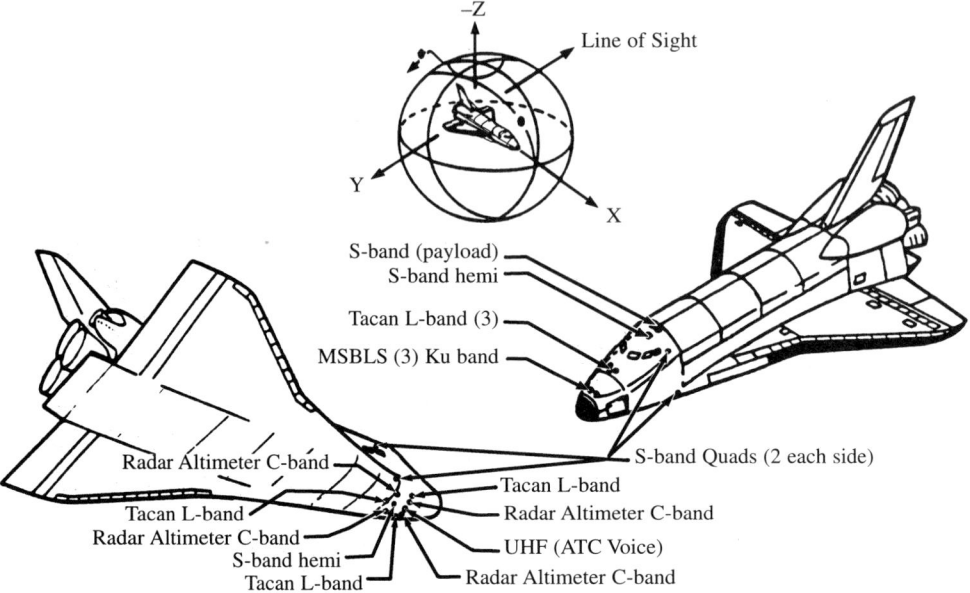

Figure 1.3   Space shuttle uses of microwave.

Figure 1.3 shows the space shuttle's use of various microwave bands. Table 1.6 shows that several bands are used for a variety of applications on the shuttle. These range from navigational sensors to communications and tracking. This list does not include all the bands aboard the shuttle, but it nonetheless demonstrates many of its microwave features.

The new U.S. military frequency designations are shown in Table 1.7. These have not gained acceptance by the industry.

Note that the range from 1 to 100 GHz, where most current microwave equipment and systems operate, is covered by these bands.

One distinct advantage of microwave is that its higher frequency yields greater bandwidth. Greater bandwidth means more room for "stuff" to be packed into the transmission.

As an example, if a typical AM carrier frequency of 1 MHz (1000 kHz) has a bandwidth of 10 kHz, the audio information is about 1% of the carrier. At 10 GHz, a 1% bandwidth is 1 MHz, which translates into as many as 100 MHz/10 kHz = 10,000 separate groups of audio information. Therefore, exploring the microwave spectrum and making it available to communications can prove to be quite productive.

1.4   MICROWAVE TECHNOLOGY

The principles of microwave technology can be taught as a separate subject apart from mainstream communications course work. This is primarily due to two unique characteristics: microwave's distributed *LC* values, rather than "lumped" values, and its relative wavelength.

Recall that the resonant frequency formula for a tuned circuit is $f_r = 1/(2\pi\sqrt{LC})$. At a resonant frequency of 10 GHz, the required values of *L* and *C* are much less than 1 µH and 1 pF,

Table 1.6 Selected Frequency Bands and Antenna Types—Space Shuttle

System	Frequency Band	Antenna Type
Navigational Sensors:		
TACAN	960–1220 MHz	Annular Slot
MSBLS	15.4–15.7 GHz	Waveguide Horn
Radar Altimeter	4.2–4.4 GHz	Waveguide Horn
Communication and Tracking:		
UHF/ATC	240–400 MHz	Annular Slot
UHF-EVA/Bio-Med	243–300 MHz (in four bands)	Microstrip Cavity Backed Slot
S-band Payload	1.75–2.3 GHz	Crossed Dipole in a Cavity
Hemi F1 Service	2200–2270 MHz	Crossed Dipole in a Cavity
Quads PM Service	1.75–2.3 GHz	Crossed Dipoles in Cavity Array
Ku-band Communications/Tracking	13.75–15.15 GHz	Parabolic Dish

respectively. With such small values it is impossible to use "lumped" circuit values (which could be done with low frequency calculations). The connecting leads of the components would probably have as much or more inductance and capacitance than the total value required. Lumped resistance would be the total amount, while the distributed resistance would be the amount of resistance associated with each meter of wire length. One must therefore use the distributed constants of $R, L,$ and C for microwave electronic components. As can be seen in the study of transmission lines (chapter 2), they have a constant $R, L,$ and C value per unit length. Distributed circuits will be used, in the form of resonant cavities (chapter 5), as the resonators to replace the familiar lower frequency "LC tank" circuit.

Table 1.7 New U.S. Military Microwave Frequency Bands

Frequency Band	Frequency Range, (GHz)
A	0.10–0.25
B	0.25–0.5
C	0.5–1.0
D	1.0–2.0
E	2.0–3.0
F	3.0–4.0
G	4.0–6.0
H	6.0–8.0
I	8.0–10.0
J	10.0–20.0
K	20.0–40.0
L	40.0–60.0
M	60.0–100.0
N	100.0–140.0

New wavelength characteristics arise at higher frequencies. This has to do with wavelength size and phase shifting. At low frequencies the wavelength is very large compared to the physical dimensions of the signal processing equipment. For example, the wavelength of the 60 Hz AC power line is 5,000 km (≈3,000 miles) and that of a 1 MHz signal is 300 meters. This large wavelength-to-equipment dimension ratio results in an extremely small phase difference between signals at various test points in a circuit. More importantly, the small equipment dimension means that a standing wave cannot be formed. A standing wave is caused by an impedance mismatch and is the result of the interference between a transmitted signal and its reflection. If a mismatch occurs at conventional lower frequencies, the worst thing that happens is that energy is not transferred at its maximum rate.

At microwave frequencies the wavelength of the signal is comparable to or smaller than the physical dimension of the signal processing equipment. For example, the wavelength of a 10 GHz signal is 3 cm (slightly more than 1 inch). Therefore, two nearby points on a circuit board could have significant phase differences. If an impedance mismatch occurs, a forward moving wave is partially reflected, and the test point signal is a superposition (a standing wave) of the forward and reflected waves. The two waves differ by phase angle and their direction of propagation. Standing waves occur because of the mismatch and a power loss results.

1.5   VELOCITY, FREQUENCY, AND WAVELENGTH

The *velocity* of an electromagnetic wave in a vacuum is 299,792,462 meters/second (and for all practical purposes in air, as well). For simplicity's sake we round this up to 300×10^6 m/s or 3.0×10^8 m/s. This also translates to 3.0×10^{10} cm/s or 3.0×10^{11} mm/s.

These values came about when Maxwell, (James Clerk Maxwell, Scottish physicist, 1831–1879), predicted that the velocity of *all* electromagnetic waves in free space (vacuum) would be given by:

$$v = 1/\sqrt{\mu_o \epsilon_o} \approx 3.0 \times 10^8 \text{ m/s},$$

where the permeability of free space is $\mu_o = 4\pi_o \times 10^{-7}$ H/m, and the permittivity of free space is $\epsilon_o = 8.85 \times 10^{-12}$ F/m. These values are simply the distributed values of L and C of free space.

The absolute permittivity of a medium is noted as ϵ and is given by the equation $\epsilon = \epsilon_r \epsilon_o$. ϵ_r is the relative permittivity (dielectric constant), which may vary from 1 to 10 for most microwave elements, and ϵ_o is the permittivity of free space as previously defined (see Table 1.2). Thus, ϵ_r becomes the ratio ϵ/ϵ_o, the form more often used in microwave equations. Likewise, the absolute permeability of a medium is noted as μ and is given by the equation $\mu = \mu_r \mu_o$. The relative permeability μ_r equals 1 for a vacuum and most nonmagnetic metals, while μ_o was previously defined (see Table 1.2). Thus, μ_r equals the ratio μ/μ_o and for most purposes we can use 1 in calculations.

The value of c is equal to the velocity of light, which is another example of an electromagnetic wave. If the wave is traveling through a medium other than free space, it can be shown that the velocity will decrease. As the new medium is virtually always nonmagnetic ($\mu = 1$), and by knowing the dielectric constant (relative permittivity) of this new medium (which is usually given), we can calculate the new velocity with:

$$v_r = c/\sqrt{\epsilon_r} \qquad \textbf{(Eq 1.1)}$$

where $\quad v_r$ = relative velocity in a new medium

c = speed of light, 3.0×10^8 m/s

ϵ_r = dielectric constant

EXAMPLE 1.1:

What is the velocity of an electromagnetic wave when the dielectric constant of the medium is 2.0 ($\epsilon_r = 2.0$)?

$$v = c/\sqrt{\epsilon_r}$$
$$= 3 \times 10^8 \text{ m/s} / \sqrt{2.0}$$
$$= 2.12 \times 10^8 \text{ m/s}$$
$$= 71\% \; c$$

When testing or calculating various microwave parameters it is often easier to work in terms of frequency, wavelength, and power rather than the voltage and current measurements commonly used with low frequency circuits.

Frequency is the number of cycles per second a signal contains and is measured in hertz. The frequency of a circuit is determined by the source generator (its oscillation frequency). Once set, it will remain constant (natural frequency drifting may occur), though the medium through which the signal travels may vary. This is not true for the velocity, and as we will see later, this is not true for the wavelength of the signal.

The period of the signal (wave) varies inversely with its frequency and is the time of one complete cycle.

$$T = 1/f \qquad \textbf{(Eq. 1.2)}$$

where $\quad T$ = period, s (of one cycle)

f = frequency, Hz

EXAMPLE 1.2:

If $f = 5.0$ GHz, what is the time of one cycle?

$$T = 1/f = 1/5 \times 10^9$$
$$= 0.2 \times 10^{-9} \text{ s}$$
$$= 0.2 \text{ ns}$$

EXAMPLE 1.3:

If $T = 500$ ps, what is the frequency of the wave? Equation 1.2 transposes to the following equation:

$$f = 1/T \qquad \textbf{(Eq. 1.3)}$$
$$f = 1/T = 1/500 \times 10^{-12}$$
$$= 2.0 \times 10^9 \text{ Hz}$$
$$= 2.0 \text{ GHz}$$

The *wavelength* (symbol λ) of an electromagnetic wave is the physical distance the wave travels in one cycle. One can make an analogy to the basic distance formula, $d = vt$ (where d equals distance, v equals velocity, and t equals time traveled) to find the distance a wave travels in a given time. The wavelength can be calculated as $\lambda = ct$, where c equals the velocity of light, and t equals the time it takes the wave to travel through one cycle (or from one peak to the next or from one trough to the next). We can modify this basic wavelength formula since frequency is the reciprocal of time. This results in the more familiar form of the wavelength equation as shown in equation 1.4.

$$\lambda = c/f \qquad \textbf{(Eq. 1.4)}$$

where
$$\lambda = \text{wavelength, m}$$
$$c = 3.0 \times 10^8 \text{ m/s}$$
$$f = \text{frequency, Hz}$$

EXAMPLE 1.4:

If the frequency of a wave is 200 MHz, the respective wavelength is:

$$\lambda = c/f$$
$$= 3.0 \times 10^8/200 \times 10^6$$
$$= 1.5 \text{ m}$$

Similarly for 2 GHz, $\lambda = 0.15$ m or 15 cm. Note that the wavelength is shorter at higher frequencies.

When the medium changes, a modification to the basic wavelength formula is required. The wavelength of an electromagnetic wave in a dielectric medium is shorter than in free space. (A similar modification was required to the velocity equation also, shown in equation 1.1). The change for wavelength is shown in equation 1.5.

$$\lambda = c/f\sqrt{\epsilon_r} \qquad \textbf{(Eq. 1.5)}$$

where
$$\lambda = \text{wavelength, m}$$
$$c = 3 \times 10^8 \text{ m/s}$$
$$f = \text{frequency, Hz}$$
$$\epsilon_r = \text{dielectric constant}$$

EXAMPLE 1.5:

Calculate the wavelength for an electromagnetic wave where the dielectric medium has $\epsilon_r = 3.0$, for a frequency of 50 GHz.

$$1 = c/f\sqrt{\epsilon_r}$$
$$= 3.0 \times 10^8/(50 \times 10^9)(\sqrt{3.0})$$
$$= 0.0035 \text{ m or } 0.35 \text{ cm or } 3.5 \text{ mm}$$

In free space $\epsilon_r = 1.0$, therefore the wavelength would be:

$$= 3 \times 10^8/(50 \times 10^9)(\sqrt{1.0})$$
$$= 0.006 \text{ m or } 0.6 \text{ cm or } 6.0 \text{ mm}$$

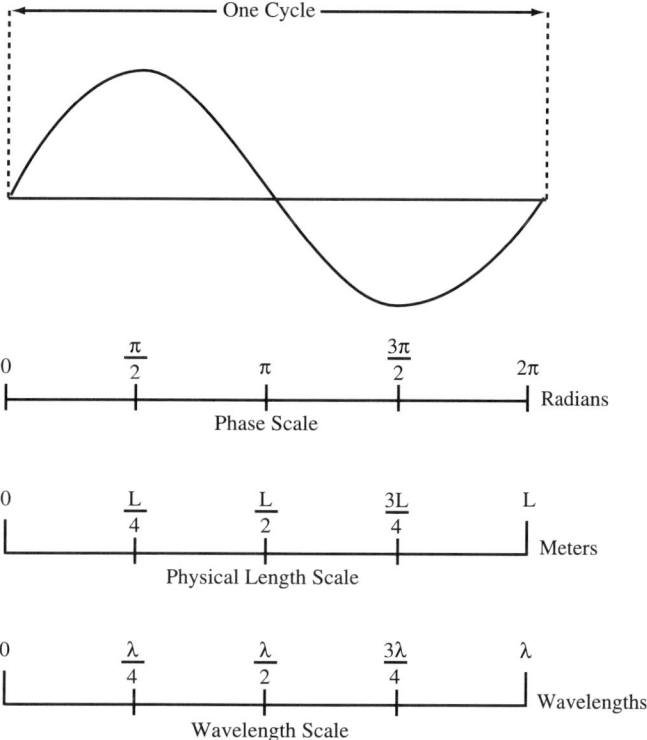

Figure 1.4 Wavelength specified by phase, physical length, and wavelength.

The wavelength of an electromagnetic wave can be expressed in different terms as the following list and Figure 1.4 demonstrate.

1. Distance where the phase change 2π radians (360°).
2. Physical length in linear units.
3. Electrical length in wavelengths.

The preceding examples for wavelength calculations used #2, above. The third term is used quite often in discussing transmission line parameters and will be used extensively throughout this text.

1.6   THE ELECTROMAGNETIC WAVE

Recall from the study of direct or alternating current (DC/AC) fundamentals that a time-varying current-carrying conductor produces an electromagnetic (EM) field surrounding the conductor. Simultaneously as the magnetic field is produced, an electric field occurs across the conductor 90° (transverse) from the magnetic field. The magnetic field (H) depends on A (amperes)/m, while the electric field (E), depends on V (volts)/m.

At low frequencies these fields follow normal Ohm's law principles and the conductor is essentially an ohmic device. As the frequency gets higher and higher, the conductor

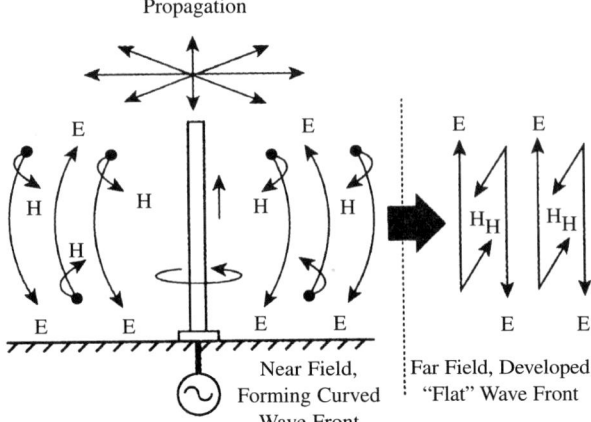

Figure 1.5 Propagation of E
and H fields from an antenna.

takes on the characteristics of a transmission line (chapter 2) and follows Maxwell's equations for electromagnetics. A line terminated by a radiating component, such as an antenna, sees the E/H mutual fields radiate into free space. These fields are at right angles (transverse) to each other and at a right angle to the direction of propagation. In other words, the electric and magnetic fields are transverse to the direction of propagation. Figure 1.5 shows the propagation of these fields from the surface of an antenna. (A more detailed analysis is covered in chapter 8.)

This wave is identified as a TEM (transverse electromagnetic) type, which occurs only in free space and on a conventional (wire) transmission line. Figure 1.6 illustrates the various features of a TEM wave. This figure shows that the direction of the electric field, the magnetic field, and propagation are mutually perpendicular in electromagnetic waves.

When an EM wave has the E field in the vertical direction with respect to Earth, the wave is identified as *vertically polarized*. (Note that the TEM wave in Figure 1.6 is shown vertically polarized.) If the E field is in the horizontal direction, the wave is identified as *horizontally polarized*. These orientations (Figure 1.7) are also known as linear polarization. Transmitting and receiving antennas must be mutually oriented with respect to the type of polarization used for maximum power transfer to occur.

Figure 1.6 Transverse electromagnetic (TEM) wave.

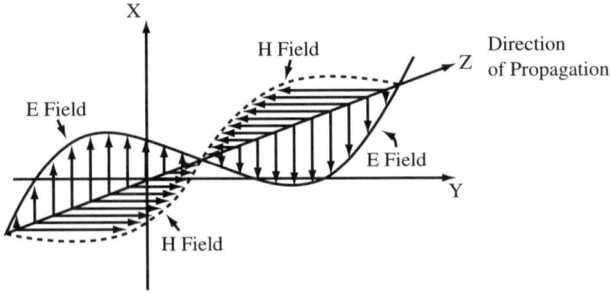

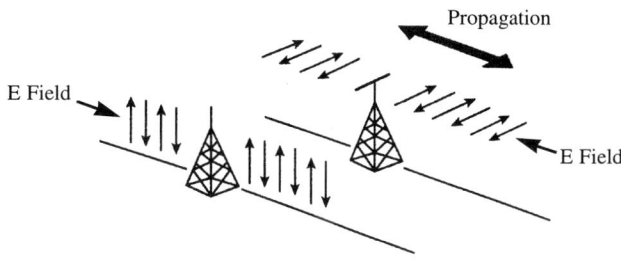

Figure 1.7 Vertical and horizontal polarization.

Amplitude modulated (AM) signals are transmitted vertically polarized (therefore, your car antenna is oriented in a vertical direction). Conventional television (TV) signals are transmitted horizontally polarized (note the orientation of a rooftop TV antenna). A C-band and Ku-band satellite transponder downlinks its signal with simultaneous vertically and horizontally polarized signals, and the new DBS service in the Ku band downlinks a circular polarized signal. A circular polarized (left-hand and right-hand) signal is shown in Figure 1.8. Here the wave is the resultant sum of two equal-amplitude E-field vectors in phase quadrature, or where one vector is 90° out of phase with the other vector. A circularly polarized wave, therefore, contains all polarizations and all information of a linearly polarized wave, although at reduced amplitude. Antenna alignment is not as critical and, further, attenuation due to rain is reduced by this form of polarization (more about this in section 1.8 and chapter 10).

1.7 PROPAGATION

The TEM wave radiates from its point source equally in all directions (assuming idealized conditions). Such a point source is said to be *isotropic*. From this point a sphere forms, much as a balloon being blown up, equally expanding in all directions. At some distance far away from the point source, the wave front would appear as a flat plane. This is comparable to looking at a map of some region on earth. The points appear as two-dimensional and the map appears to show a flat area (plane).

Figure 1.8 Circular polarization.

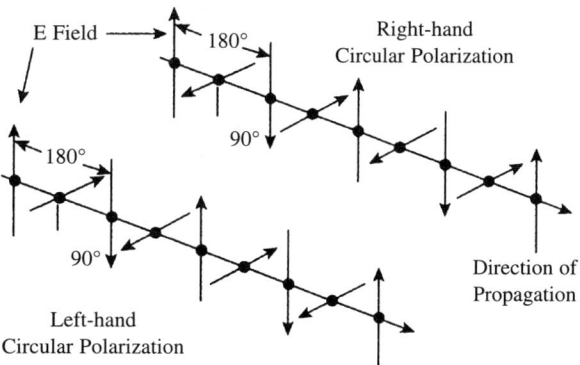

The spherical TEM wave is known as a plane wave since the wave front appears to be a two-dimensional shape. This terminology is useful in characterizing the optical effects and behavior of TEM waves under certain conditions.

It should also follow that energy contained in the wave front at any given distance (r) from the source would distribute itself equally in all directions over the sphere. The area of this sphere is found from $A = 4\pi r^2$. Determining the power density (P_D) at any point in free space as a function of the total power radiated (P_T) divided by the area (A) over which it is distributed yields equation 1.6.

$$P_D = P_T / 4\pi r^2, \qquad \textbf{(Eq. 1.6)}$$

where P_D is called the power density, P_T is in watts, and the area (denominator) is given in square meters.

EXAMPLE 1.6:

If the total radiated power (P_T) is 1,000 watts and the distance from the source (r) is 1,000 km, then the power density (P_D) at any point on the sphere is:

$$P_D = P_T/4\pi r^2$$
$$= 1000W/4\pi(1,000 \text{ km})^2$$
$$= 7.96 \times 10^{-11} \text{ W/m}^2$$
$$\approx 80 \text{ pW/m}^2$$

Note that P_D is expressed in watts per square meter, not simply in watts. As the power spreads out, the sphere continues to expand.

What is important about this relationship is that power density varies with the *inverse* of the distance (r) squared. Therefore, if the distance doubles, the power density is 1/4 as great. This phenomenon follows the familiar inverse square law relationship.

1.8   PROPAGATION PATHS

The preceding section described electromagnetic waves as leaving an isotropic antenna uniformly in all directions. We can describe the propagation of the wave using three terms: *ground waves, space waves,* and *sky waves* (Figure 1.9). Ground waves occur with higher power, low-frequency transmissions (example: AM radio). These waves travel over and near the surface of the earth. They suffer a frequency-dependent attenuation due to absorption into the ground. A wave induces currents in the ground over which it passes and thus loses some energy by absorption. This is especially true for horizontally polarized waves; thus, they effectively "short circuit" the electric component. This is the main reason why conventional AM radio is broadcast vertically polarized. There is also an absorption loss due to the fact that as the wave propagates over the earth, it tilts over more and more, and the increasing tilt causes the vertically polarized wave at some point to effectively lie in a horizontal position with respect to the earth and short circuit itself. It is important to realize this, since it shows that the maximum range of such a transmitter depends on its frequency as well as

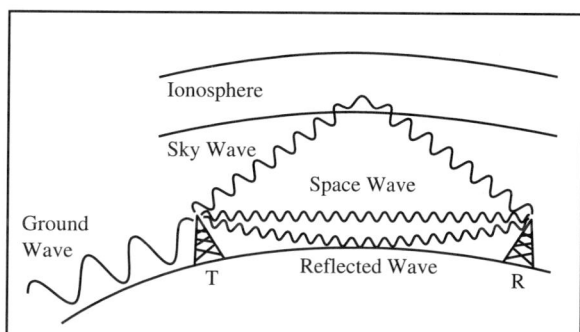

Figure 1.9 Propagation paths.

its power. Because of the aforementioned, lower frequency assignments are preferred by broadcasters in this frequency band.

Sky waves (typically in the HF band) move outward and upward into the ionosphere. The ionosphere is a layer of ionized gas that extends from 35 to 250 miles above the earth's surface. Refraction (signal bending) is the mechanism for most sky wave propagation phenomena. It occurs when the wave passes between mediums of differing density. There are many types of propagation anomalies associated with refraction; however, they are beyond the scope of this text because we are more concerned with the line-of-sight propagation paths that microwaves employ.

The space wave is also a ground wave phenomenon, but it is radiated several wavelengths above the surface and does not actually touch the ground. VHF, UHF, and microwave band signals are usually propagated as space waves. They travel in (more or less) straight lines. There are two components of the space wave to be considered: *direct* and *reflected* waves. Direct waves travel line-of-sight (LOS) to the receiving antenna, while reflected waves, by bouncing off the earth (or some obstacle) arrive at the antenna sometime later. Phase shifting occurs between these waves and causes the signals to add algebraically to either increase or decrease signal strength.

A category of reception problems, called *multipath* phenomena, exists from the interference between the direct and reflected components of the space wave. The form of multipath phenomena that is, perhaps, familiar to most is *ghosting* in television reception.

Another consideration to take in account with space waves (LOS) is the physical location of the transmitting and receiving antennas. If the antennas are at ground level, the curvature of the earth (Figure 1.10) limits the range of the space wave. The Atlantic Ocean has a 200-mile-high "hump" between Europe and the United States. Between these continents, microwave communication can normally occur only via satellite transmission.

The earth's horizon line is what affects the optical range (line-of-sight path) between the tops of the transmitting and receiving antennas. The optical range in statute miles between these antennas can be approximated by equation 1.7.

$$R = 1.42 \left(\sqrt{h_1} + \sqrt{h_2} \right) \qquad \textbf{(Eq. 1.7)}$$

R = range in statute miles

h_1 = transmitting antenna height, ft

h_2 = receiving antenna height, ft

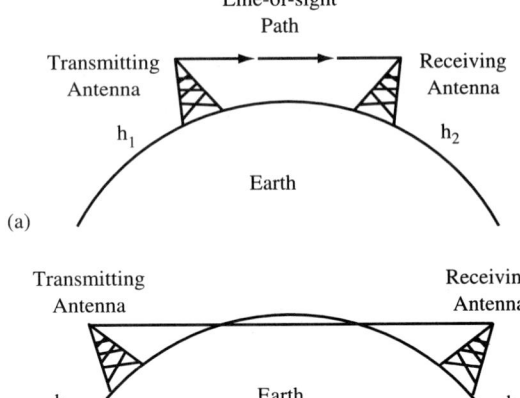

Figure 1.10 Space wave propagation and physical limitation. (a) Good space wave propagation. (b) Propagation restricted by Earth's curvature.

EXAMPLE 1.7:

Let both the transmitting and receiving antennas be 25 feet above the surface of the ground. What is the optical range?

$$R = 1.42 \, (\sqrt{25} + \sqrt{25})$$
$$= 14.2 \text{ miles.}$$

In terms of MKS units, equation 1.7 can be modified to:

$$R = 4.12 \, (\sqrt{h_1} + \sqrt{h_2}) \qquad \textbf{(Eq. 1.8)}$$

R = range in km

h_1 = transmitting antenna height, m

h_2 = receiving antenna height, m

EXAMPLE 1.8:

Let the transmitting antenna be at the top of a 15 m tower, while the receiving antenna is 1.5 m above ground. What is the optical range?

$$R = 4.12 \, (\sqrt{15} + \sqrt{1.5})$$
$$= 21.0 \text{ km}$$

The optical range can actually be extended an average of 4/3 longer because of signal path bending due to atmospheric diffraction. (Because of weather conditions, this may extend to only 10% to 30% longer.) The direct wave may actually bend beyond the horizon. This phenomenon is described as the *radio horizon*.

Note that wherever possible FM and TV stations locate their antennas atop nearby mountains, towers, or buildings. This increases the range and service area.

Attenuation of free space transmission is primarily caused by water vapor and oxygen molecules in the atmosphere. The degree of attenuation depends on the frequency transmitted. A 10 GHz radar system may have a range of 75 km in dry air, 68 km in light drizzle, 55 km in

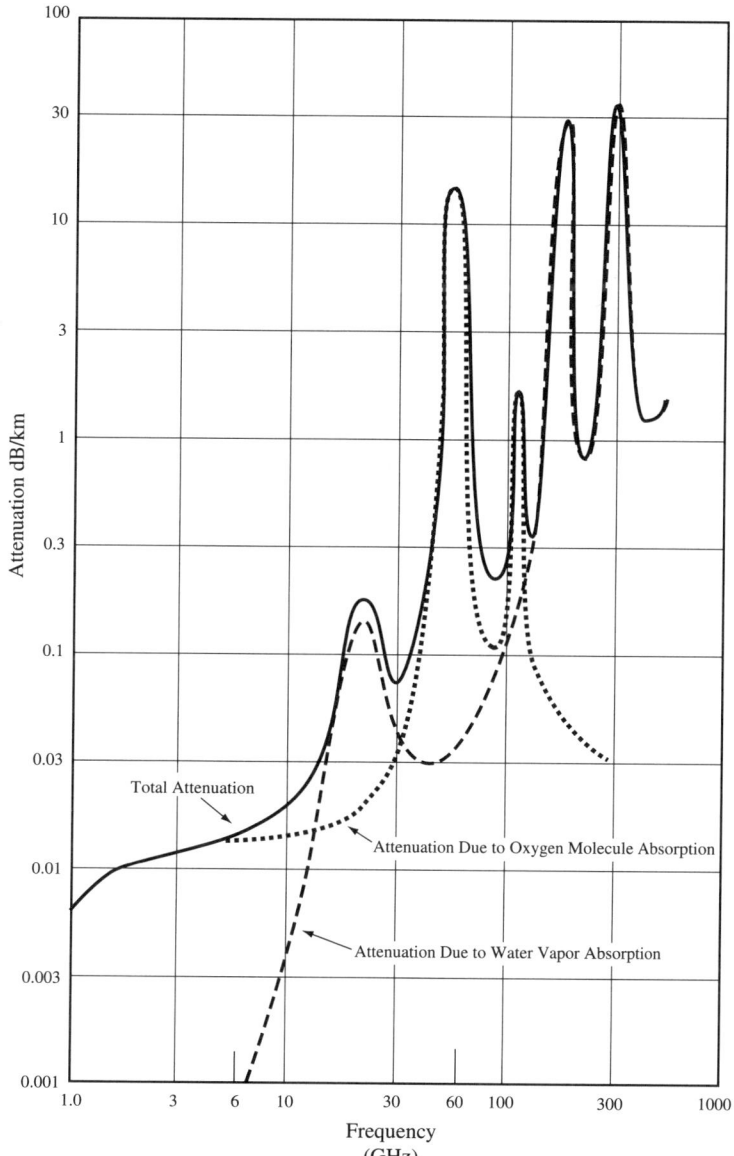

Figure 1.11 Microwaves vs. atmospheric absorption.

light rain, 22 km in moderate rain, and 8 km in heavy rain, showing effectively how precipitation causes severe absorption at microwave frequencies. Figure 1.11 shows where the peaks and valleys of attenuation occur with respect to frequency. The peaks occur around 23, 60, 120, 180, and 320 GHz. Some of this attenuation is caused by water vapor absorption and some is caused by oxygen molecule absorption of the microwave signal. Obviously these frequencies should be avoided. However, satellites parked in geosynchronous orbit (22,300 miles) can "talk" to each other in their orbits, since there is no atmosphere between them. There are a couple of good "windows" above 12 GHz, occurring at 33 and 110 GHz, respectively. They will no doubt be used sometime in the future, as transmitting frequencies go higher.

1.9   SUMMARY

1. Microwave technology expanded with the development of radar and mushroomed during World War II.
2. Microwaves are those electromagnetic (EM) waves with frequencies from 500 MHz to 300 GHz. More specifically, most microwave equipment operates from 1 GHz to 100 GHz.
3. Conventional electronic components do not work well at microwave frequencies because wavelengths approximate component sizes. "Lumping" *LC* values is not feasible at microwave frequencies. Instead we use the distributed constants of *L* and *C*.
4. EM waves propagate at the speed of light, 3.0×10^8 m/s, in free space (and air as well).
5. Because of microwave's higher frequency, greater bandwidth is available, and therefore information-carrying capacity is greater.
6. Both velocity and wavelength decrease when the medium is other than free space.
7. Polarization of an electromagnetic wave is determined by the orientation of the electric (E) field with respect to Earth.
8. The signal strength of a propagated wave decreases as a function of the distance squared.
9. Propagation paths are in the form of ground waves, space waves, and sky waves. Microwaves usually travel as space waves (line-of-sight).
10. Refraction (wave bending) occurs in the ionosphere.

Key Equations:

$v_r = c/\sqrt{\epsilon_r}$	(Eq. 1.1)
$T = 1/f$	(Eq. 1.2)
$f = 1/T$	(Eq. 1.3)
$\lambda = c/f$	(Eq. 1.4)
$\lambda = c/f\sqrt{\epsilon_r}$	(Eq. 1.5)
$P_D = P_T/4\pi r^2$	(Eq. 1.6)
$R = 1.42 \left(\sqrt{h_1} + \sqrt{h_2}\right)$	(Eq. 1.7)
$R = 4.12 \left(\sqrt{h_1} + \sqrt{h_2}\right)$	(Eq. 1.8)

PROBLEMS

1. Convert the velocity to cm/s for an EM wave in a certain medium that has a velocity of 2.75×10^8 m/s.

2. Determine the velocity of an EM wave in a medium where the dielectric constant is 2.25.

3. Determine the dielectric constant of a medium, if the velocity of an EM wave is reduced to 81% of *c*.

4. Determine the time of one cycle when $f = 100$ MHz, 1.0 GHz, and 10 GHz, respectively.

5. Determine the frequency when the time of one cycle is 20 ms, 200 μs, 5.0 ns, and 12 ps, respectively.

6. Determine the wavelength when the frequency is 50 kHz, 900 kHz, 25 MHz, 2.4 GHz, and 33 GHz, respectively.

7. An EM wave has a wavelength in free space of 125 μm. What is its frequency in terahertz?

8. The dielectric constant of a certain medium propagating a TEM wave is 1.55. The frequency of the signal is 10 GHz. Determine the wavelength of the signal.

9. How many inches long is the wavelength for problem 8?

10. How many seconds are required for a microwave signal to travel 56 km through the air?

11. A microwave signal radiated into free space travels _____ miles in 52 μs.

12. An isotropic transmitter in free space sends a signal whose power is 50 watts. The power density at a point 18 km away is _____.

13. The power density at a certain distance from a 1 kW isotropic source in free space is 3.18×10^{-14} W/m². How far away is the source?

14. Calculate the optical range in miles between two antennas. The height of one is 100 feet, while the height of the other is 200 feet.

15. The optical range of problem 14 could be extended to an average _____ miles, via "radio horizon" effects.

16. Determine the optical range in km between two antennas when one is at a height of 50 m while the other is at a height of 100 m.

QUESTIONS

1. Define the term *microwave*.

2. What factors hampered the early developments of microwave devices?

3. Describe the evolution of microwave from World War II to the present.

4. Most microwave devices and equipment fall into what CCIR bands of operation?

5. Describe the specific principles of microwave technology that are different from ordinary high-frequency course work.

6. Describe the term *standing wave*.

7. Define the terms *frequency, period,* and *wavelength*.

8. The velocity and wavelength of an EM wave (increases, decreases) in a medium other than free space.

9. Describe the features of a TEM wave.

10. Describe the term *polarization*.

11. Describe the propagation paths of EM waves in free space.

12. Describe the term *multipath phenomena*.

13. What extends the optical range of a transmitted signal?

14. Describe what factors attenuate the free space transmission of microwave signals.

15. Do geosynchronous satellites "talking" to each other suffer the same maladies described in question 14? Why or why not?

2 TRANSMISSION LINE FUNDAMENTALS

OBJECTIVES

1. To define a transmission line.
2. To describe the losses associated with a transmission line.
3. To describe the "distributed" constants of a transmission line.
4. To calculate characteristic impedances of various transmission lines.
5. To define standing wave, reflection coefficient, return loss, and VSWR.
6. To calculate various parameters of reflection coefficient, return loss, and VSWR.
7. To compare and contrast various characteristics associated with transmission lines.
8. To compare and contrast characteristics of various line terminations and their effects on standing waves.
9. To describe the uses of time-domain reflectometry (TDR).

2.1 INTRODUCTION

A simple transmission line is a pair of conductors linking two electrical systems, components, or devices. Transmission lines are also used to connect a source component to a load component (example: a generator to an antenna). All transmission lines have a *characteristic impedance* (also known as *surge impedance*) value, which is labeled Z_O. This impedance is typically constant regardless of transmission line length. Figure 2.1 shows examples of some of the common transmission lines.

Transmission lines in the form of parallel wires have been in use for some time. Parallel wires for radio wave applications were first introduced by Lecher in 1890. This idea eventually evolved into microwave striplines, which are forms of single wire lines. Single-wire transmission lines were designed to simplify transmission line construction. As miniaturization of electronic circuits has evolved, microstrip and stripline transmission lines have become popular because they lend themselves to printed circuit board fabrication techniques.

All transmission lines suffer from some sort of attenuation or power losses. Either singly or in combinations, these include copper losses, dielectric losses, and radiation losses.

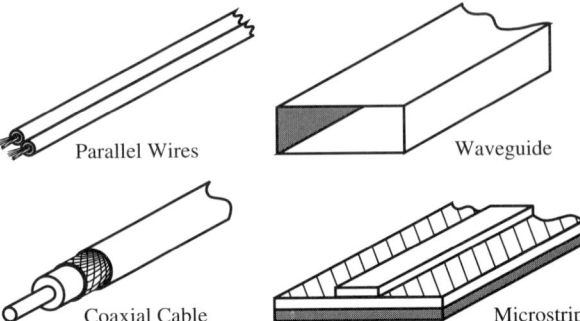

Figure 2.1   Common transmission lines.

Copper losses are caused by the resistance of conductors that make up the transmission line. Current flowing in the conductors causes the conductors to dissipate energy in the form of heat, so this energy will not be available at the output of the transmission line. This loss is proportional to the length of the line.

Skin effect is another form of copper loss and is sometimes referred to as AC resistance. Skin effect is the tendency of the signal to flow near the outer surface of the conductor at high frequencies. It is caused by self-inductance occuring at the center of the line. This effectively reduces the cross-sectional area of the line and increases its resistance. This resistance varies directly with frequency. Some conductors are constructed with hollow centers to create a transmission line that is lightweight and just as efficient as solid conductors.

Dielectric losses are primarily caused by dielectric "hysteresis." The dielectric does not allow the electrical flux to reverse as quickly as the polarity of the signal changes. The net effect is an energy (heat) loss that is also frequency dependent.

Radiation loss occurs when the transmission line acts like a radiator (antenna). Loss occurs as the signal propagates into space instead of being transferred to the load.

Coaxial lines were developed for high-frequency signal transmission to prevent losses due to radiation and thereby replace parallel lines. They are limited in power handling capability. Waveguides, which are hollow-tube transmission devices, show even better characteristics in most microwave frequency ranges. They can handle higher power as well. Microstrip and stripline transmission lines also suffer from the same maladies, but because the length of their lines is relatively small, their losses are minimal. All types of transmission lines have had their characteristics employed in a variety of applications as microwave technology evolved. Their shortcomings meant that another type would be found to replace the unsatisfactory one.

The emphasis in this chapter is on the conventional two-wire line. Other types of transmission lines are covered in detail in chapter 4.

2.2   CONCEPTS OF TWO-WIRE LINES

Consider two parallel conductors as shown in Figure 2.2. Between the conductors is a dielectric material consisting of air; the insulation coatings are dielectric as well. All conductors have some ohmic resistance R directly proportional to length. When the source energy is AC, a time-varying change in voltage or current exists, therefore electromagnetic and electrostatic fields form in the vicinity of the wires. The dielectric medium is polarized

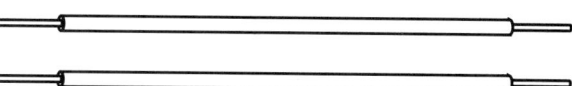

Figure 2.2 Parallel conductors.

with positive charges on one side and negative charges on the opposite side and alternates this charge as the signal varies. This phenomenon is the capacitor effect of the line. The line can be labeled C, to represent this capacitance.

The electromagnetic field produced causes the wire to act like an inductor. This inductance effect, labeled L, is proportional to the rate of change of the incoming current. Note that C is a parallel quantity, while L is a series quantity of the line.

All dielectric media allow some leakage current to flow between the conductors, especially at high voltages. The current leakage through it can be represented by a shunt conductance G.

The transmission line circuit shown in Figure 2.3a is the schematic equivalent of a general transmission line. As energy is transmitted over the line, the energy must overcome the losses caused by R and G, overcome local inductance opposition due to L, and charge the capacitance effect, C. Assuming an infinitely long line, then the bottom wire's characteristics can be added with the upper wire's values. This simplifies the line to that in Figure 2.3b. The quantities of R, G, L, and C are specified as quantities per unit length, e.g., per meter, because they occur periodically along the line. They are thus distributed throughout the length of the line.

At radio frequencies, the inductive reactance is much larger than the resistance, while the capacitive susceptance is much larger than the shunt conductance. Therefore, both R and G can be ignored, resulting in a line that is considered lossless (as a very good approximation for radio frequency [RF] calculations). The equivalent circuit is simplified as shown in Figure 2.3c.

Figure 2.3 Equivalent transmission lines. (a) The general equivalent transmission line circuit. (b) A simplified line created by adding bottom wire characteristics with the upper wire's values. (c) A further simplified line that is considered lossless.

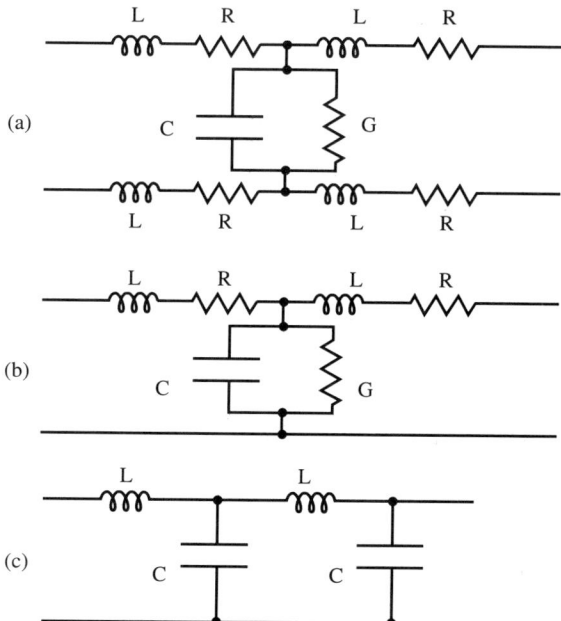

Any circuit that consists of series and shunt impedances must have an input impedance. For the transmission line, this input impedance depends on the type of line, its length, and the termination at the far end. The input impedance taken as the reference is known as the characteristic impedance of that line. By definition, a transmission line's characteristic impedance, Z_O, is the impedance measured at the input of this line when its length is infinite. Under these conditions the type of termination at the far end has no effect, and consequently is not mentioned in the definition.

It can be shown that the characteristic impedance of a line will be measured at its input when the line is terminated at the far end in an impedance equal to Z_O, no matter how long the line is. It follows that the characteristic impedance of an iterative circuit consisting of series and shunt elements is given by:

$$Z_O = \sqrt{(Z/Y)}$$

where
$$Z = \text{series impedance per section}$$
$$= R + j\omega L$$
$$Y = \text{shunt admittance per section}$$
$$= G + j\omega C$$

Therefore:

$$Z_O = \sqrt{(R + j\omega L/G + j\omega C)}$$

As was shown earlier, R and G can be ignored, so the expression for Z_O reduces to:

$$Z_O = \sqrt{(j\omega L/j\omega C)}$$

and further reduces to:

$$Z_O = \sqrt{(L/C)} \qquad \textbf{(Eq. 2.1)}$$

where
$$Z_O = \text{characteristic impedance, } \Omega$$
$$L = \text{inductance per unit length, H/m}$$
$$C = \text{capacitance per unit length, F/m}$$

Quantities L and C are distributed constants and are specified on a manufacturer's data sheet. (Note that the impedance of the line is in ohms, which shows that this characteristic impedance is resistive at radio frequencies.)

EXAMPLE 2.1:

Using RG59 cable, where $L = 370$ nH/m and $C = 67$ pF/m, the characteristic impedance, Z_O, is calculated as follows:

$$Z_O = \sqrt{(370 \times 10^{-9}/67 \times 10^{-12})}$$
$$= 74.3 \ \Omega$$

2.3   FREE SPACE CHARACTERISTIC IMPEDANCE

An antenna signal propagates through free space (or air) to a receiving antenna; therefore, free space can be considered a form of a transmission line. It can be shown to have a characteristic impedance as well.

The characteristic impedance for *any* medium in which an EM wave travels can be calculated using equation 2.2.

$$Z_O = \sqrt{(\mu/\in)}$$ **(Eq. 2.2)**

where Z_O = characteristic impedance of any medium, Ω

μ = permeability, H/m

$\in$ = permittivity, F/m

Note that this equation is consistent with equation 2.1 in that the ratio of μ (henries) to $\in$ (farads) is equivalent to the ratio of L to C.

We can now calculate the characteristic impedance of free space using equation 2.2. For free space, $\mu = 4\pi \times 10^{-7}$ H/m and $\in = 8.85 \times 10^{-12}$ F/m, with Z_O calculated as follows:

$$Z_O = \sqrt{(4\pi \times 10^{-7}/8.85 \times 10^{-12})}$$
$$= 120\pi \ \Omega$$
$$= 377 \ \Omega$$

Terminations on a transmission line, such as an antenna, must therefore have an impedance that matches that of free space for maximum power transfer to occur.

2.4   MATCHED TERMINATIONS

From low-frequency electronics, recall that maximum power is transferred to the load when the source impedance matches the load impedance. If a generator is connected to a lossless cable of characteristic impedance Z_O, then Z_L or R_L must equal Z_O to achieve maximum power transfer.

At microwave frequencies, the maximum power transfer theorem is equally important. The idea of an infinitely long transmission line is a very useful analysis tool to describe what happens. If we start with an infinitely long line whose characteristic impedance is Z_O, then we can keep one segment (unit length) of the cable as our model. We then can replace the rest of the infinitely long line with a load whose impedance is also equal to Z_O. This is shown in Figure 2.4. As this figure shows, a finite length of transmission line terminated by an impedance Z_O is electrically equivalent to an infinitely long transmission line of characteristic impedance Z_O.

The importance of this concept is that by using an infinitely long transmission line, any signal leaving the generator keeps going down the line and never gets to the end. Therefore, the signal *never* returns, and no reflection can occur. Reflections inside a transmission line mean that we have inefficient transfer of power from the source to the receiving end. Reflections also cause distortion of the signal.

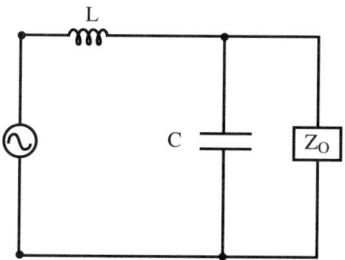

Figure 2.4   Matched termination. A finite length of transmission line terminated by an impedance Z_O is electronically equivalent to an infinitely long transmission line of characteristic impedance Z_O.

If we have a lossless line of finite length, no self-absorption occurs, therefore, the transmitted signal must be absorbed by the terminating load. This happens only when Z_O of the line matches the load impedance, Z_L. The source generator cannot tell the difference between an infinitely long line and a finite line terminated by a matched load. No reflections occur in either case.

If the impedances of the line and the load don't match, then reflections occur causing standing waves to form. These are described in section 2.6 of this chapter.

2.5   PROPAGATION VELOCITY, VELOCITY FACTOR, AND ATTENUATION CONSTANT

Recall that in free space an EM wave travels at a velocity equal to c, which is approximately equal to 3×10^8 m/s. Inside a transmission line, the voltage (E field) and current (H field) propagations are delayed somewhat due to the effects of inductance and capacitance along the line. Thus, the *propagation velocity* is reduced. The velocity of propagation in a transmission line of known inductance and capacitance per unit length is given by equation 2.3.

$$v = 1/\sqrt{LC} \qquad \text{(Eq. 2.3)}$$

where
$$v = \text{velocity, m/s}$$
$$L = \text{inductance, H/m}$$
$$C = \text{capacitance, F/m}$$

EXAMPLE 2.2:

Using the RG59 cable again, the velocity within the line is:

$$v = 1/\sqrt{(370 \times 10^{-9})(67 \times 10^{-12})}$$
$$= 2.0 \times 10^8 \text{ m/s}$$
$$\approx 67\% \ c$$

The 67% c denotes a reduction in the velocity of the wave in the transmission line. This reduction is called the *velocity factor*, v_f. It is calculated using the dielectric constant of the associated transmission line. The dielectric constants of materials commonly used in transmission lines range from about 1.2 to 2.8, giving corresponding velocity factors from 0.9 to 0.6. The equation is given by:

$$v_f = 1/\sqrt{\in_r} \qquad \text{(Eq. 2.4)}$$

where
$$v_f = \text{velocity factor}$$
$$\in_r = \text{dielectric constant of material}$$

EXAMPLE 2.3:

Using the RG59 cable again, with a dielectric of polyethylene, $\in_r = 2.25$, determine the velocity factor.

$$v_f = 1/\sqrt{2.25}$$
$$= .667$$

Certain practical considerations regarding velocity factor result from the fact that the physical and electrical lengths of a transmission line are not equal. The physical length of the line is longer than the equivalent electrical length. To achieve a desired electrical length, the physical length of line must be shortened.

Consider a certain type of phased array antenna with radiating elements spaced half-wavelengths apart and to be fed 180° (half-wave) out of phase with each other. The simplest interconnect is a half-wave transmission line between the appropriate elements. Because of the velocity factor, the physical length for a one-half wavelength cable is shorter than the free-space half-wave distance between the elements. The cable is too short to reach between the radiating elements by the amount of the velocity factor.

Another characteristic associated with transmission lines is the *attenuation constant*, α. Practical transmission lines suffer losses like those described in section 2.1, particularly at higher frequencies. The attenuation constant, α, can be found from:

$$\alpha = (R/2Z_O + GZ_O/2) \text{ Neper/m}$$

The Neper is a dimensionless quantity based on the natural log (1 Np = 8.689 dB). The attenuation constant, α, is usually given in manufacturers' catalogs in dB/100 m or dB/100 ft. In dB/m,

$$\alpha = 8.689 (R/2Z_O + GZ_O/2)$$

A partial listing from a manufacturer's catalog shows many important characteristics of standard transmission lines, as shown in Table 2.1. Note that the attenuation increases with frequency.

2.6   STANDING WAVES

Recall from section 1.4 of chapter 1 that a transmission line at low frequencies is physically only an infinitesimal percent of wavelength size. The wavelength for all practical purposes "looks" infinitely long, therefore, no reflections occur. This is comparable to what happens on an infinitely long transmission line. At microwave frequencies the physical length of a transmission line can contain many multiples of wavelengths, and when mismatches occur reflected waves and their associated phase shifts can and do take place.

Table 2.1   Manufacturer's Catalog Entries for Selected Transmission Lines

JAN Type No.	Outside Diameter (mm)	Maximum Operating Frequency (MHz)	Attenuation (dB/100m) @ 450 MHz	Attenuation (dB/100m) @ 1500 MHz	Attenuation (dB/100m) @ 6000 MHz	Velocity Factor
RG-213/U	10.29	1000	16.4	–	–	0.659
RG-213/U Commercial	10.29	–	9.2	19.0	48.2	0.84
RG-214/U	10.79	11,000	18.0	37.1	93.8	0.659
RG-142B/U	4.95	12,400	27.6	55.4	135	0.695
RG-393/U	9.91	11,000	16.1	32.8	79.4	0.695
RG-6/U	8.43	300	22.0	45.9	–	0.659
RG-11/U	10.29	1000	16.7	–	–	0.659

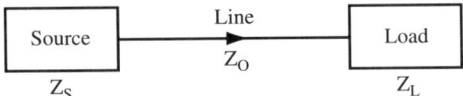

Figure 2.5   A simple
microwave system.

Consider a simple microwave building block consisting of a source, a line, and a load. The source could be a signal generator. The line could be some length of coaxial cable or waveguide. The load could be a simple resistive termination or an antenna. These building blocks are shown in Figure 2.5.

Each block exhibits some characteristic impedance. Assuming a lossless line, energy produced by the source travels down the line toward the load. If Z_S matches Z_O, we have a match between source and line, and if Z_L matches Z_O, we have achieved a match between line and load. Maximum power transfer occurs at the load, which ensures the most efficient use of the power delivered by the source. The match must occur at both ends of the line. The source-to-line impedance match is relatively simple to accomplish in actual practice during the manufacturing process.

In many applications the source is a Gunn diode or klystron oscillator (or some other form of a signal generator), with a built-in adapter to match the impedance to a waveguide or coaxial transmission line. Therefore, the line-to-load match becomes the one to be concerned about (particularly for practical applications). In chapter 3 we will look at several possible ways to make a line-to-load match from an existing unmatched condition.

Any impedance mismatch manifests itself by the formation of reflected waves. The amplitude of the reflected wave depends on the degree of mismatch. The consequence of this impedance mismatch is that whenever the reflected wave meets the incident wave, a new wave, called a *standing wave,* is produced. Algebraic addition of the incident and reflected waves occurs to create a standing wave, that is, a wave that appears to stand still rather than travel on the line as do the incident and reflected waves. The formation of standing waves occurs as time varies from t_1 to t_3 as shown in Figures 2.6 a–c. Because of the phase differences between the incident and reflected waves, the standing wave has both a stationary node (minimum) and a stationary antinode (maximum) on the transmission line. This forms a standing wave envelope as shown in Figure 2.6d.

Both voltage and current wavefronts from the source can be reflected. It is more common to represent the standing waves as a voltage variation. This is because voltage is easier to detect with instrumentation when doing measurements or lab experiments.

The nodes represent a stationary position of zero voltage on the line. Antinodes (maximums) are also stationary in position and midway between two adjacent nodes. The distance between the nodes is one-half wavelength. Of course there is also $\lambda/2$ distance between antinodes, but the nodes provide a sharper deflection so that a detection device can "see" them more easily.

With this understanding of standing waves and their formation, we can now derive a quantitative means of expressing the degree of mismatch for a given load-line impedance condition. These include the voltage reflection coefficient, the power reflection coefficient, return loss, and the voltage standing-wave ratio. We will look at these parameters in the sections that follow.

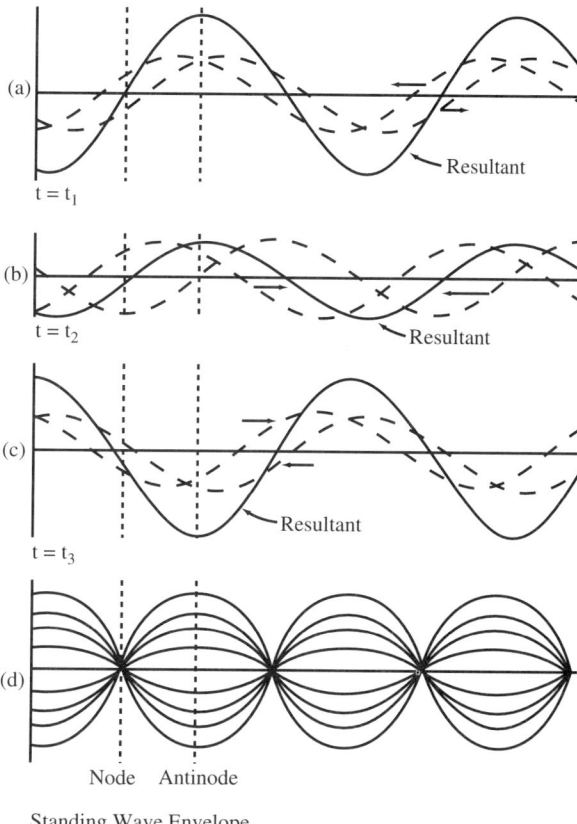

Figure 2.6 Instantaneous values of incident and reflected waves on a transmission line varying with time.

Standing Wave Envelope

Incident Wave ⟶
Reflected Wave ⟵

2.7   REFLECTION COEFFICIENT AND RETURN LOSS

The *reflection coefficient* is a unitless numerical ratio that represents the amount of incident energy reflected from a load. The voltage reflection coefficient of the load impedance Z_L, Γ (gamma), is equal to the ratio of reflected voltage to incident voltage. Γ is a phasor quantity, having both magnitude and phase angle. The magnitude can vary from 0 to 1. The phase angle represents the fact that the reflected wave is not necessarily reflected in phase with the incident wave. The phase angle varies at points other than the load (along the line), but the voltage reflection magnitude remains the same anywhere along the line. Equation 2.5 shows this ratio.

$$\Gamma = V_r / V_i \qquad \textbf{(Eq. 2.5)}$$

where Γ = voltage reflection coefficient, no units

V_r = reflected voltage

V_i = incident voltage

Since the specific reflected and incident voltages are difficult to separate and measure, we can solve for Γ much more easily by using known impedances. This is shown in equation 2.6.

$$\Gamma = (Z_L - Z_O)/(Z_L + Z_O) \tag{Eq. 2.6}$$

where

Γ = voltage reflection coefficient, no units

Z_L = load impedance, Ω

Z_O = characteristic impedance of the line, Ω

For resistive terminations on a lossless line, $Z_L = R_L$, $Z_O = R_O$ and equation 2.6 simplifies to:

$$\Gamma = (R_L - R_O)/(R_L + R_O) \text{ (Resistive load)} \tag{Eq. 2.7}$$

If R_L is greater than R_O, then Γ is positive, and $\theta = 0°$. Γ is negative when R_L is less than R_O and $\theta = -\pi$ ($-180°$).

EXAMPLE 2.4:

Determine Γ when $R_L = 100 \ \Omega$ and $R_O = 50 \ \Omega$.

$$\Gamma = (100 - 50)/(100 + 50)$$
$$= 50/150$$
$$= 0.333 \text{ (positive value denotes } \theta = 0°)$$

EXAMPLE 2.5:

Determine Γ when $R_L = 25 \ \Omega$ and $R_O = 50 \ \Omega$.

$$\Gamma = (25 - 50)/(25 + 50)$$
$$= -25/75$$
$$= -0.333 \text{ (negative value denotes } \theta = -180°)$$

When the load is complex ($R \pm jX$), equation 2.6 solves for the voltage reflection coefficient and the associated phase angle.

EXAMPLE 2.6:

Determine Γ when $Z_O = 50 \ \Omega$, and $Z_L = 40 - j20 \ \Omega$.

$$\Gamma = (Z_L - Z_O)/(Z_L + Z_O)$$
$$= [(40 - j20) - 50/(40 - j20) + 50]$$
$$= (-10 - j20)/(90 - j20)$$
$$= 22.4\angle - 116.1°/92.2\angle - 12.5°$$
$$= 0.24\angle - 103.6°$$

Rho, ρ, is used in this text to also indicate the voltage reflection coefficient, but only the magnitude of Γ, or $|\Gamma|$. It has no phase association. Rho (ρ) for the preceding example is 0.24.

Power reflection coefficient is another term that describes reflection on the line and can be found rather easily. Since power is a function of the voltage squared divided by Z_O, then the power reflection coefficient equals the square of the voltage reflection coefficient. This is shown by equation 2.8.

$$\text{Power reflection coefficient} = \rho^2 \qquad \textbf{(Eq. 2.8)}$$

EXAMPLE 2.7:

Determine the power reflection coefficient when $\rho = 0.333$.

$$\rho^2 = (0.333)^2$$
$$= 0.111$$

To express the voltage or power coefficients in percent, simply multiply the respective coefficient by 100.

Return loss (RL) is another parameter used to analyze reflection on a transmission line. Return loss is defined as the ratio of the incident power to the reflected power. This is shown in equation 2.9.

$$RL = P_{in}/P_{ref} \qquad \textbf{(Eq. 2.9)}$$

where
RL = return loss, no units

P_{in} = incident power, W

P_{ref} = reflected power, W

EXAMPLE 2.8:

Determine RL when P_{in} = 30 mW, and P_{ref} = 1 μW.

$$RL = 30 \text{ mW}/1 \text{ } \mu\text{W}$$
$$= 3.0 \times 10^4$$

For a perfect match, $RL = \infty$. Therefore, the larger the return loss the less reflection occurs. Return loss can also be expressed in dB. The corresponding equation is:

$$RL_{(dB)} = 10 \log RL \qquad \textbf{(Eq. 2.10)}$$

Applying this equation to the value of RL above:

$$RL_{(dB)} = 10 \log (3.0 \times 10^4)$$
$$= 44.8 \text{ dB}$$

Again, a perfect match has infinite return loss when expressed in dB, and the poorest match equals 0 dB return loss.

An alternate method used to calculate return loss is found by using the voltage reflection coefficient ρ, (remember $\rho = |\Gamma|$). This is shown in equation 2.11.

$$RL = 1/\rho^2 \qquad \textbf{(Eq. 2.11)}$$

where
RL = return loss, no units

ρ = magnitude of voltage reflection coefficient

EXAMPLE 2.9:

If $\rho = 0.2$, determine the return loss.

$$RL = 1/(0.2)^2$$
$$= 25$$

and, $$= 14 \text{ dB (from equation 2.10)}$$

Equation 2.11 can be converted to one that yields the return loss in dB of power. This is given in equation 2.12.

$$RL_{(dB)} = -20 \log \rho \qquad \textbf{(Eq. 2.12)}$$

EXAMPLE 2.10:

Using the same ρ as in example 2.9, determine return loss.

$$RL_{(dB)} = -20 \log 0.2$$
$$= 14 \text{ dB}$$

2.8   VOLTAGE STANDING-WAVE RATIO

We can now derive a numerical expression to describe the load-line impedance mismatch condition in terms of the standing wave. This term is the *voltage standing-wave ratio* (VSWR). It is defined as the ratio of maximum to minimum standing wave voltages present at the load. (Note: Some microwave literature refers to VSWR as simply *S*, while others refer to it as sigma, σ.) VSWR ranges from 1 to ∞. The larger this value, the greater the mismatch. On a practical matched system, a VSWR of 1.2 or less is usually regarded as acceptable. In a laboratory setting, even better matches are easily obtainable relative to the application being done.

The ratio of V_{max} to V_{min} is defined as the VSWR, as shown in equation 2.13.

$$\text{VSWR} = V_{max}/V_{min} \qquad \textbf{(Eq. 2.13)}$$

The voltage standing-wave ratio expressed in decibels is called the standing-wave ratio ($\text{SWR}_{(dB)}$). This term can be found using the familiar log voltage equation.

$$\text{SWR}_{(dB)} = 20 \log \text{VSWR} \qquad \textbf{(Eq. 2.14)}$$

Maximum and minimum voltages are difficult to measure unless one has a slotted line available, so an easier method is to determine VSWR from the voltage reflection coefficient. Equation 2.15 shows how this can be calculated, while equation 2.16 shows how rho (ρ) can be calculated from the VSWR.

$$\text{VSWR} = (1 + \rho)/(1 - \rho) \qquad \textbf{(Eq. 2.15)}$$

and, $$\rho = (\text{VSWR} - 1)/(\text{VSWR} + 1) \qquad \textbf{(Eq. 2.16)}$$

EXAMPLE 2.11:

Determine the VSWR when $\rho = 0.2$.

$$VSWR = (1 + 0.2)/(1 - 0.2)$$
$$= 1.2/0.8$$
$$= 1.5$$

EXAMPLE 2.12:

Determine ρ when the VSWR = 3.0.

$$\rho = (3.0 - 1)/(3.0 + 1)$$
$$= 2/4$$
$$= 0.5$$

EXAMPLE 2.13:

Determine $SWR_{(dB)}$ when VSWR = 1.5.

$$SWR_{(dB)} = 20 \log 1.5$$
$$= 20(0.176)$$
$$= 3.52 \text{ dB}$$

The VSWR can also serve to determine the minimum and maximum impedances that will occur on a line under mismatched conditions. (Remember when $Z_O = Z_L$, the line reflects a constant impedance equal to Z_O anywhere on the line). Maximum impedance occurs at a point of maximum voltage and can be found from the following:

$$Z_{max(\Omega)} = (VSWR)Z_O \qquad \textbf{(Eq. 2.17)}$$

Minimum impedance occurs at a point of minimum voltage and can be found from the following:

$$Z_{min(\Omega)} = Z_O/VSWR \qquad \textbf{(Eq. 2.18)}$$

EXAMPLE 2.14:

Determine Z_{max} and Z_{min} on a transmission line when the VSWR = 4.0 and $Z_O = 50\ \Omega$.

$$Z_{max(\Omega)} = 4(50)$$
$$= 200\ \Omega$$
$$Z_{min(\Omega)} = 50/4$$
$$= 12.5\ \Omega$$

For convenience, Table 2.2 shows the conversions for some common values of VSWR, return loss, and voltage reflection coefficient. (Note: Return loss is given for power and is in dB).

Table 2.2   Conversion Table
Between VSWR and Return Loss

VSWR	Return Loss	Reflection Coefficient (ρ)
1.01	46.06	0.0050
1.02	40.08	0.0099
1.03	36.60	0.0148
1.04	34.15	0.0196
1.05	32.25	0.0244
1.06	30.71	0.0291
1.07	29.41	0.0338
1.08	28.29	0.0385
1.09	27.31	0.0431
1.10	26.44	0.0476
1.11	25.65	0.0521
1.12	24.94	0.0566
1.13	24.28	0.0611
1.14	23.68	0.0654
1.15	23.12	0.0698
1.20	20.82	0.0909
1.25	19.08	0.1111
1.30	17.69	0.1304
1.40	15.56	0.166
1.50	13.97	0.20
2.0	9.54	0.33
3.0	6.021	0.50
4.0	4.437	0.6
5.0	3.522	0.666
10.0	1.743	0.818
20.0	0.869	0.904
30.0	0.579	0.935
40.0	0.434	0.9512
50.0	0.347	0.960
60.0	0.290	0.967
70.0	0.248	0.971
80.0	0.217	0.975
90.0	0.193	0.978
100.0	0.174	0.980

2.9   OPEN AND SHORTED TERMINATIONS

If a line is terminated by an open, then the incident and reflected voltages are equal and in phase at the open circuit. This is shown in Figure 2.7, drawings 1 through 8, which provide more detail than the standing waves shown in earlier figures. The voltage is maximum at the open and every half-wavelength thereafter. The incident and reflected current waves are 180° out of phase and cancel at the open. This is shown in Figure 2.7, drawings a through h. As you check each picture individually, you will see that the bold curve is the algebraic sum of the other two curves. One curve moves to the right, while the other moves to the left. The curve moving to the right is the incident wave; the one moving left is the reflected wave. Note that VSWR equals ∞, while Γ equals 1 for this situation.

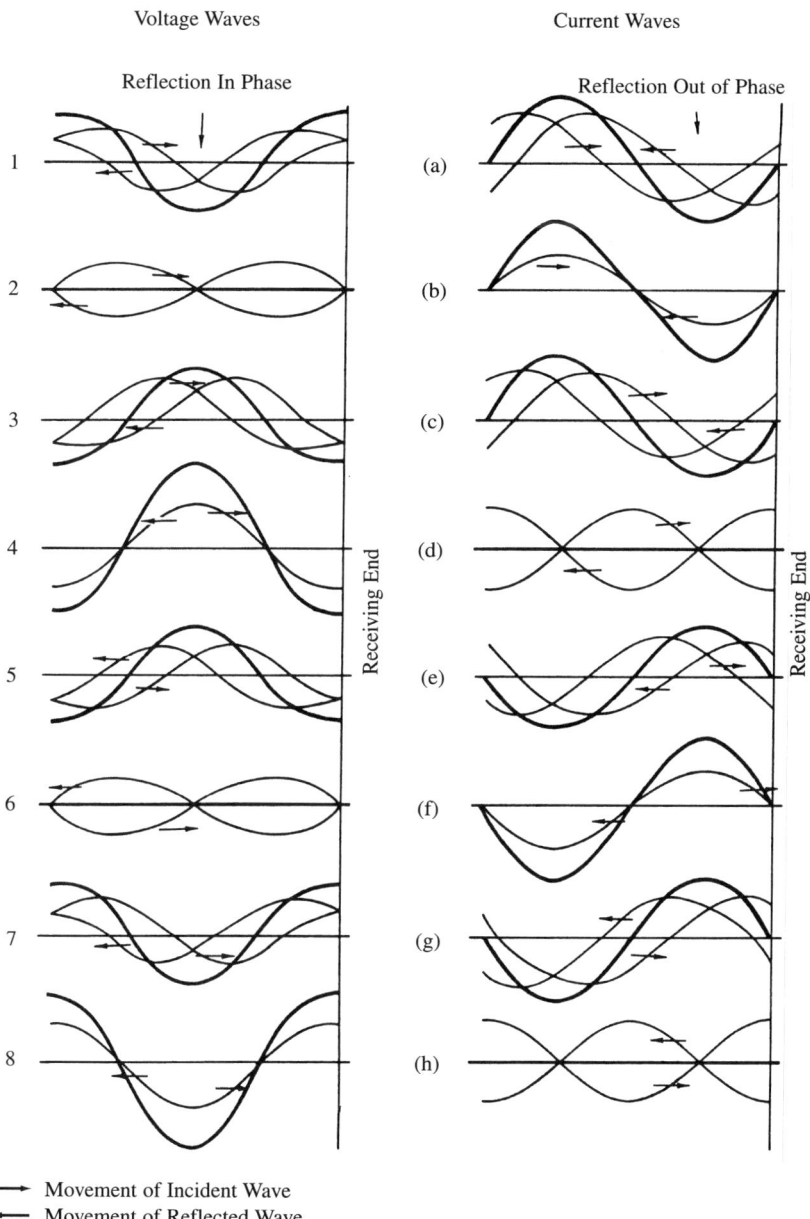

Figure 2.7 Instantaneous values of incident and reflected waves on a line terminated by an open.

If a line is terminated by a short circuit, accomplished by attaching a piece of heavy metal across the termination, then the incident and reflected voltages are equal and 180° out of phase and, thereby, cancel. Voltage is zero at the short and every half-wavelength thereafter. Incident and reflected current waves are in phase and maximum at the short. VSWR is ∞, while Γ is −1.

Figure 2.8 summarizes the effects of various terminations on standing waves. Terminations in Z_O cause a constant voltage or current to be measured anywhere along the line, providing a lossless line is used. This is shown in Figure 2.8a. With a lossy line, the voltage and current diminish as they move down the line toward the load. This is shown in Figure 2.8b. In an open-circuited line (Figure 2.8c), the voltage is maximum at the load, but the current is minimum. When the load is a short (Figure 2.8d), the voltage is zero at the load and the current is maximum. When the load is resistive but not equal to Z_O (Figures 2.8e and 2.8f), some energy is absorbed and some is reflected. If the line is terminated in capacitance (Figure 2.8g), the capacitor does not absorb energy but returns it all to the circuit. This means

Figure 2.8   Effects of various terminations on standing waves.

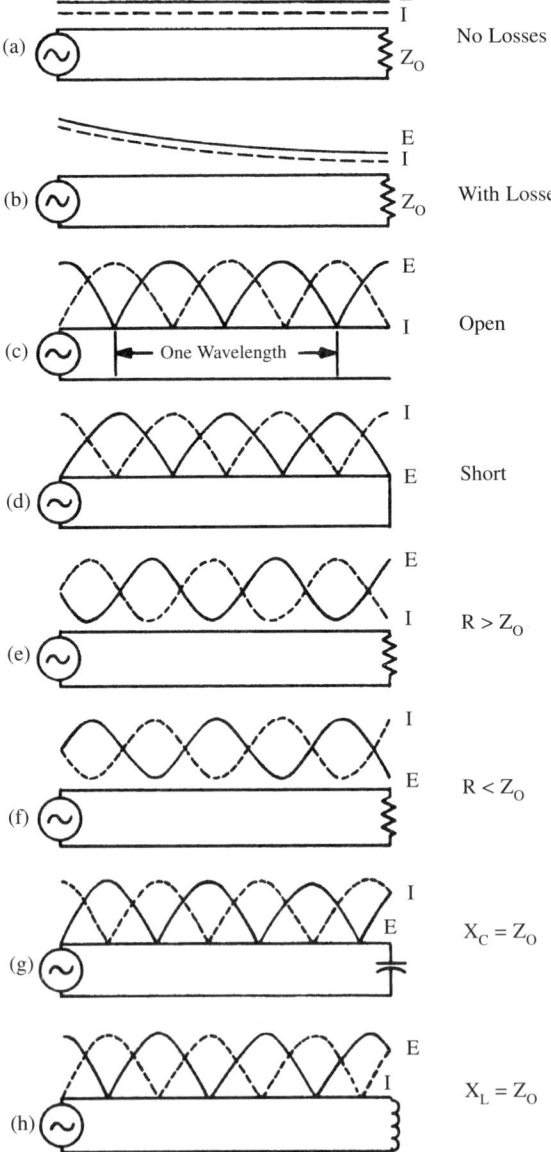

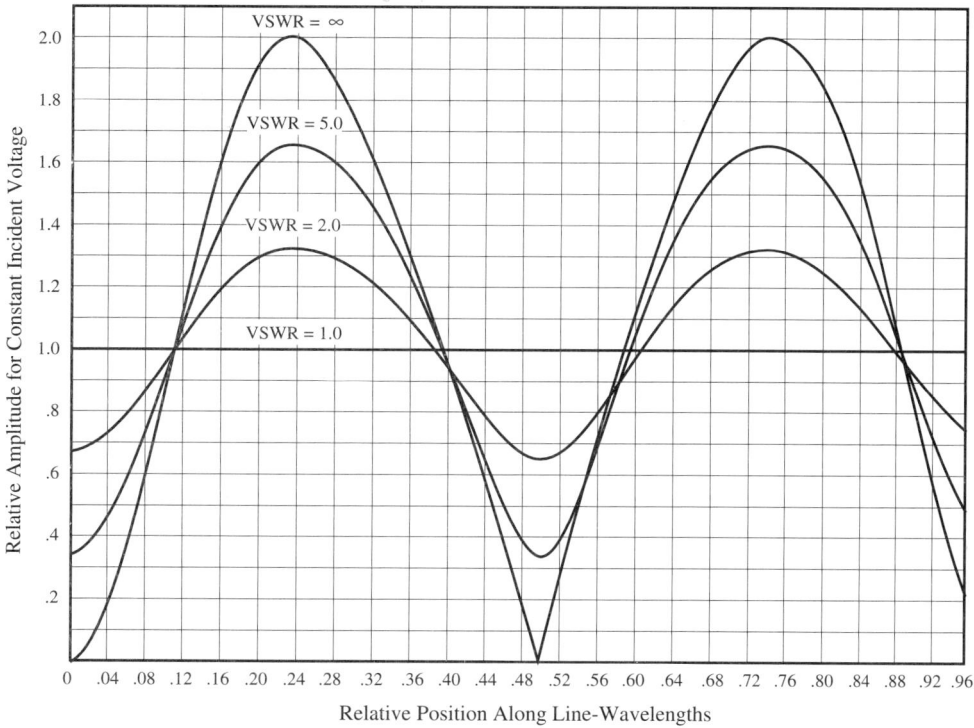

Figure 2.9   Relative amplitudes of various standing waves.

100% reflection. The voltage and current are phase shifted at the load due to the capacitance and Z_O connected in series with it. Figure 2.8h shows an inductive load. When X_L and Z_O are equal, the resulting standing waves are as shown, exactly the opposite of those of the capacitive load.

Figure 2.9 shows the relative amplitudes of various standing-wave ratios, from a perfect match (VSWR = 1) to a shorted termination (VSWR = ∞).

You may be asked to replicate this graph when doing lab exercises. Shorted terminations are often done in the lab to do various types of measurements. These include determining standing waves and the VSWR and determining an unknown load impedance.

2.10   TRANSMISSION LINE SECTIONS

When a quarter-wave ($\lambda/4$) section of a transmission line is open at the end, the voltage at the end is high and the current is low. At the sending end the voltage is low and the current is high. The quarter-wave section effectively inverts the voltage and the current. The impedance at each end of the line is opposite (inverted) with the receiving end acting like a short and the sending end acting like an open.

When a quarter-wave section is terminated in a resistance less than Z_O, the section inverts it to resemble a resistance greater than Z_O. Conversely, when terminated in a resistance greater than Z_O, the section inverts it to resemble a resistance less than Z_O. Both the voltage and current values at the terminations are inverted as well. This is shown in Figures 2.10 a and 2.10 b.

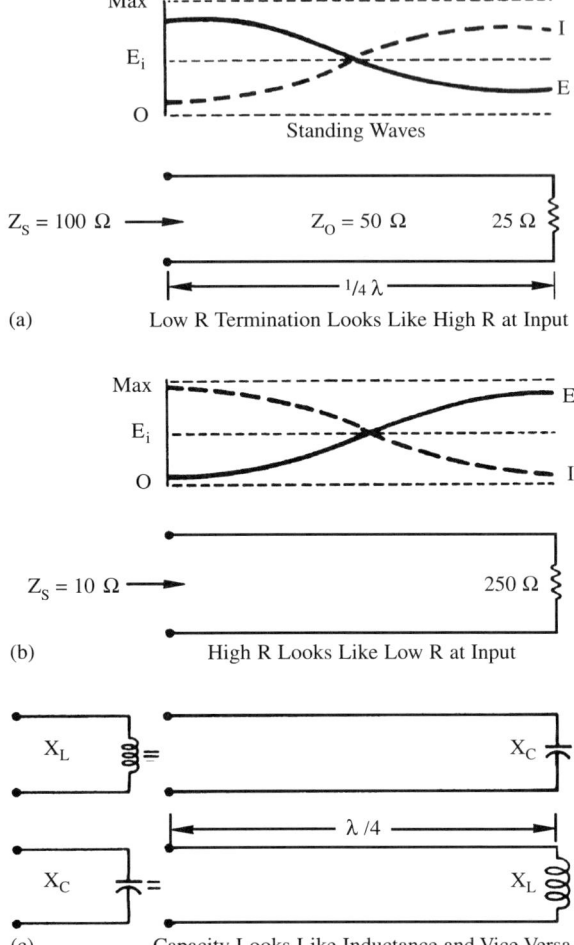

Figure 2.10 Impedance inversion with quarter-wave (λ/4) section.

Mathematically, the quarter-wave section can be expressed by:

$$Z = Z_O^2/Z_L$$ **(Eq. 2.19)**

where

Z = impedance looking into the section

Z_L = load impedance

Z_O = characteristic impedance of the section

To avoid confusion when the section is inserted into the main transmission line, let us replace Z with the term Z_O, since the idea is to have the input impedance of the section equal to the characteristic impedance of the main line. The term Z_O in the above equation can be replaced by Z'_O to emphasize that this is the impedance of the λ/4 section. Equation 2.19 then becomes:

$$Z_O = Z'^2_O/Z_L$$

To match a load impedance Z_L to a line impedance Z_O, the λ/4 section must have an impedance given by:

$$Z'_O = \sqrt{Z_O Z_L}$$ **(Eq. 2.20)**

EXAMPLE 2.15:

Find the impedance and length of a λ/4 section to match a 100 Ω to a 50 Ω line at a frequency of 500 MHz.

$$Z'_O = \sqrt{(50 \times 100)}$$
$$= 70.7 \ \Omega$$

The length of line depends on the dielectric. Assuming a dielectric with a velocity factor of 0.66:

$$\lambda = .66 c/f$$
$$= (0.66 \times 3.0 \times 10^8)/(500 \times 10^6)$$
$$= 0.396 \ \text{m}$$

The line must be a λ/4 long, so its length is:

$$L = 0.25 \times 0.396$$
$$= .099 \ \text{m}$$

Effectively the λ/4 section is an impedance-matching transformer at *one frequency only*. This inverting property also holds for reactance. If the section is terminated in capacitive reactance, the input is inductive reactance. If the section is terminated in inductance, the input displays all the characteristics of a capacitor, as illustrated in Figure 2.10c.

When the section of a transmission line is a half-wave long, the situation resembles two connected λ/4 sections, as shown in Figure 2.11a. The first λ/4 section inverts the terminal impedance and the second inverts it again, making the input impedance the same as the terminating impedance.

In a short-circuited λ/2 section (Figure 2.11b), the input is shorted also. This is equivalent to a series resonant circuit. If open circuited, the λ/2 section input is also open. This is equivalent to a parallel resonant circuit. This is shown in Figure 2.11c. When the λ/2 section is terminated in resistance (other than Z_O), in capacitance, or in inductance, it always repeats this impedance at the sending end (Figures 2.11d–f).

2.11   TIME-DOMAIN REFLECTOMETRY

With all the transmission lines that are laid underground or under water for a variety of communications applications, what happens when a loose connection or break occurs and one attempts to determine where the fault is located? The techniques of time-domain reflectometry (TDR) come to the rescue.

TDR utilizes a step or pulse input applied to the input of a transmission line. By examining the reflected signal, a good deal of information can be gained about such things as faulty splices or connectors, water leaking into the cable, kinks, breaks, and so forth.

A pulse generator and a high-speed oscilloscope can be connected as shown in Figure 2.12. However, dedicated TDR units are available, as well as optical TDR units for fiber-optic cable.

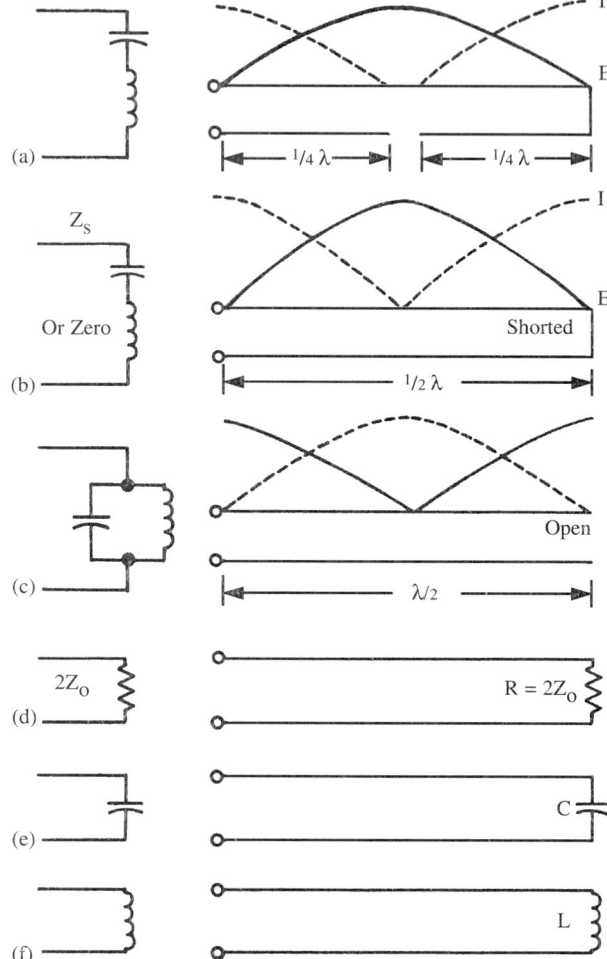

Figure 2.11   Half-wave ($\lambda/2$) section reflects same impedance as termination.

The most important use of TDR is to determine the position and types of defects on a line. With TDR the distance to the defect is easy to determine from the time the reflection takes to return to the source. The type of defect can be gauged to some extent from the nature of the reflected signal.

Figure 2.13 shows several types of reflections. In Figure 2.13a, the line is open-circuited, and the voltage rises to double the initial value. In Figure 2.13b, the termination is a short circuit, causing the final voltage to fall to zero. Reactive terminations can be distin-

Figure 2.12   TDR setup.

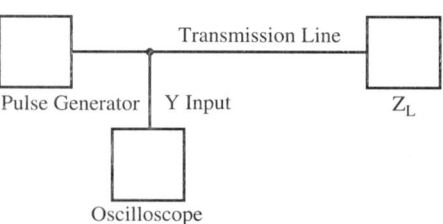

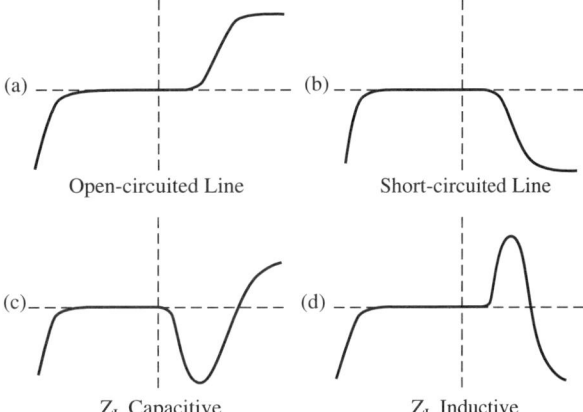

(a)

Open-circuited Line

(b)

Short-circuited Line

(c)

Z_L Capacitive

(d)

Z_L Inductive

Figure 2.13   TDR reflections.

guished as well. A capacitance initially appears as a short circuit, then becomes more like an open circuit as it charges, as can be seen in Figure 2.13c. On the other hand, an inductance initially appears as an open circuit, then gradually draws more current until it eventually appears as a short (Figure 2.13d).

If there is no fault on the line, the signal is absorbed by the termination, no signal returns to the instrument, and no reading is displayed.

A dedicated waveform type TDR is shown in Figure 2.14. The 1502B metallic cable tester has a fast rise time step pulse so that identification of multiple faults as close together as 0.6 inches can be determined. Figure 2.15 shows this unit being used in a field application of cable testing.

Until the 1980s the waveform TDR was the only type available. The advent of digital numeric instruments has had an impact on the popularity and usage of the TDR. A digital TDR is easier to operate and is less costly, but it is limited in both range and capability. It also displays only the distance to the fault.

Lengths of cable that can be tested may vary from as little as a few feet to more than 50,000 feet. However, in reality, most tests are from pedestal to pedestal, amplifier to amplifier, tap to tap, or alley to house.

Figure 2.14   Time-domain reflectometery (TDR) for metallic cables. (Photo courtesy of Tektronix, Inc.)

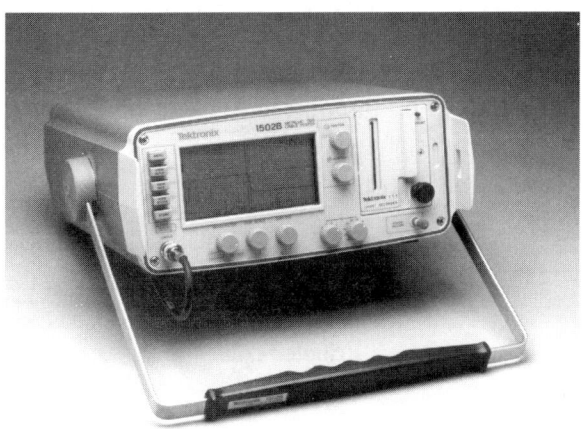

Figure 2.15 Field application using the TDR. (Photo courtesy of Tektronix, Inc.)

Some practical considerations that help reduce errors when using a TDR are:

1. Maintain a constant impedance by using connectors and adapters designed for the same impedance as the cable you are testing.
2. Know the velocity factor (v_f) of the cable and adjust the TDR for that value.
3. Vary the pulse width as necessary for the type and length of cable being tested.
4. Start with the shortest pulse width or range. If a fault is not seen, switch to the next larger pulse width and retest.
5. If the fault is located in the first 5 to 15 feet of cable, the fault is sometimes difficult to locate. Adding a length of cable (properly connected) between the TDR and the cable under test can help uncover these hidden faults.
6. Testing a cable from *both* ends helps reduce error.

2.12   SUMMARY

1. A simple transmission line couples a source component to a load component.
2. Transmission line losses are due to copper losses, dielectric losses, and radiation losses.
3. The characteristic impedance of a two-wire transmission line can be found from its distributed constants.
4. Free space is a form of a transmission line and has a characteristic impedance of 377 Ω.
5. Matched terminations result in maximum transfer of power between the line and the load.

6. Standing waves result when an incident wave is reflected from a load termination and algebraically adds to the incident wave. This happens because the respective impedances don't match and phase shifting occurs.

7. Reflection coefficient Γ indicates both the magnitude and phase angle. The range for Γ is from 0 to 1.

8. Return loss is the ratio of incident power to reflected power. A perfect match has a return loss of infinity.

9. The VSWR for a perfect match is 1.0. On a practical system, a VSWR of 1.2 or less is considered acceptable.

10. An open termination results in a VSWR $= \infty$, $\Gamma = 1$, with the voltage wave reflected in phase.

11. A shorted termination results in a VSWR $= \infty$, $\Gamma = -1$, with the voltage wave reflected 180° out of phase.

12. A $\lambda/4$ section is an impedance-matching transformer at one frequency only.

13. A $\lambda/2$ section repeats the terminating impedance at the sending end.

14. TDR is a technique to determine faulty splices, connectors, water leakage, kinks, breaks, and so forth, in transmission lines.

15. The most important use of TDR is to determine the position and types of defects on a line.

Key Equations:

$Z_O = \sqrt{L/C}$	**(Eq. 2.1)**
$Z_O = \sqrt{\mu/\epsilon}$	**(Eq. 2.2)**
$v = 1/\sqrt{LC}$	**(Eq. 2.3)**
$v_f = 1/\sqrt{\epsilon_r}$	**(Eq. 2.4)**
$\Gamma = V_r/V_i$	**(Eq. 2.5)**
$\Gamma = (Z_L - Z_O)/(Z_L + Z_O)$	**(Eq. 2.6)**
$\Gamma = (R_L - R_O)/(R_L + R_O)$	**(Eq. 2.7)** (Resistive load)
Power reflection coefficient $= \rho^2$	**(Eq. 2.8)**
$RL = P_{in}/P_{ref}$	**(Eq. 2.9)**
$RL_{(dB)} = 10 \log RL$	**(Eq. 2.10)**
$RL = 1/\rho^2$	**(Eq. 2.11)**
$RL_{(dB)} = -20 \log \rho$	**(Eq. 2.12)**
$VSWR = V_{max}/V_{min}$	**(Eq. 2.13)**
$SWR_{(dB)} = 20 \log VSWR$	**(Eq. 2.14)**
$VSWR = (1 + \rho)/(1 - \rho)$	**(Eq. 2.15)**
$\rho = (VSWR - 1)/(VSWR + 1)$	**(Eq. 2.16)**
$Z_{max(\Omega)} = (VSWR)Z_O$	**(Eq. 2.17)**
$Z_{min(\Omega)} = Z_O/VSWR$	**(Eq. 2.18)**
$Z = Z_O^2/Z_L$	**(Eq. 2.19)**
$Z'_O = \sqrt{Z_O Z_L}$	**(Eq. 2.20)**

PROBLEMS

1. Determine the characteristic impedance of a two-wire line when $L = 6.5\ \mu H/m$ and $C = 8.7\ pF/m$.

2. A transmission line has a distributed inductance of $11.3\ \mu H/m$. Its characteristic impedance is $250\ \Omega$. What is the value of the line's distributed capacitance?

3. Determine the velocity of a wave in the transmission line of problem 1.

4. Calculate Γ if the incident voltage is 0.9 and the reflected voltage is 0.1.

5. Calculate Γ if $R_L = 12\ \Omega$ and $R_O = 50\ \Omega$.

6. Determine Γ when $R_O = 50\ \Omega$ and $R_L = 95\ \Omega$.

7. Determine Γ when $Z_O = 50\ \Omega$ and $Z_L = 33 - j21\ \Omega$.

8. Determine the power reflection coefficient for problem 6.

9. Determine RL when the incident power is 10 mW and the reflected power is 0.3 mW. Convert this return loss into dB also.

10. Calculate RL from the Γ calculated in problem 7. Convert this return loss to dB using equation 2.8 also.

11. Convert the Γ of problem 7 directly into $RL_{(dB)}$.

12. Determine the VSWR when $V_{max} = 84.84$ V and $V_{min} = 56.56$ V. Convert this VSWR into dB also.

13. Determine the VSWR from the given values in problem 7.

14. Determine ρ when the VSWR = 1.5.

15. Determine Z_{max} and Z_{min} on a transmission line when $Z_O = 75\ \Omega$ and the VSWR is the same as in problem 13.

16. Determine Z'_O for a $\lambda/4$ section when $Z_O = 50\ \Omega$ and $Z_L = 100\ \Omega$.

QUESTIONS

1. Describe the term *transmission line*.

2. What are the commonly used types of transmission lines?

3. Discuss the advantages and disadvantages of the common transmission lines.

4. Describe the common losses associated with transmission lines.

5. Define the term *characteristic impedance*.

6. Describe how to determine the Z_O of a two-wire line.

7. What are the *distributed constants* associated with the two-wire line.

8. Discuss the importance of matched terminations.

9. What is the significance of an infinitely long transmission line?

10. How does the propagation velocity of an EM wave in a transmission line compare to the velocity of an EM wave in a medium other than free space?

11. Discuss the concept of *standing waves*.

12. What are typical sources for a signal generator on a line?

13. Describe the effects on Γ when $R_L < R_O$ and when $R_L > R_O$.

14. What is a perfect VSWR? An acceptable VSWR?

15. Why are shorts placed on a transmission line?

16. Describe the effects on standing waves for different terminations.

17. Describe the effects of the $\lambda/4$ section; the $\lambda/2$ section.

18. Describe TDR techniques and uses.

19. What practical considerations should be implemented when using a TDR?

3 SMITH CHART ANALYSIS

OBJECTIVES

1. To describe the functions of the Smith chart.
2. To solve transmission line problems using the Smith chart.
3. To demonstrate the use of the radial scales to solve for unknown transmission line parameters.
4. To describe matching techniques used to match a transmission line to a load.
5. To compare and contrast methods of matching techniques.

3.1 INTRODUCTION

Before continuing with the study of the other common forms of microwave transmission lines, we'll take a look at an instrument that can be used to simplify transmission line calculations. The formulas used in chapter 2 provide relatively easy solutions to transmission line problems. But when more complex problems are posed, such as determining impedance matches or determining the value of unknown load impedances, this instrument can offer solutions without the need for advanced mathematics such as trigonometry or calculus.

The instrument is called the Smith®[1] chart, named after and first published by P. H. Smith in 1939. The Smith chart shown in Figure 3.1 was introduced in 1944 and is a conformal transformation of a reactance chart. It pulls the infinite points of reactance around until they join at a common "infinity" point on the right-hand side of the chart. Described another way, the chart is a special kind of impedance coordinate system, mechanically arranged so that the relationships of impedances can be determined at any point along a transmission line. All of the solutions are found through graphical analysis.

---

[1]Smith is a registered trademark of the Analog Instruments Co., P.O. Box 808, New Providence, NJ 07974.

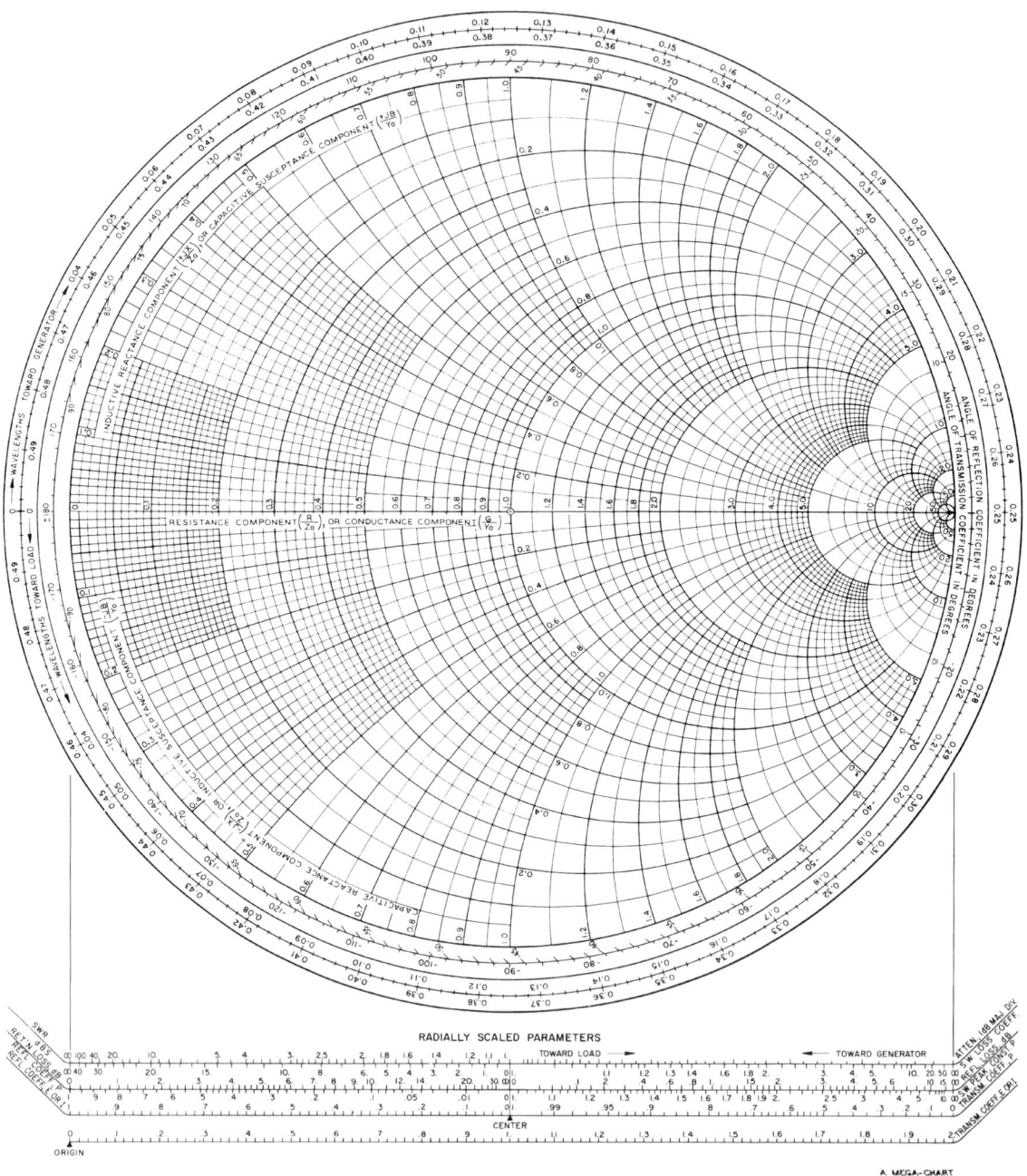

NAME	TITLE	DWG. NO.
SMITH CHART FORM 82-BSPR (9-66)	KAY ELECTRIC COMPANY, PINE BROOK, N.J., ©1966. PRINTED IN U.S.A.	DATE

IMPEDANCE OR ADMITTANCE COORDINATES

Figure 3.1   The Smith chart. (Courtesy of Analog Instruments Co., P.O. Box 808, New Providence, NJ 07974.)

Blank Smith charts and a plastic transmission line calculator are available from:

Analog Instruments Company
P.O. Box 808
New Providence, NJ 07974
(908) 464-4214

All values of resistance and reactance on a Smith chart are normalized, meaning the load impedance to be plotted is divided by the characteristic impedance of the line, Z_L or R_L/Z_O. This effectively converts the chart to a "universal chart" that can be used with any combination of Z_O and Z_L. Multiplication by the characteristic impedance of the line converts these normalized values back to actual values. The Smith chart shown in Figure 3.1 is a universal chart.

Circles tangent to the right (infinity) side of the chart are circles of constant resistance. The constant resistance circles for normalized resistances of 0, 0.3, 1.0, 3.0, and 10.0 are shown in Figure 3.2.

The horizontal centerline represents the resistive part of a load. Loads can be purely resistive ($R\pm j0$), purely reactive ($0\pm jX$), or complex ($R\pm jX$). The centerline (diameter) also represents conductance values. On this line, $R = 0$ or $G = 0$, is at the left end, while $R = \infty$ or $G = \infty$, is at the right end.

Curved lines starting from the right (infinity) side of the chart and going above and below the centerline are the reactance coordinates. They form constant reactance circles. The horizontal centerline represents zero reactance to these values. Curved lines going above the centerline represent positive series reactance (inductance); lines going below the centerline represent negative series reactance (capacitance). Note that the normalized-reactance values are labeled along the outer limits of the reactance lines. Positive and negative reactance coordinates are emphasized in Figure 3.3 for values of 0, ± 0.3, ± 1.0, ± 3.0, and ± 10.0.

Any specific value of impedance can be located on the Smith chart by locating the proper coordinate positions. For example, the normalized impedance of a line terminated in its characteristic impedance is $1.0+j0$. On the chart this point is at 1.0 on the horizontal centerline. This indicates a perfect match and represents a single point.

To plot other normalized values you can go back to the familiar rectangular coordinate graph from your AC electronics background as a reference. Recall that resistive values are plotted to the right on the x-axis labeled $+R$, while inductive values (X_L) are plotted on the y-axis labeled $+j$, and capacitive values (X_C) are plotted on the y-axis labeled $-j$. Some examples are shown in Figure 3.4a. The corresponding values are also shown in Figure 3.4b plotted on the Smith chart.

As a further guide, several points have been plotted on a Smith chart (Figure 3.5) to show relative impedance values. Point A indicates a perfect match, where $Z_O = R_L$. Point B represents a zero (shorted) load impedance, while point C represents an infinite (open) load impedance. Points D and E represent loads of pure resistance. Point D shows a load resistance less than the characteristic impedance of the line ($R_L < Z_O$), while point E represents a load resistance that is greater than Z_O, ($R_L > Z_O$). Point F represents a complex load having both resistance and inductive reactance ($R+jX_L$), while point G shows a complex load having both resistance and capacitive reactance ($R-jX_C$). (Note: Complex loads are always shown in the rectangular coordinate form for use on a Smith chart.)

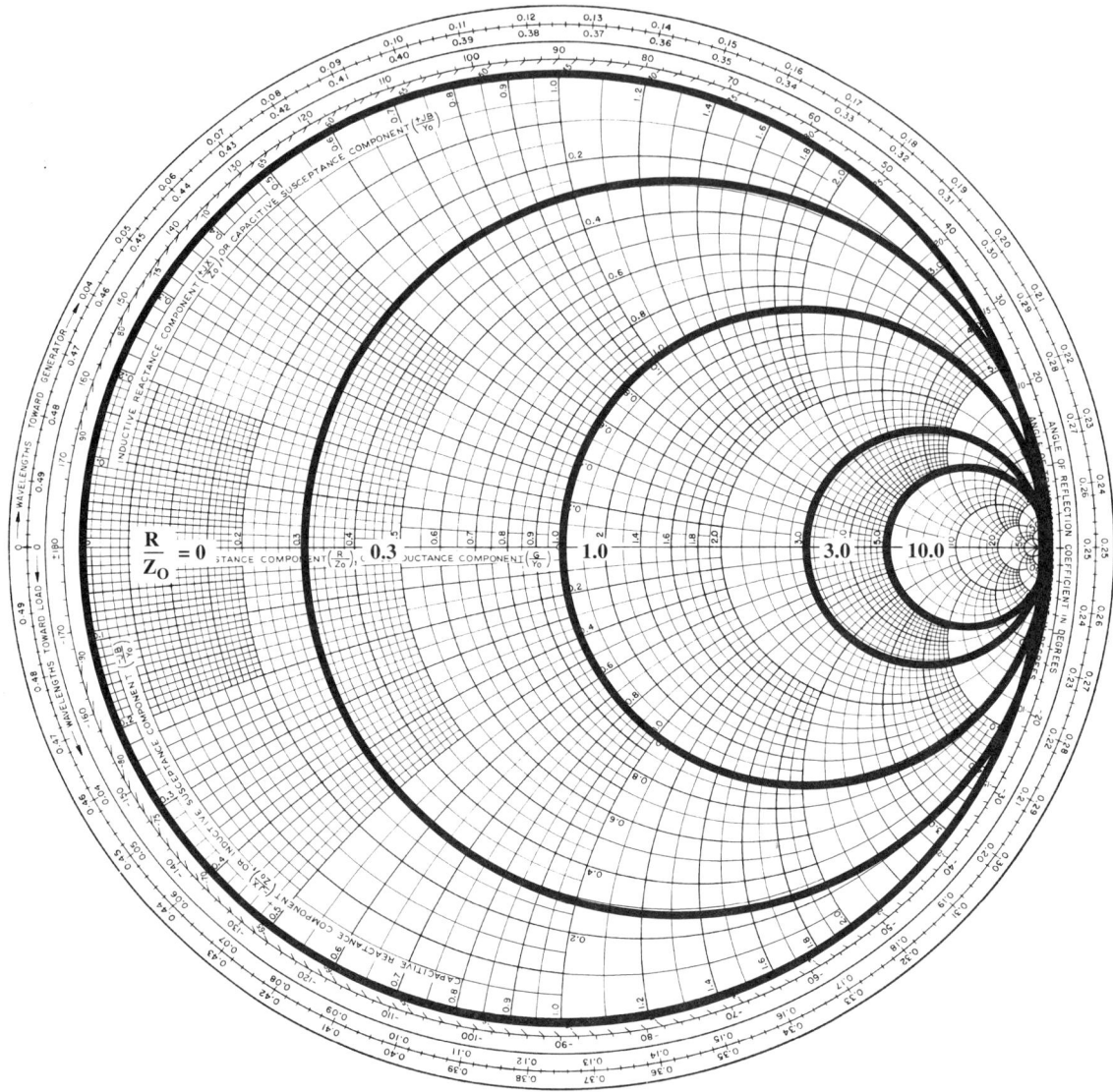

Figure 3.2 Constant resistance circles. (Courtesy of Analog Instruments Co., P.O. Box 808, New Providence, NJ 07974.)

One of the inherent characteristics of the Smith chart is that any lossless uniform line is plotted as a circle about the point 1+j0, where the radius of the circle is determined by the known amount of impedance at any point in the line. Thus, a single impedance value is all that is required to enter the chart. This circle is also known as the *constant VSWR* circle, since any impedance indicated by this circle produces the same VSWR. The phase relationship along a transmission line is represented by movement around the chart on one of the concentric circles centered at 1+j0, and the electrical distance on the line is also directly related to the distance around the circle. To allow convenient measurement of circular

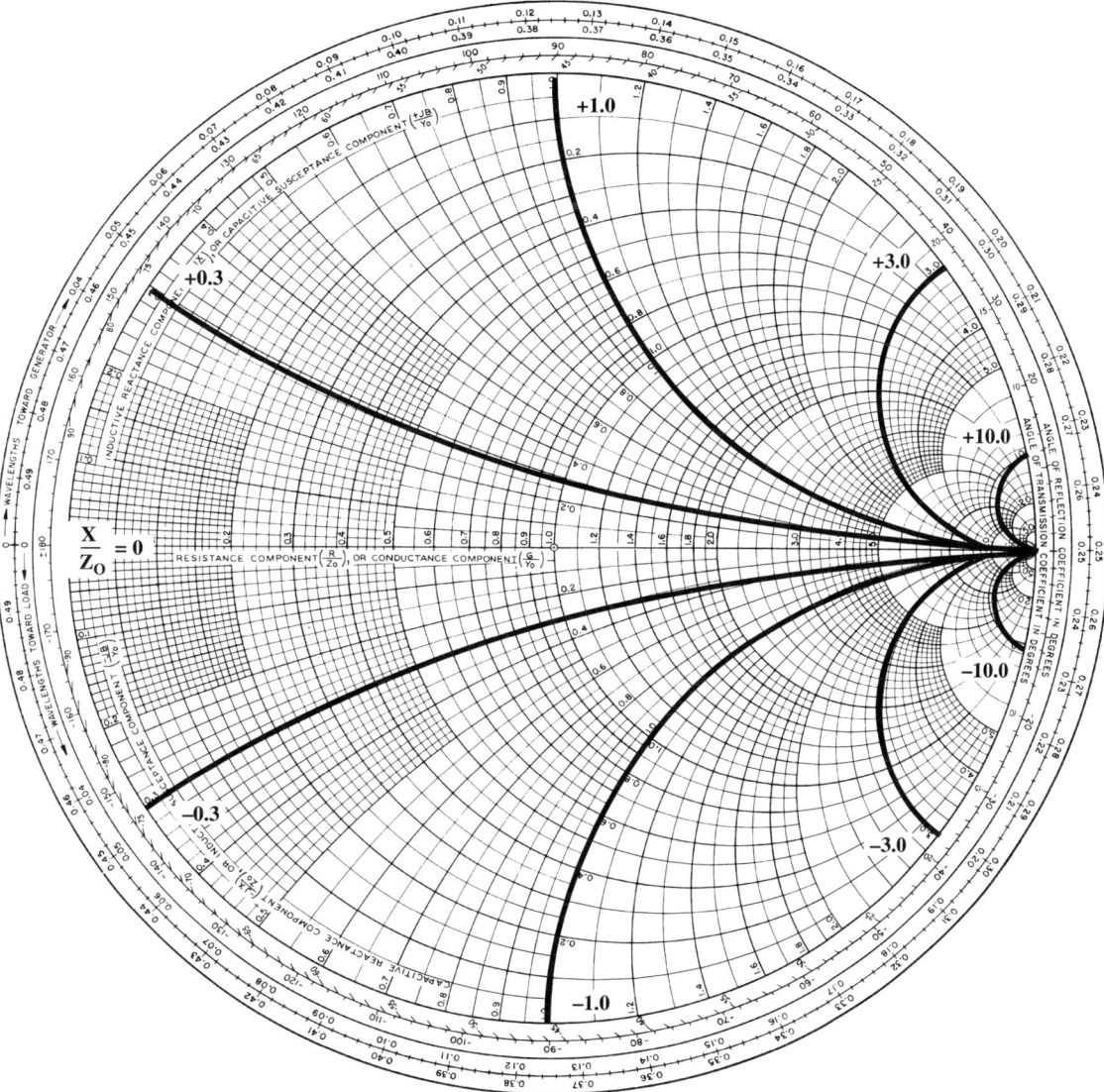

Figure 3.3 Positive and negative reactance coordinates. (Courtesy of Analog Instruments Co., P.O. Box 808, New Providence, NJ 07974.)

movement, the chart has two circular scales on its outer edge. One is calibrated in fractional wavelength; the other is in degrees.

The wavelength scale shows that a complete revolution on the chart is equivalent to a half wavelength. Thus, 180 electrical degrees on the transmission line are represented by 360 degrees of revolution on the chart. The degree scale shows that in a complete revolution of the chart the reflection coefficient goes through a complete cycle of 180 degrees positive and 180 degrees negative. These scales are important, because the impedance varies cyclically along a line terminated by a value different from its characteristic impedance. They are

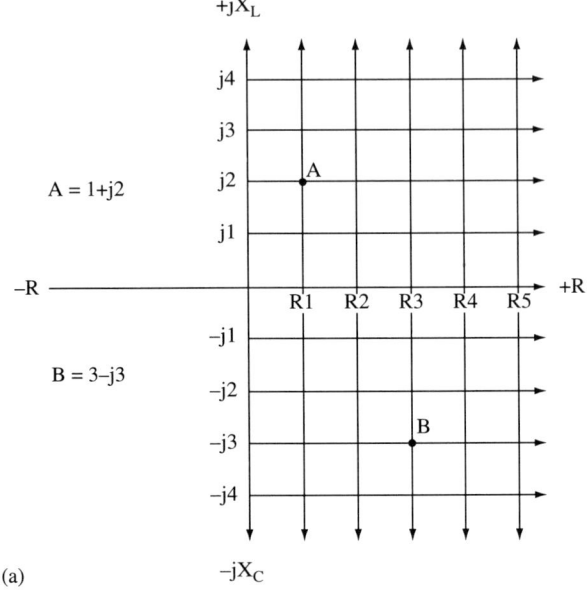

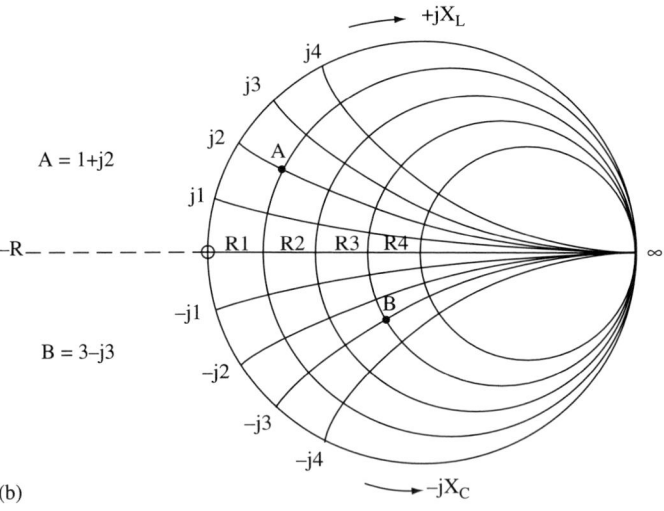

Figure 3.4 Plotted normalized impedance values. (a) Rectangular coordinator graph. (b) Smith chart (chart not to scale).

used to determine the impedance at various points along a line after the impedance is determined for any one specific point. The entire variation in impedance on a line is repeated cyclically in each half wavelength along the line.

Some example problems are in order to clarify the usefulness of the Smith chart. These are covered in the next section.

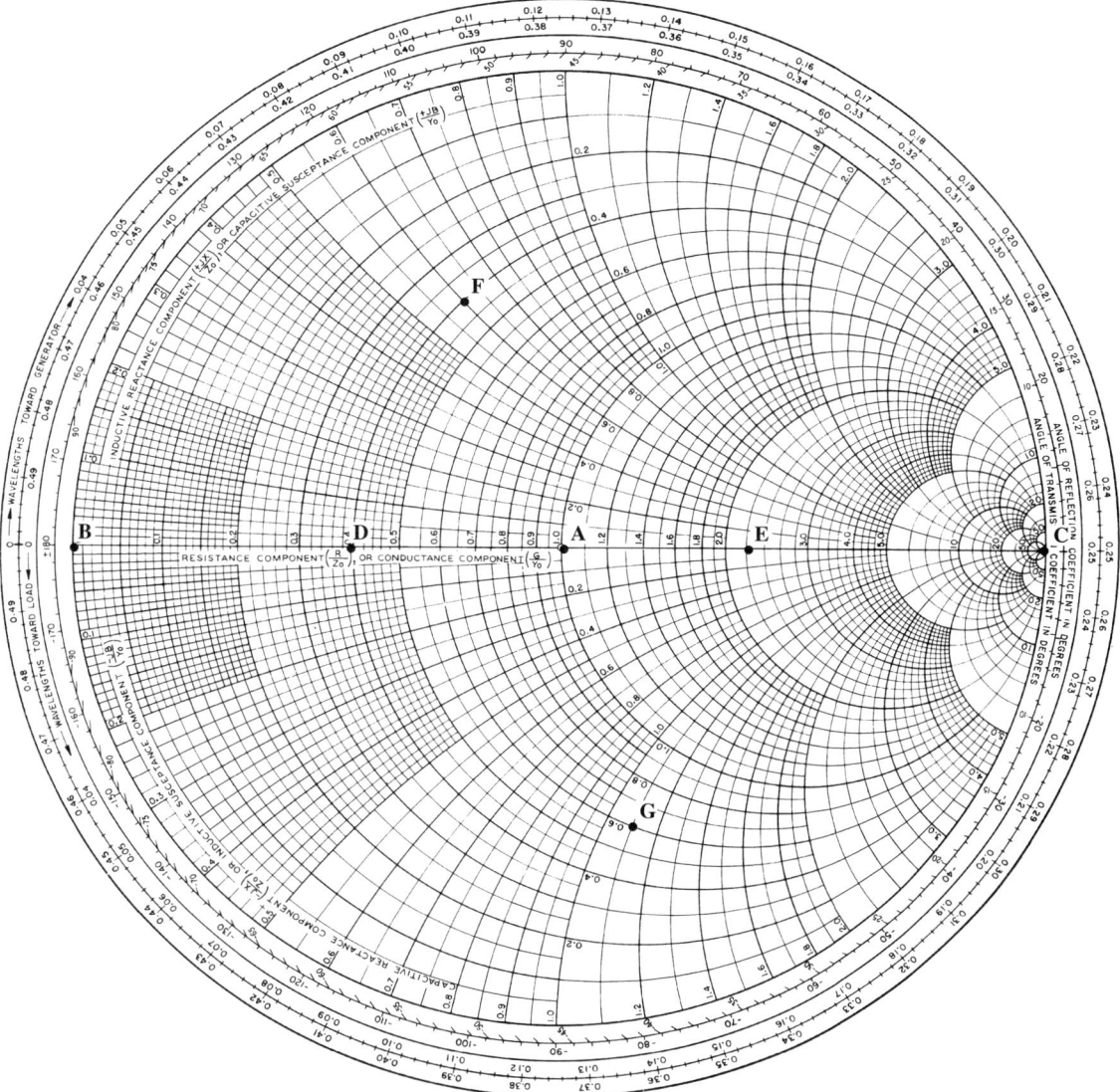

Figure 3.5 Various loads plotted on a Smith chart. A: Perfect match; B: Zero (shorted) load impedance; C: Infinite (open) load impedance; D: Pure resistance ($R_L < Z_O$); E: Pure resistance ($R_L > Z_O$); F: Complex load with resistance and inductive reactance; G: Complex load with resistance and capacitive reactance. (Courtesy of Analog Instruments Co., P.O. Box 808, New Providence, NJ 07974.)

3.2   USING THE SMITH CHART

The Smith chart can be used to solve a plethora of transmission line problems. These include:

1. Plotting resistive, reactive, and complex loads.
2. Finding Γ for a given Z_L.

3. Finding VSWR for a given Z_L.
4. Finding the input impedance to a terminated load that may be shorted, open, or complex.
5. Locating the distance to the minimum and maximum points of a standing wave at a distance from any termination.
6. Determining the value of an unknown impedance from a given or measured VSWR. (This requires a lab exercise utilizing the short-circuit minima shift method to facilitate the solution).
7. Finding transmission coefficients for a given Z_L.
8. Finding return loss for a given Z_L.
9. Finding an admittance value for a given Z_L.
10. Finding Z_{max} and Z_{min} for a given Z_L.
11. Determining the relative λ position of a given load.
12. Determining the impedance on a line from a given distance from the load.
13. Matching line terminations to the line using a $\lambda/4$ transformer.
14. Matching line terminations to the line using single- and double-stub tuners or slide-screw tuners.

This list is not inclusive of all possible transmission line problems that can be solved on the Smith chart. For practical as well as time constraints, *not* all of those on the list will be shown by example.

EXAMPLE 3.1:

Plot the impedance of $Z_L = 25+j50\ \Omega$ when $Z_O = 50\ \Omega$.

1. Normalize the impedance as follows:

$$Z_{L(norm)} = Z_L/Z_O = (25/50+j50/50)$$
$$= 0.5+j1.0$$

2. Start at the left end of the centerline ($R = 0$) and move to the right to the resistance circle labeled 0.5.
3. Follow the resistance circle up until it crosses the reactance arc marked 1.0. This point represents $Z_L = 0.5+j1.0$ and is marked A in Figure 3.6.

EXAMPLE 3.2:

Determine the VSWR that results when $Z_L = 50+j50\ \Omega$ if $Z_O = 50\ \Omega$.

1. Plot the normalized impedance. (Remember, just divide Z_L by Z_O.) $Z_{L(norm)} = 1.0+j1.0$. This is point A in Figure 3.7.
2. With a compass, draw a circle centered at $1+j0$ having a radius equal to the distance from the center to point A. This circle is called the constant VSWR circle; all impedances on this circle produce the same VSWR.
3. Read the value of the VSWR where the circle crosses the diameter line to the right of center. (VSWR = 2.6)

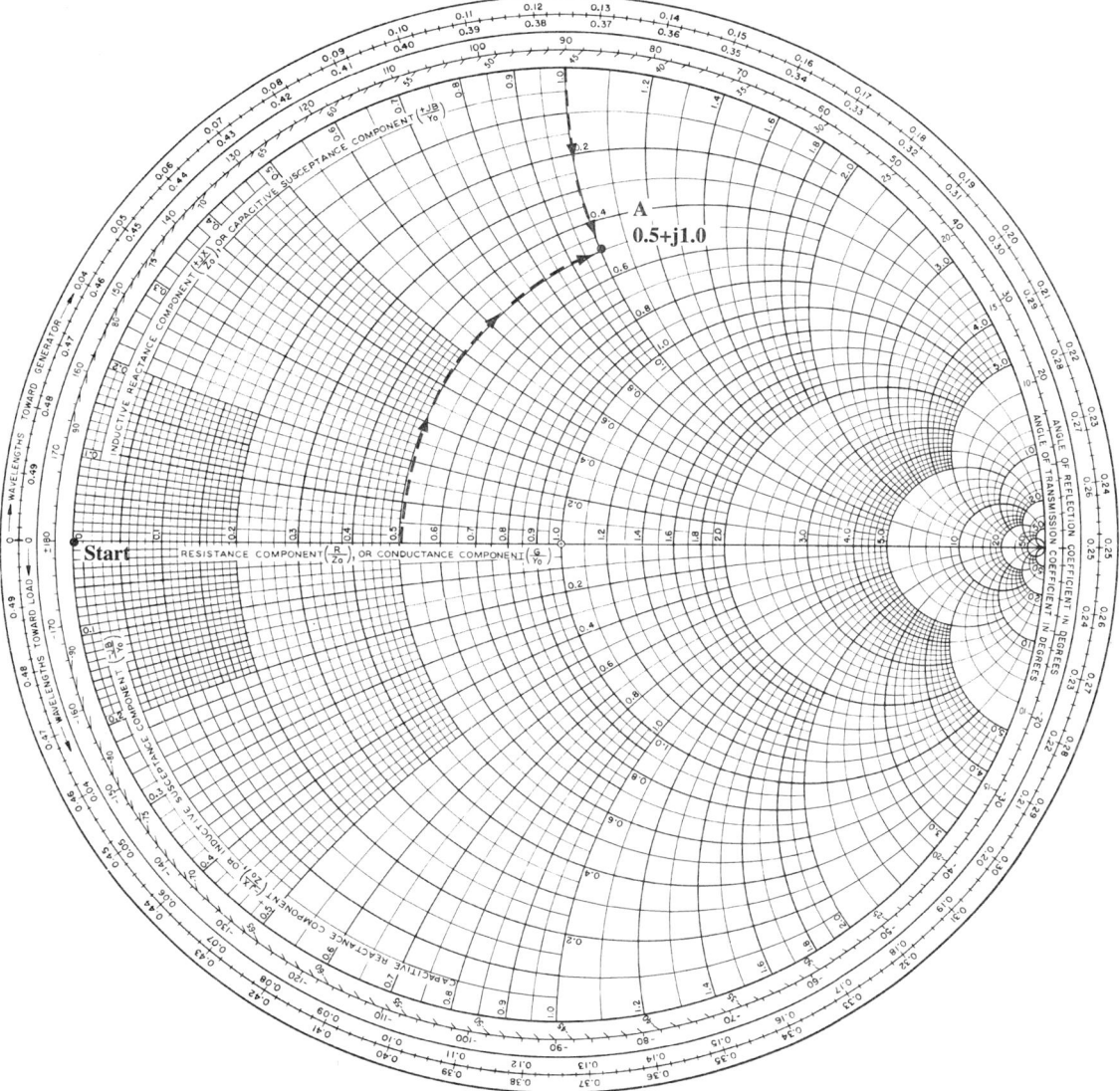

Figure 3.6 Plotting an impedance. (Courtesy of Analog Instruments Co., P.O. Box 808, New Providence, NJ 07974.)

An alternate method, yielding a more precise value of the VSWR, is to set a compass for the distance of the radius OA (the center of the VSWR circle to point A). Then transfer this distance to the radially scaled parameters directly below the chart. Place one end of the compass on the center point of the SWR scale and the other end of the compass to the left on the SWR scale. The numerals above the SWR line indicate the VSWR, while the numerals below the line indicate the SWR in dB. Read the VSWR directly off the scale. (VSWR = 2.6 or SWR_{dB} = 8.3 dB)

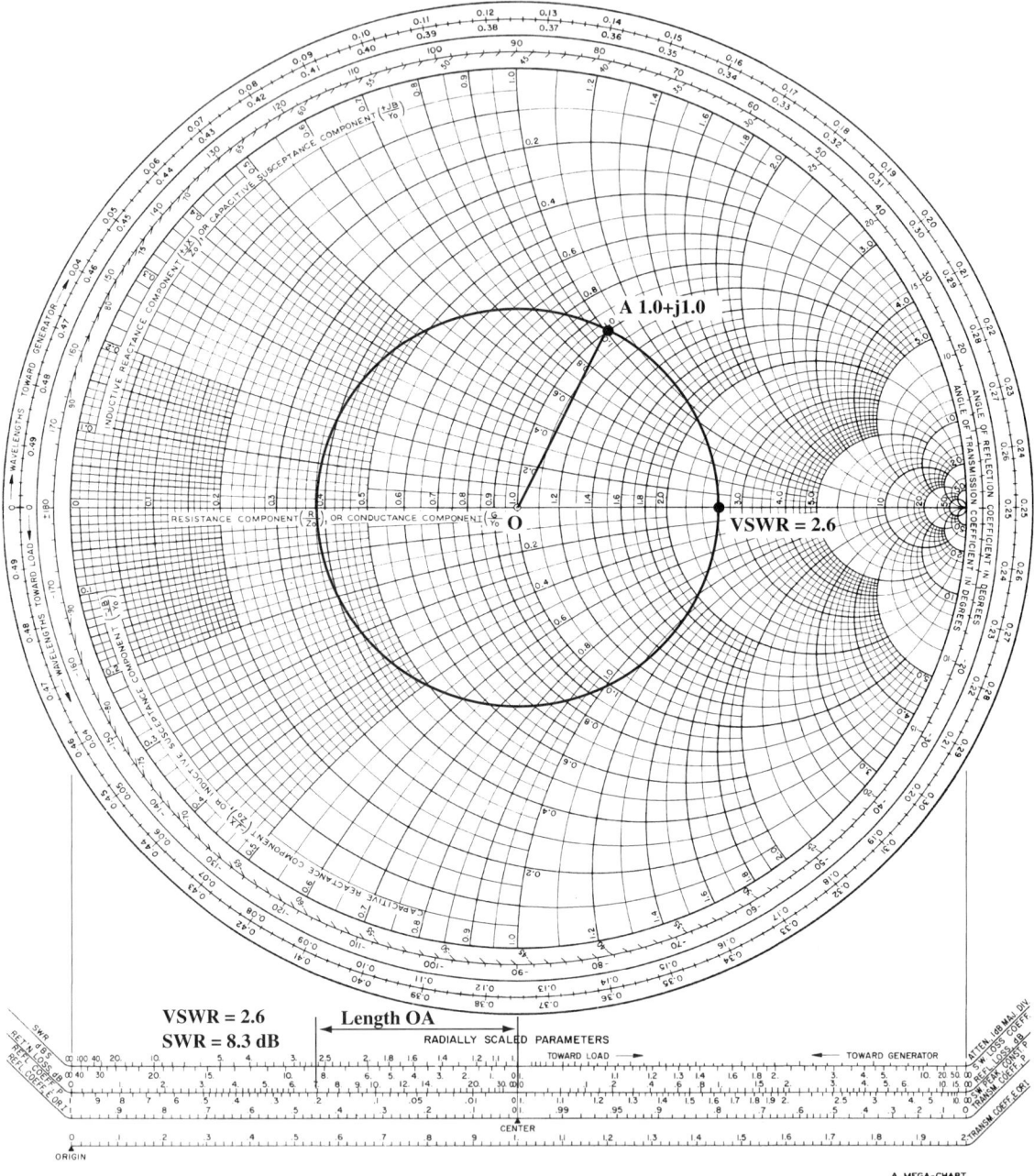

Figure 3.7 Determining VSWR. (Courtesy of Analog Instruments Co., P.O. Box 808, New Providence, NJ 07974.)

EXAMPLE 3.3:

Determine Γ for a transmission line with $Z_O = 100\ \Omega$ terminated by $Z_L = 100 - j200\ \Omega$.

1. Normalize and plot the load. $Z_{L(norm)} = 1.0 - j2.0$. This is point A in Figure 3.8. Draw the VSWR circle for this load.
2. Set the compass for the radius of the VSWR circle (length OA). Transfer this distance to the Voltage Reflection Coefficient scale, on the radially scaled parameters. (This is labeled Refl. Coeff., E or I.) One end of the compass is put at the center point; the other is placed to the left on the reflection coefficient scale. Read off the magnitude of Γ. ($\rho = 0.71$)
3. To get the phase angle for Γ, draw a line that starts from the center of the VSWR circle ($1+j0$) and passes through point A extending outward to the Angle of Reflection Coefficient in Degrees scale at the edge of the chart. Read the phase angle of reflection coefficient where this line crosses the angle value. ($\phi \approx -45°$)
4. Note that the power reflection coefficient ρ^2 can be found by transferring length OA to the Refl. Coeff., P scale just above the voltage reflection scale. Read the power reflection coefficient directly off the scale. ($\rho^2 \approx 0.5$)

EXAMPLE 3.4:

Determine the Z_{max} and Z_{min} on a transmission line when $Z_O = 50\ \Omega$ and $Z_L = 100 - j100\ \Omega$.

1. Normalize and plot the load. $Z_{L(norm)} = 2.0 - j2.0$. Draw the VSWR circle for this load. Refer to Figure 3.9.
2. Note the value where the VSWR circle crosses the diameter line on both the left and right sides.
3. At the right side of the circle is the location of Z_{max}, which is a function of this impedance value (which equals the VSWR) times $50\ \Omega$. You have to denormalize, therefore, $4.2(50) = 210\ \Omega$. ($Z_{max} = 210\ \Omega$)
4. At the left side is the location of Z_{min}, which is a function of this impedance value times $50\ \Omega$. You have to denormalize; therefore, $0.24(50) = 12\ \Omega$. ($Z_{min} = 12\ \Omega$)
5. Note that 4.2 is the VSWR for the line and 0.24 is the reciprocal of VSWR. This yields the same results for Z_{max} and Z_{min} as equations 2.17 and 2.18 of chapter 2.

EXAMPLE 3.5:

(Many Smith chart solutions require that a plotted point be in terms of admittance rather than impedance.)

Determine $Y_{L(norm)}$ for a given $Z_L = 50 + j100\ \Omega$ when $Z_O = 50\ \Omega$.

1. Plot the normalized impedance on the chart. ($Z_{L(norm)} = 1.0 + j2.0$) This is point A in Figure 3.10.
2. Plot the constant VSWR circle.
3. Diametrically opposite point A is point B, which is the normalized admittance $Y_{L(norm)}$. Note: This point is $\lambda/4$ away from point A.

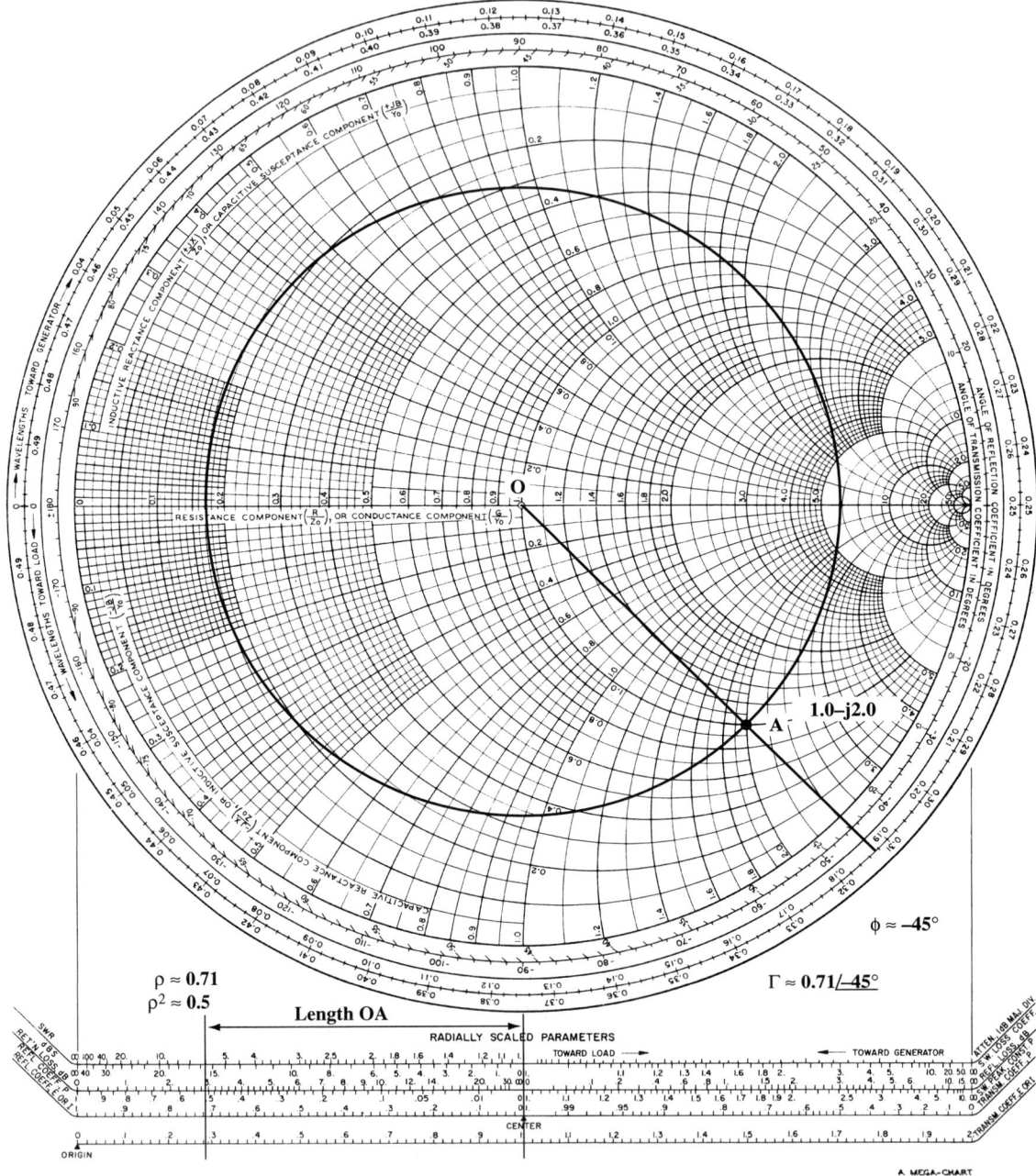

Figure 3.8 Determining voltage reflection coefficient. (Courtesy of Analog Instruments Co., P.O. Box 808, New Providence, NJ 07974.)

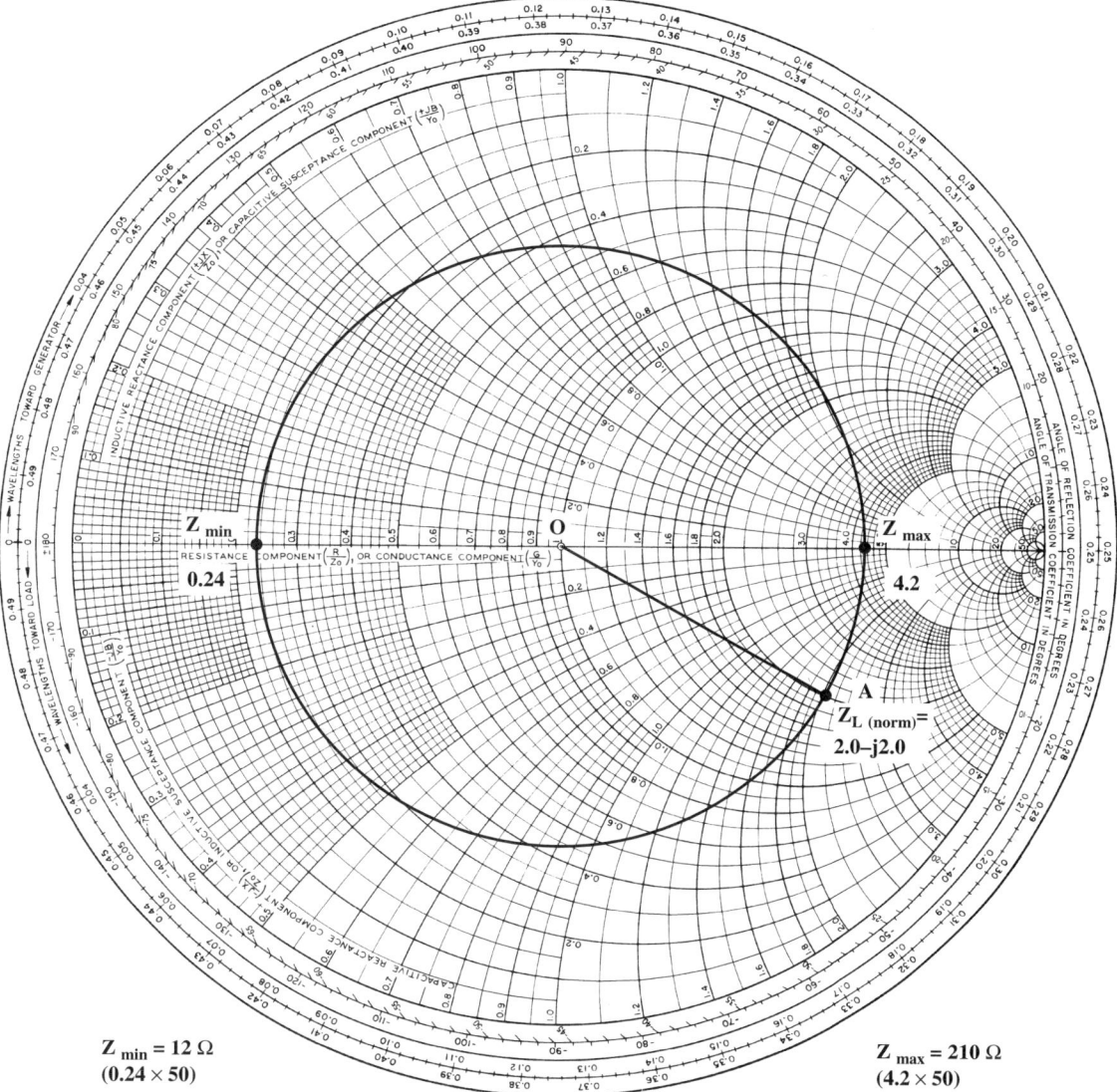

Figure 3.9 Determining Z_{max} and Z_{min}. (Courtesy of Analog Instruments Co., P.O. Box 808, New Providence, NJ 07974.)

4. Read the $Y_{L(norm)}$ at point B.

$$Y_{L(norm)} = 0.2 - j0.4$$

5. Denormalize Y_L. (Note: You divide by 50 rather than multiply.)

$$Y_L = 0.2/50 - j0.4/50$$
$$= 0.004 - j0.008 \text{ S}$$
$$= 4.0 - j8.0 \text{ mS}$$

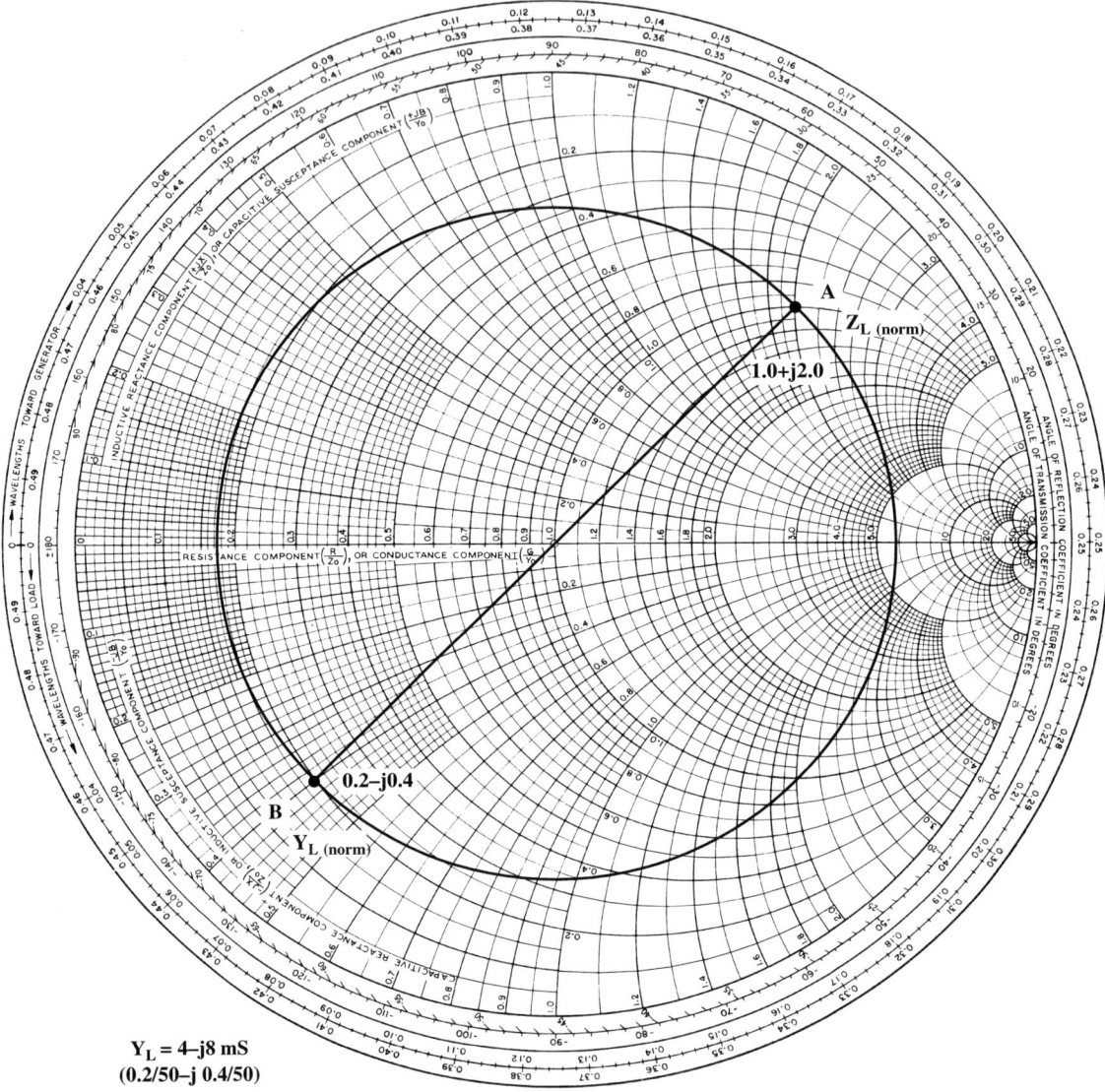

Figure 3.10   Determining Y_L. (Courtesy of Analog Instruments Co., P.O. Box 808, New Providence, NJ 07974.)

EXAMPLE 3.6:

Determine the relative location in λ of Z_L when $Z_L = 75+j75\ \Omega$ and $Z_O = 50\ \Omega$.

1. Plot the normalized impedance on the chart. ($Z_{L(norm)} = 1.5+j1.5$) This is point A in Figure 3.11.
2. Draw a line through point A from the center of the chart outward toward the outer scales. Note that the outer scales are labeled λ toward the load (counterclockwise labeling), and λ toward the generator (clockwise labeling). Since we are already at the load, we will read the clockwise labeling, λ toward the generator.

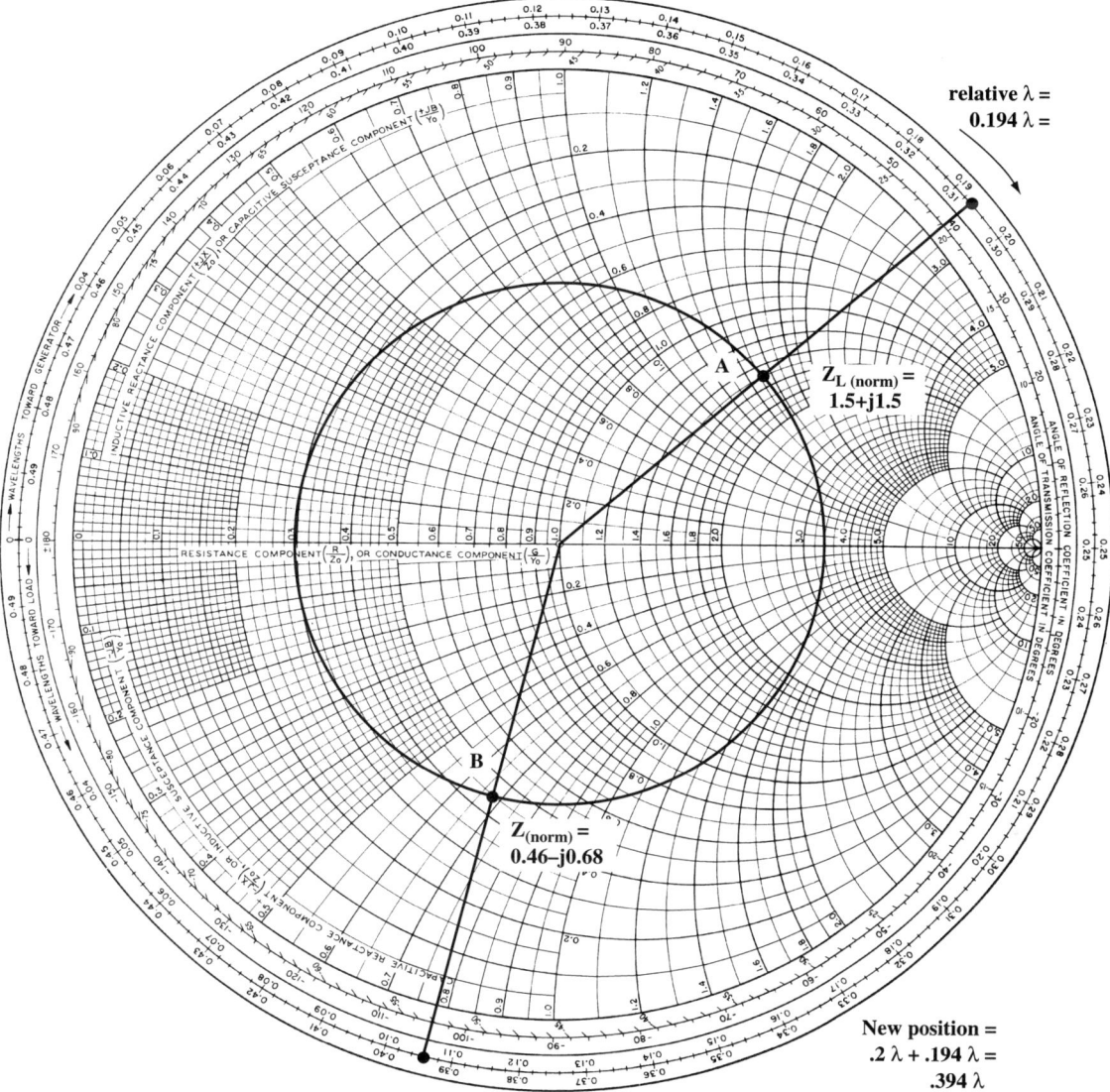

relative λ = 0.194 λ =

$Z_{L (norm)} = 1.5 + j1.5$

$Z_{(norm)} = 0.46 - j0.68$

New position = .2 λ + .194 λ = .394 λ

Figure 3.11 Determining relative wavelength position. (Courtesy of Analog Instruments Co., P.O. Box 808, New Providence, NJ 07974.)

3. Read the relative λ value off this scale.

relative λ = 0.194λ toward generator

By completing the drawing of the VSWR circle, any impedance can be found on the line in relation to the location of the load. This is because the circle not only reflects constant

VSWR (all impedances on the circle produce the same VSWR), the points on the circle represent any and all impedances that occur on the line. Just rotate around the chart the number of λs toward the generator to any point. That point's impedance is found by drawing a line from this point back to the center of the chart. Where this line crosses the VSWR circle locates the normalized impedance at that point. Note that one revolution around the chart is equal to one-half wavelength. To rotate values greater than one-half wavelength, subtract multiples of one-half wavelength until a value less than one-half wavelength is obtained.

EXAMPLE 3.7:

Determine the impedance at a point on the line 0.2λ from the load. This is also shown on Figure 3.11.

1. From the relative position of the load rotate clockwise (toward generator) 0.2λ.
2. Draw a line from the center of the chart through this point (point B) crossing the VSWR circle.
3. Read the normalized impedance at point B.

$$Z_{(norm)} \text{ 0.2λ from load} = 0.46 – j0.68$$

3.3   SHORT-CIRCUIT MINIMA SHIFT METHOD

The short-circuit minima shift method is a procedure for determining the value of an unknown impedance. A short circuit is placed on the line to provide a large standing wave. A slotted line is used to take measurements on the line. (The slotted line is a common microwave component that is described in chapter 5.) The detector within the slotted line is then used to locate the minima. The minima are very sharp voltage deflections and easy to "see." The distance in millimeters between successive minima is one-half wavelength (this value is used to determine the guide wavelength, λ_g) as indicated on the slotted line rule. Next, the unknown impedance is placed on the line. Again, the slotted line is used to determine the location of minima. The minimum shifts and is found between the locations of the short-circuit minima. The distance the minimum shifted and the one-half wavelength dimension are input to equation 3.1 to solve for the phase angle of the reflection coefficient.

$$\phi = 180° \, [1 – (4d/\lambda_g)] \qquad \textbf{(Eq. 3.1)}$$

where

ϕ = phase angle in degrees

d = distance minimum shifted, mm

λ_g = guide wavelength, mm

The angle of reflection and the given VSWR are then plotted on the Smith chart. A line starting from the center of the chart and drawn toward the calculated value of the angle of reflection crosses the VSWR circle at one point. This point represents the unknown impedance.

EXAMPLE 3.8:

Determine the value of an unknown load impedance that produces a VSWR of 2.0. (Assume the VSWR is given or has already been measured with a slotted line.)

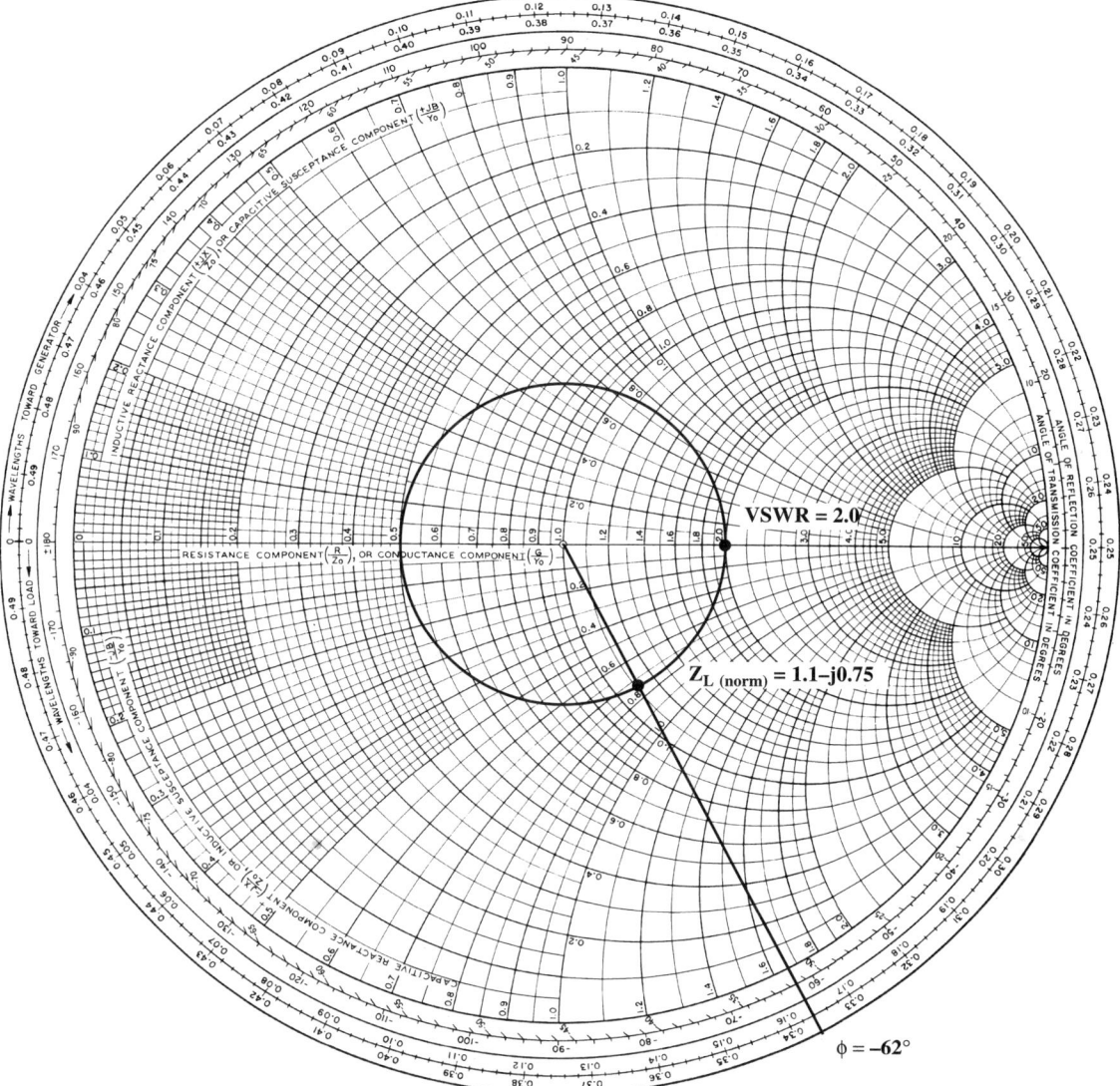

Figure 3.12   Determining unknown load impedance. (Courtesy of Analog Instruments Co., P.O. Box 808, New Providence, NJ 07974.)

1. Plot the constant VSWR circle on the chart as shown in Figure 3.12.
2. With the slotted line, determine the phase angle of the reflection coefficient ϕ as follows:

    a. With a short circuit replacing the unknown load, determine $NULL_1$, the location of the minimum of the standing-wave pattern that is nearest the short circuit. The results used in this example, and shown in Figure 3.13, were obtained with Lab-Volt X-band microwave equipment. $NULL_1$ represents the closest measurable minimum to the short circuit.

    Location of $NULL_1 = 36.9$ mm

b. Determine the location of the next minimum toward the source.

Location of $NULL_2 = 55.2$ mm

c. Replace the short circuit with the load in question. Determine the minimum located between the positions of $NULL_1$ and $NULL_2$.

Location of minimum = 42.9 mm

d. Determine the distance this minimum shifted.

$$d = NULL_2 - \text{minimum}$$
$$= 55.2 - 42.9 \text{ mm}$$
$$= 12.3 \text{ mm}$$

e. Determine the guide wavelength from $NULL_1$ and $NULL_2$. (Note: You must multiply this value by 2, since this value is for one-half wavelength.)

$$\lambda_g = 2(NULL_2 - NULL_1)$$
$$= 2(18.3) \text{ mm}$$
$$= 36.6 \text{ mm}$$

f. Use equation 3.1 to calculate the angle of the reflection coefficient ϕ.

$$\phi = 180° [1 - (4(12.3)/36.6)]$$
$$= 180°(-0.3444)$$
$$\approx -62°$$

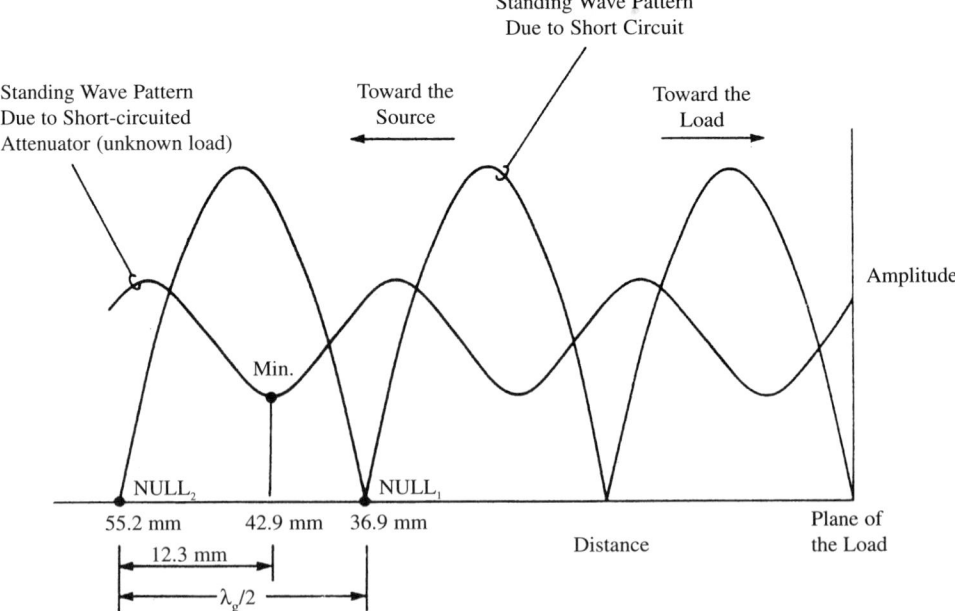

Figure 3.13   Short-circuit minima shift method.

3. Draw a line from the center of the VSWR circle (1+j0) to the outer edge where the Angle of the Reflection Coefficient in Degrees scale indicates –62°. (This is shown in Figure 3.12.)
4. Read the normalized impedance of the unknown load where the line crosses the VSWR circle. $Z_{L(norm)} = 1.1–j0.75$
5. Denormalize $Z_{L(norm)}$ (multiply by 50). This value represents the unknown load impedance.

$$Z_L = 50(1.1)–j50(.75)$$
$$= 55–j37.5 \ \Omega$$

3.4   IMPEDANCE MATCHING TECHNIQUES

In an ideal microwave system, all components have impedances equal to the characteristic impedance of the transmission line. This ensures that no reflected waves occur anywhere on the line. Such a system is called *matched*. Another way to describe a matched system is to say the line is *flat*, indicating there are no standing waves (VSWR = 1.0).

The importance of a matched system is that it eliminates problems that arise when reflected waves are present. If there is a high reflected voltage, then useful signal power is obviously diminished. This is particularly noticeable in "active" components such as diodes and transistors because of their impedance characteristics.

Problems also can occur with measurements. The accuracy of power measurements is in jeopardy when mismatches occur. The desired signal to be measured may contain an unknown component caused by the reflections.

The term *matching* refers to techniques employed, or circuit modifications made, that eliminate unwanted reflections. (Z_L must equal Z_O for a matched condition to occur.)

Another parameter affected by impedance matching techniques is circuit bandwidth. Many systems must operate over a fairly large bandwidth. Not all matching techniques work equally well over the complete operating frequency range.

Two basic categories describe matching techniques. The first of these is called *lossless matching*, because the components used do not dissipate energy. No power is unnecessarily lost due to the matching techniques employed. However, these techniques tend to work only over a limited range of frequencies (narrow bandwidth). These techniques lend themselves to CAD/CAE (computer-aided design/computer-aided engineering) programs. Many software programs are available to assist in circuit simulation and to provide computer solutions to matching problems. Software is available to do these matching techniques on a computer-generated Smith chart. One example is SmithMatch™ from Microwave Software, Laguna Hills, California. This software uses reactive circuit elements to match the line. Another software program is Smith Chart for Windows from Antenna Design Associates, Inc., Leverett, Massachusetts. Many of these programs are being used for design work with the higher frequency (millimeter wave) bands using microwave monolithic integrated circuits on microstrip lines. It is beyond the scope of this text to describe these CAD/CAE methods.

The other basic matching technique employs the use of *lossy matching* elements, such as resistive pads. Results include very broadband frequency ranges, but this is accomplished at a cost. Some power is lost that can degrade circuit efficiency.

A simple matching technique is to add shunt inductance or capacitance to the line to effectively cancel out unwanted reactances. These are normally lossless elements. Series and parallel *LC* tuned elements, or their transmission line equivalents, are added to transform the load to the center of the Smith chart. The center of the chart, 1+j0, represents a matched impedance (50 Ω for a 50 Ω Z_O line). This means the reflection coefficient Γ is zero and VSWR = 1.0. The actual impedance match occurs for *only a single frequency*.

This technique requires that Z_L be converted to Y_L. We work with admittances because admittances are additive in parallel. Our matching elements are shunt devices, which makes the problem easier to do. The only drawback to this method is that it is inherently a narrow-band solution.

The lumped *LC* values inherent in a transmission line can cause problems due to self-resonance in circuits that use microstrip lines and in other forms of microwave monolithic integrated circuits. "Chip" capacitors and printed circuit inductors are added to the line to provide a means of matching.

Commercially available tuners can be used when the matching circuit cannot be built into the load itself. They are available in waveguide and coaxial designs. The waveguide tuner is called a slide-screw tuner. (For a picture of a waveguide tuner, see Figure 5.9 in chapter 5.)

This tuner has a variable post that can extend into a slotted line opening that causes either a capacitive or an inductive effect depending on the depth of penetration. By moving the post along the slot, and through adjustment of depth, a wide range of mismatches can be tuned.

A quarter-wave transformer (section 2.10) uses the properties of a length of transmission line that is one-quarter wavelength long and is placed between the line and the load. The characteristic impedance of this section of line must be chosen to match the load impedance.

EXAMPLE 3.9:

Determine the point, nearest the load, at which a λ/4 transformer may be inserted to provide matching between a Z_L = 250+j450 Ω and Z_O = 300 Ω. Find the characteristic impedance (Z'_O) of the transmission line to be used for the transformer.

1. Normalize the load impedance with respect to the line: $Z_{L(norm)}$ = 0.83+j1.5. Plot this as point A, shown in Figure 3.14.
2. Draw a circle whose center lies at the center of the chart, passing through the plotted point A. This circle is referred to as the constant VSWR circle.
3. Move toward the generator (clockwise) from point A to find the nearest point at which the line impedance is purely resistive. This is on the diameter line where this line crosses the VSWR circle on the right side, labeled point B.
4. Measure the distance between points A and B in wavelengths. (d = 0.250 – 0.17 = 0.08 λ)
5. Read the normalized resistance at point B, r = 4.75, and convert this normalized resistance into an actual resistance by multiplying by the Z_O of the line. Here R = 4.75 × 300 = 1425 Ω.
6. This resistance is that which the λ/4 transformer must have to match the 300 Ω line. Z'_O is found by:

$$Z'_O = \sqrt{Z_O Z_L}$$

(Eq. 2.19)

$$= \sqrt{(300 \times 1425)}$$

$$= 654 \ \Omega$$

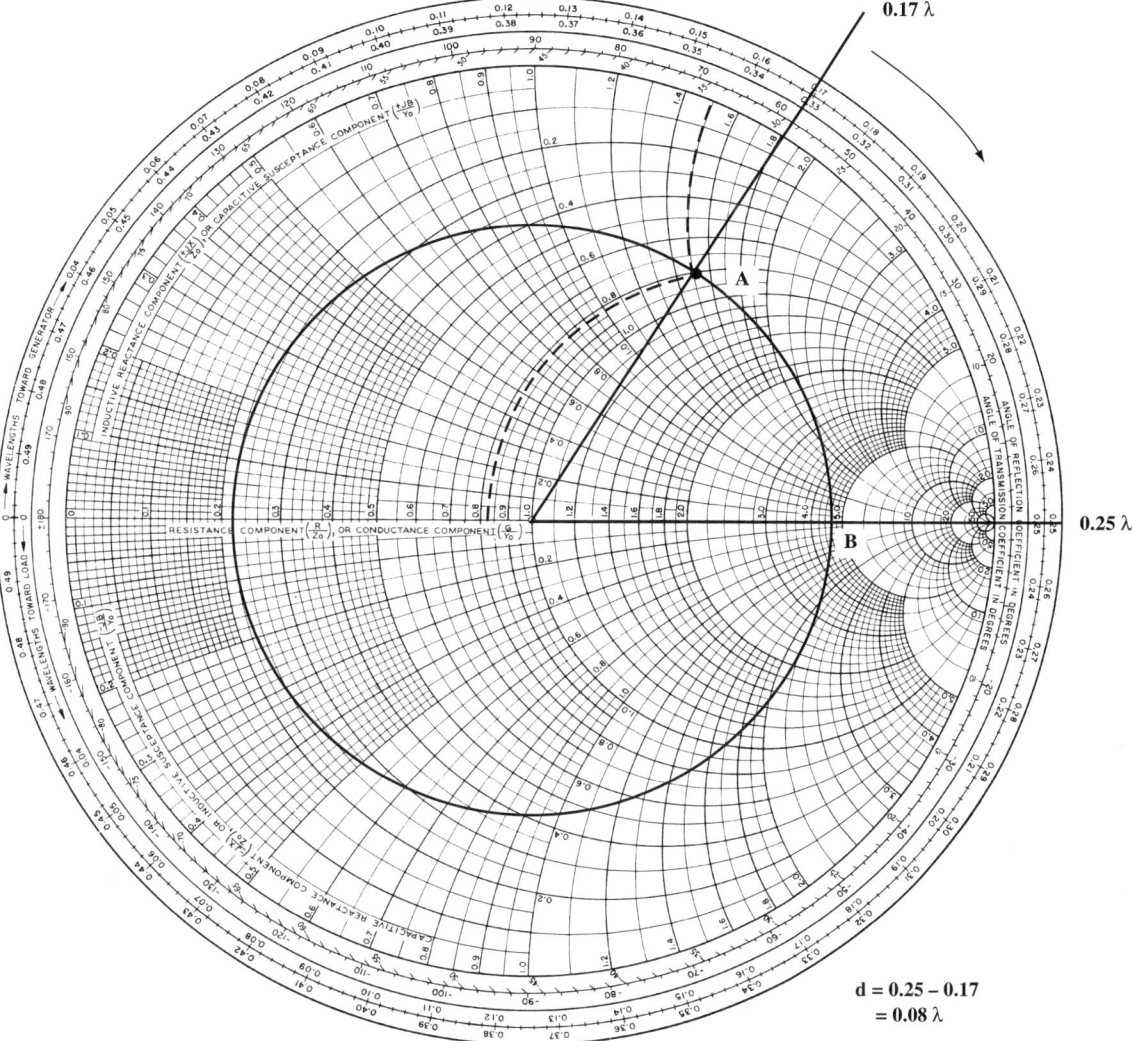

0.17 λ

0.25 λ

d = 0.25 − 0.17
  = 0.08 λ

Figure 3.14   Matching via λ/4 transformer. (Courtesy of Analog Instruments Co., P.O. Box 808, New Providence, NJ 07974.)

Another technique employs a coaxial line single- or double-stub tuner. Electrically these stubs act as reactive components. They have sliding shorts that can be moved in perpendicular arms, thus reflections can be tuned out.

The single-stub method demonstrates how versatile the Smith chart is in solving transmission line matching problems. A short-circuit stub is placed across the line at a point such that the impedance to the right of this point effectively matches that of Z_O. The stub attachment, with dimensions taken from example 3.10, is shown in Figure 3.15. Short-circuit stubs are preferred to open-circuit stubs because an infinite terminating impedance is more difficult to realize than a zero impedance. One reason for this is that radiation can

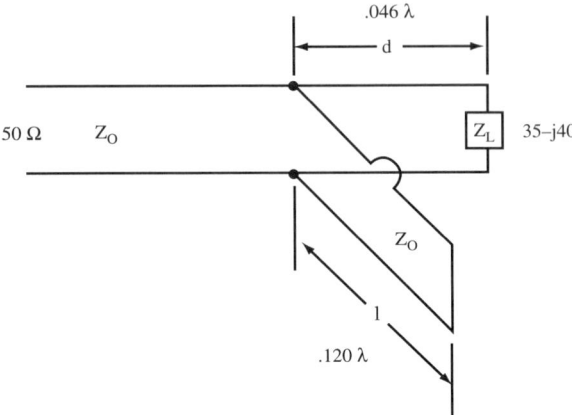

Figure 3.15 Matching via a short-circuit stub.

occur from an open end. Moreover, a short-circuit stub of adjustable length and a constant characteristic impedance is much easier to construct. Remember, the stub represents a reactive component.

It is desirable to analyze the matching requirement in terms of admittance since we are working with a parallel connected matching device. Distance d is found as a point on the line away from the load where we find a normalized admittance, $Y = 1.0 \pm jB$. The stub is attached to the line at this point. The stub must have a length such that its input impedance is reactive and cancels the line reactance at this point. Thus, the line admittance is $Y = 1.0 \pm j0$. The net result is that reflected energy from the load is canceled by the stub, leaving the line matched. The following examples demonstrate this process. They are shown in Figures 3.16 and 3.17.

EXAMPLE 3.10:

Determine d and l dimensions, in wavelengths, of a short-circuit single stub in order to effect a matched line. $Z_O = 50\ \Omega$ and $Z_L = 35 - j40\ \Omega$.

1. Normalize and plot Z_L on the Smith chart as point A, as shown in Figure 3.16. ($Z_{L(norm)}$ = 0.7–j0.8)
2. Draw the VSWR circle with the center at 1+j0.
3. Locate $Y_{L(norm)}$ opposite $Z_{L(norm)}$ as point B. Note that this point is located at relative position 0.118 λ toward the generator.
4. From point B move clockwise (λ toward the generator) around the VSWR circle, and note that there are two points on this circle where the Y has a $G = 1.0$. One of these occurs above the centerline and the other below it. These points are labeled P_1 and P_2 respectively. The distance in λ from B to either of these points represents the length for distance d. (This is the point where the stub will be attached to the line.) Only one of these two points needs to be considered, and usually the one that is the shorter distance from the load is preferred. However, for the sake of uniformity in doing this type of problem and those at the end of the chapter, we will use P_1 for dimensional purposes, whether or not it is the shorter distance of the two. The relative position of P_1 is 0.164 λ toward the generator.

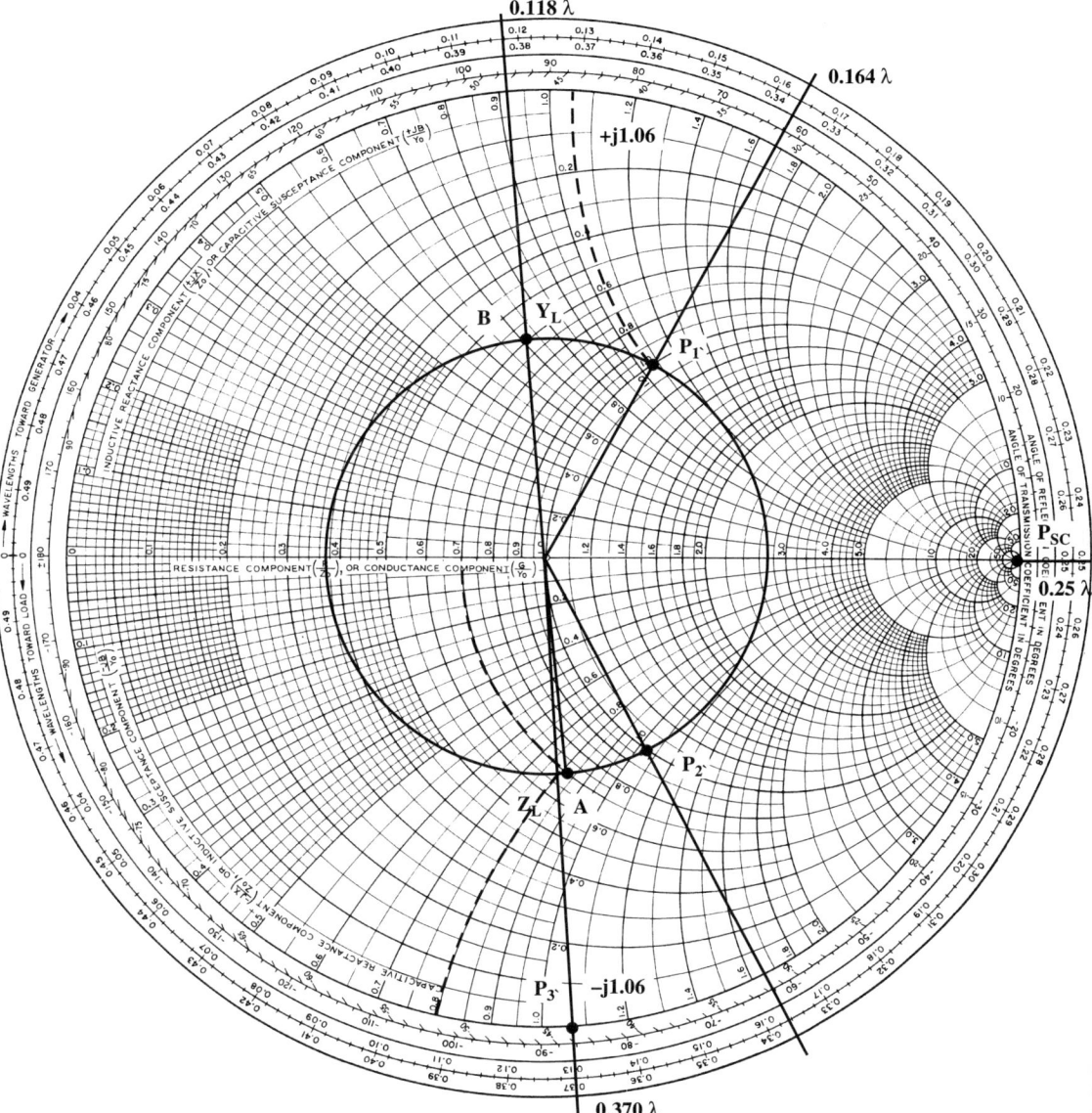

Figure 3.16 Matching via a short-circuit stub. (Courtesy of Analog Instruments Co., P.O. Box 808, New Providence, NJ 07974.)

5. Calculate the length for *d* as:

$$d = P_1 - B$$
$$= 0.164 - 0.118 \, \lambda$$
$$= 0.046 \, \lambda$$

6. Note: Point P_1 has an admittance of 1.0+j1.06. The +j1.06 represents the susceptance that needs to be cancelled out.

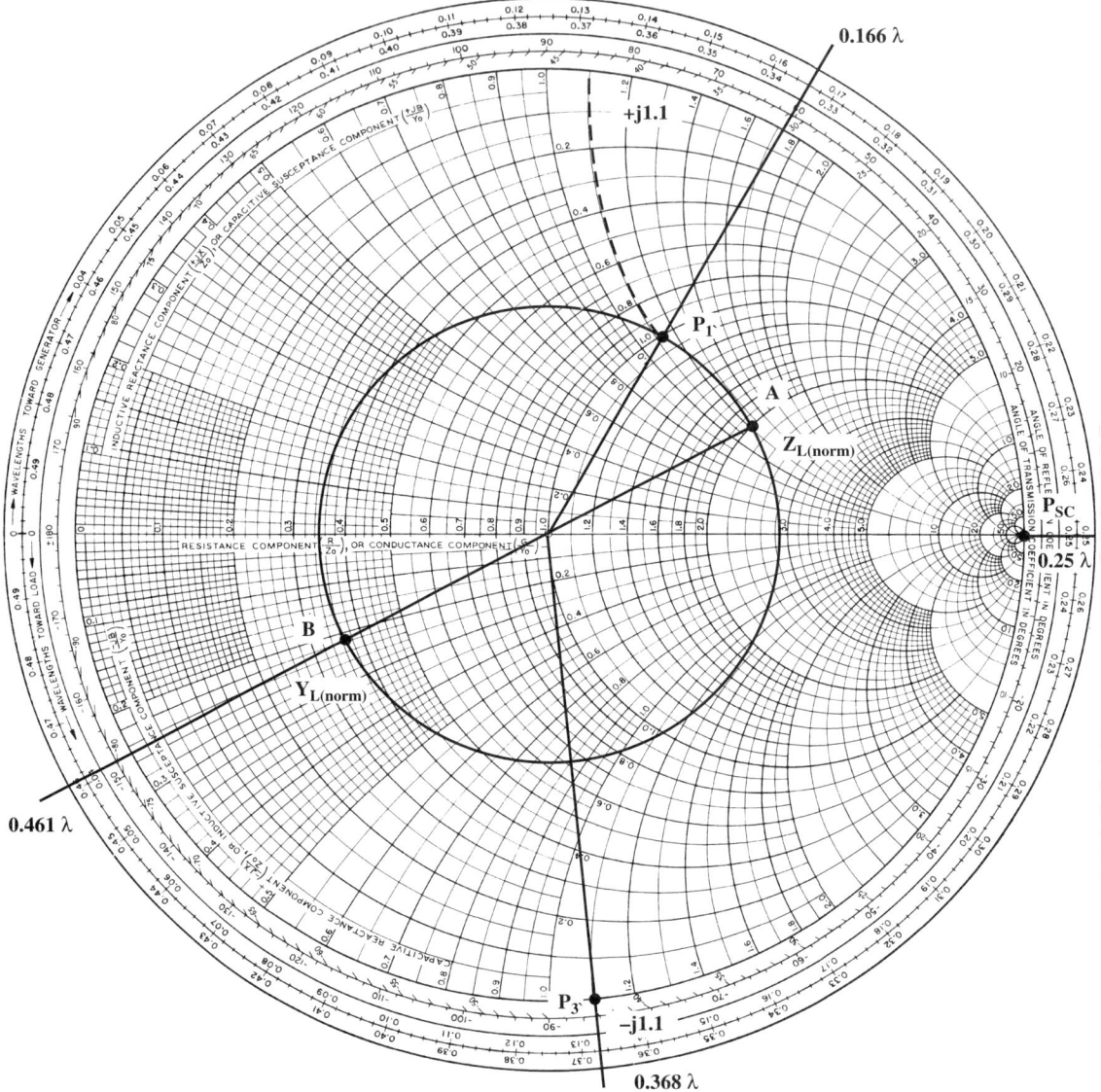

Figure 3.17   Matching via a short-circuit stub. (Courtesy of Analog Instruments Co., P.O. Box 808, New Providence, NJ 07974.)

7. Locate and label the point of equal and opposite susceptance, –j1.06, as P_3, the conjugate of the value at P_1.
8. Locate and label the point of infinite short-circuit conductance, which is also the point where the VSWR = ∞. The stub acts as a reactance, thus its VSWR = ∞. This is point P_{sc} (0.25 λ) located to the extreme right end of the diameter line.
9. Calculate the length of the short-circuited stub to provide –j1.06 susceptance as:

$$l = P_3 - P_{sc}$$
$$= 0.37\,\lambda - 0.250\,\lambda$$
$$= 0.120\,\lambda$$

10. $d = 0.046\,\lambda$ and $l = 0.120\,\lambda$. These dimensions can be converted to a physical dimension in millimeters or centimeters if the source frequency is known, as $\lambda = c/f$, and if the velocity factor is known. The next example shows this.

EXAMPLE 3.11:

Determine d and l dimensions, in cm, of a short-circuit single stub in order to effect a matched line. $F = 1.0$ GHz, $v_f = 0.95$, $Z_O = 50\ \Omega$, and $Z_L = 100+j60\ \Omega$.

1. Normalize and plot Z_L on the Smith chart as point A, as shown in Figure 3.17. ($Z_{L(norm)} = 2.0+j1.2$)
2. Draw the VSWR circle with the center at $1+j0$.
3. Locate $Y_{L(norm)}$ opposite $Z_{L(norm)}$ as point B. Note that this point is located at relative position $0.461\,\lambda$ toward the generator.
4. From point B move clockwise around the VSWR circle to point P_1 ($G = 1.0$). The relative position of P_1 is $0.166\,\lambda$ toward the generator.
5. Calculate the length for d as:

$$d = (0.500 - B) + (P_1 - 0.000)$$
$$= (0.500 - 0.461) + (0.166 - 0.000)$$
$$= 0.039 + 0.166$$
$$= 0.205\,\lambda$$

6. Note: P_1 has an admittance of $1.0+j1.1$. The $+j1.1$ represents the susceptance that needs to be cancelled out.
7. Locate and label the point of equal and opposite susceptance, $-j1.1$, as P_3, the conjugate of the value at P_1.
8. Locate the point of infinite short-circuit conductance as P_{sc}.
9. Calculate the length of the short-circuited stub to provide $-j1.1$ susceptance as:

$$l = P_3 - P_{sc}$$
$$= 0.368 - 0.250$$
$$= 0.118\,\lambda$$

10. $d = 0.205\,\lambda$ and $l = 0.118\,\lambda$. Convert to cm as follows:

$$\lambda = (0.95 \times 3.0 \times 10^{10}\ \text{cm/s})/(1.0 \times 10^9\ \text{GHz})$$
$$= (2.85 \times 10^{10})/(1.0 \times 10^9)$$
$$= 28.5\ \text{cm}$$
$$d = 0.205 \times 28.5\ \text{cm}$$
$$= 5.8\ \text{cm}$$
$$l = 0.118 \times 28.5\ \text{cm}$$
$$= 3.36\ \text{cm}$$

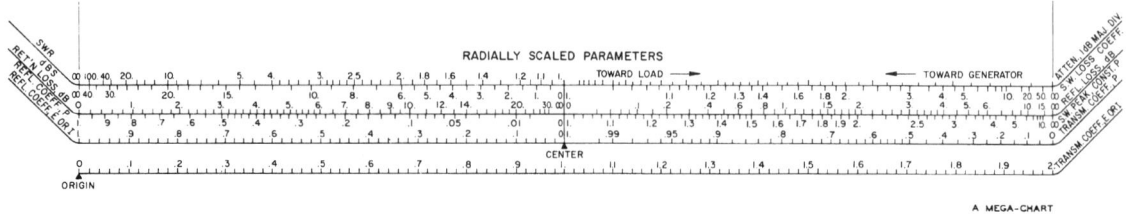

Figure 3.18   Smith chart scale for radially scaled parameters. (Courtesy of Analog Instruments Co., P.O. Box 808, New Providence, NJ 07974.)

3.5   SMITH CHART RADIALLY SCALED PARAMETERS

The radially scaled parameters are found on scales located directly below the circular part of the Smith chart and are shown in Figure 3.18. They are very useful in solving for many numerical answers. These include reflection and transmission coefficients, reflection loss, return loss, conversion from coefficient ratios into dB, and the like.

Examples in section 3.2 used the scales to find the VSWR, Γ, and ρ^2. Once the radius of a VSWR circle is known and transferred onto the scales below, many other parameters can be determined about the problem in question. Some parameters that are convenient to know are SWR in dB, return loss in dB, and the transmission coefficient. Note that the most frequently used parameters are found to the left of center on the scales and thus were demonstrated.

3.6   SUMMARY

1. The Smith chart shows that the relationships of impedances can be determined at any point along a transmission line.

2. The Smith chart is constructed of (1) circles of resistance, (2) arcs of inductive reactance, and (3) arcs of capacitive reactance. These circles and arcs are all tangent at the right side (infinity) of the chart.

3. All values of impedance must be normalized before plotting on the Smith chart and denormalized when coming from the Smith chart. To normalize impedance values, divide the actual impedance by the Z_O of the transmission line.

4. Voltage standing wave ratio (VSWR) circles are drawn concentric to the center of the chart, and all values of impedance contained in one-half wavelength of a given transmission line are on the VSWR circle.

5. Electrical distances on the Smith chart are expressed in terms of wavelength toward the generator or toward the load, and also in terms of electrical degrees.

6. A matched system is also known as a flat line.

7. Impedance matching techniques employ either lossy, or broadband, solutions or lossless, or narrow band, solutions.

8. Common matching techniques can employ shunt reactive components, commercial tuners, a quarter-wave transformer, and either double or single short-circuit stubs.

Key Equation:

$$\phi = 180° \, [1 - (4d/\lambda_g)] \qquad\qquad \textbf{(Eq. 3.1)}$$

PROBLEMS

1. Determine the normalized values of the following load and characteristic impedances:

   a. $Z_L = 75-j25\ \Omega$ $Z_O = 50\ \Omega$
   b. $Z_L = 50+j100\ \Omega$ $Z_O = 75\ \Omega$
   c. $Z_L = 15-j40\ \Omega$ $Z_O = 50\ \Omega$
   d. $Z_L = 100+j75\ \Omega$ $Z_O = 100\ \Omega$
   e. $Z_L = 35-j75\ \Omega$ $Z_O = 50\ \Omega$

2. Plot the normalized values of problem 1 on a Smith chart and label a through e, respectively.

3. Using the normalized value of problem 1a., determine the VSWR, voltage reflection coefficient (Γ) including the angle of reflection, and the power reflection coefficient (ρ^2) using a Smith chart.

4. Repeat problem 3 using the normalized value of problem 1b.

5. Find Z_{max} and Z_{min} using the values of problem 1c.

6. Determine Y_L using the values of problem 1d.

7. Find the relative λ position of the load using the values of problem 1e.

8. Determine the impedance on the line 0.3 λ in front of the load using the values of problem 7.

9. Having determined that $\lambda_g = 32.6$ mm, and that the minima shifted 9.8 mm, calculate the angle of reflection.

10. Given a VSWR of 2.0 and the angle of reflection from problem 9, determine the value of an unknown normalized Z_L.

11. Determine the stub distance d and length l required to obtain a matched load-line condition between a $Z_L = 41.25-j22.5\ \Omega$ and a $Z_O = 75\ \Omega$.

12. Repeat problem 11 for a $Z_L = 31.25+j10\ \Omega$ and $Z_O = 50\ \Omega$.

13. Determine d and l in problem 11 in cm, when the frequency is 2.0 GHz and the velocity factor is 0.95.

14. Determine d and l in problem 12 in cm, when the frequency is 1.2 GHz and the velocity factor is 0.66.

QUESTIONS

1. Describe what the Smith chart represents.

2. What can be determined from using a Smith chart?

3. What is meant by a *complex impedance*?

4. What is meant by the term *normalized impedance*?

5. What is a VSWR circle and what does it represent?

6. What is meant by the reactive properties of a transmission line?

7. What does movement around the VSWR circle represent?

8. Where are Z_{max} and Z_{min} located on a Smith chart?

9. Why is the intersection of the VSWR circle with the $r = 1$ circle such an important point?

10. What is a short-circuit stub?

11. Why do we work with admittances when doing short-circuit stub problems?

12. Why do we need to have a matched line?

13. Describe both *lossy* and *lossless* matching techniques.

14. What are the pros and cons of lossy vs. lossless matching techniques?

15. Describe what the radially scaled parameters can be used for.

4 MICROWAVE TRANSMISSION LINES

OBJECTIVES

1. To describe common characteristics of basic transmission lines.
2. To calculate characteristic impedance of two-wire and coaxial lines.
3. To describe propagation and propagation modes in transmission lines.
4. To describe the "higher-order" modes of transmission lines.
5. To define cutoff wavelength, cutoff frequency, and guide wavelength.

4.1   INTRODUCTION

Transmission lines are devices that guide electromagnetic waves from one place to another. Recall from chapter 2 that transmission lines fall into four general categories:

1. Two-wire (parallel)
2. Coaxial
3. Waveguide
4. Stripline and microstrip

Two-wire lines are used only at the low end of the microwave band. Their primary limitation is due to radiation loss, but they also exhibit dielectric losses and skin effect losses. Above 200 MHz, coaxial lines are more efficient.

Coaxial lines overcome the radiation loss problems due to the inherent shielding provided by the outer conductor that surrounds the center conductor. However, they have limitations due to dielectric loss and skin effect loss. Their main drawback is limited power-handling capability.

Waveguides have better all-around characteristics than most other transmission lines. They are used almost exclusively in radar with its high power requirements.

As frequencies are pushed up even higher, especially near the millimeter wave bands, the microstrip and stripline have distinct advantages. Many higher-frequency systems are used in low-power applications. Low-power circuits employ monolithic microwave integrated circuit technology (MMIC) or hybrid microwave integrated circuit technology

(HMIC) with the copper trace on the circuit board becoming the transmission line. This allows stripline and microstrip designs to be the choice for these requirements since these lines lend themselves to printed circuit board techniques.

This chapter elaborates on the characteristics of these four general types of transmission lines. Concepts of cutoff wavelength and frequency, velocities, higher-order modes, characteristic impedance, and practical aspects are covered for these transmission lines.

4.2   TWO-WIRE LINES

Most of the general characteristics of the two-wire line are covered in chapter 2. This provided the necessary foundation for understanding how the Smith chart (chapter 3) can solve various types of transmission line problems.

When the inductance and capacitance per unit length are known, the characteristic impedance of any line can be found, as was shown in chapter 2 by equation 2.1.

$$Z_O = \sqrt{L/C} \qquad \textbf{(Eq. 2.1)}$$

Values L and C depend on geometric configurations on the line. Specific two-wire lines have finite spacing between the conductors and a definitive diameter of the conductors. These factors can be used rather than values of L and C per unit length. Using these factors, we can calculate the characteristic impedance for the two-wire line shown in Figure 4.1 by using equation 4.1.

$$Z_O = (276/\sqrt{\epsilon_r})(\log_{10} 2D/d) \qquad \textbf{(Eq. 4.1)}$$

where Z_O = characteristic impedance, Ω

D = distance between the conductors

d = diameter of the conductors

ϵ_r = relative dielectric constant (ϵ_r = 1.0 for air)

Figure 4.1   Two-wire line.

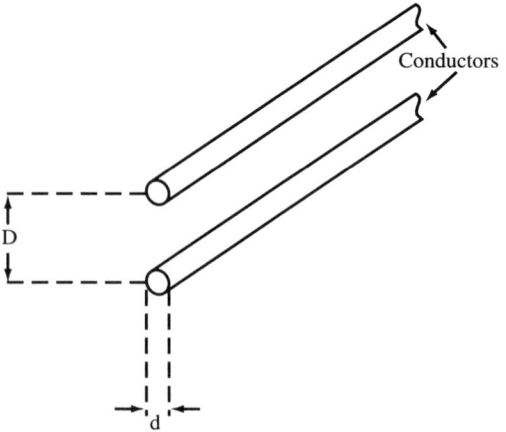

EXAMPLE 4.1:

Determine the characteristic impedance of a two-wire transmission line when the diameter of the conductors is 0.2 cm separated by a distance of 0.8 cm. The dielectric is air.

$$Z_O = (276/\sqrt{1})[\log_{10}(2[0.8]/0.2)]$$
$$= 276 \log_{10} 8$$
$$= 249.3 \ \Omega$$

Note that it is the ratio of $2D$ to d that determines Z_O. Therefore, unequal diameters and appropriate spacing could produce the same Z_O. The usual range of characteristic impedances for parallel lines is 150 to 600 Ω.

4.3 COAXIAL LINES

The second conductor of the two-wire line eventually became the shield surrounding the center conductor, thus evolving into the coaxial line as shown in Figure 4.2a. The center wire can be supported by dielectric spacers when air is the dielectric as shown in Figure 4.2b. The dielectric can actually support the center conductor via polyethylene or a similar dielectric. This is shown in Figure 4.2c.

Coaxial cable is the familiar "feeder" connecting a domestic television receiver to a rooftop antenna. The electrical characteristics of various types of coaxial lines can differ significantly, particularly at microwave frequencies, due to the dimensions of the cable and the choice and construction of the dielectric material.

Figure 4.2 Coaxial lines.

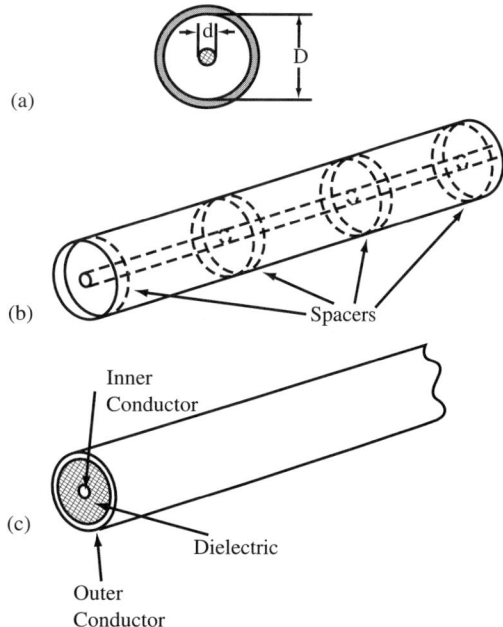

The outer connector (shield) of a coaxial line is connected to ground, so a coaxial line is referred to as an *unbalanced* line. This may be compared to a *balanced* line, such as the familiar TV twin-lead feeder, where the two signal-carrying conductors are equally isolated from ground. The matching transformer used to match a coaxial line to a TV twin-lead or antenna is referred to as a *balun* transformer. This is indicative of a match occurring between an unbalanced and a balanced line. The balun also serves as an impedance matching device, matching the 300 Ω twin-lead to the 75 Ω coaxial line.

Coaxial cables intended for microwave systems are considerably larger in diameter than those for conventional television reception. In microwave radio systems, 7/8-inch and 1 5/8-inch cables (diameter of the outer conductor) with foam or air dielectrics are common.

The usual range of characteristic impedances for coaxial lines is 40 to 150 Ω. The characteristic impedance of a coaxial cable employing a solid, uniform dielectric spacing material is given by:

$$Z_O = (138/\sqrt{\epsilon_r})(\log_{10} D/d) \tag{Eq. 4.2}$$

where
Z_O = characteristic impedance, Ω

D = diameter of the outer conductor

d = diameter of the inner conductor

ϵ_r = dielectric constant (ϵ_r = 1.0 for air)

EXAMPLE 4.2:

A 7/8-inch diameter coaxial cable employs a 1/4-inch diameter inner conductor with solid insulating material yielding a dielectric constant of 2.25. Calculate Z_O for the cable.

$$Z_O = (138/\sqrt{2.25})(\log_{10} 0.875/0.25)$$
$$= 92 \log_{10} 3.5$$
$$= 50 \ \Omega$$

Many companies have proprietary brand names for their transmission line products. Andrew Corporation uses the HELIAX® system for its coaxial lines, while Gore and Associates uses the GORE-TEX® dielectric for its coaxial lines. Coaxial lines can be used up to 40 GHz, with aspirations of going even higher. However, more high-frequency work is being done on the MMIC circuit board without the need for additional cabling.

Propagation through a coaxial line is done via the TEM mode (see chapter 1), since a coaxial line is a two-wire transmission line. In a coaxial line, the TEM wave is called the principal or dominant wave mode and is the preferred mode of operation. When operated in the dominant mode, frequencies from DC to several GHz can be propagated. Because of this, the coaxial line is considered a broadband device.

As the frequency gets higher, the relative wavelength of the signal and the physical dimensions of the line come into play. When a high-frequency wavelength equals or is smaller than the physical dimension of the line, *higher-order* modes of wave propagation exist. The critical dimension of the coaxial line is a function of the circumference of the dielectric that exists between the inner and outer conductors. These higher-order modes distribute their energy fields in the dielectric between the conductors.

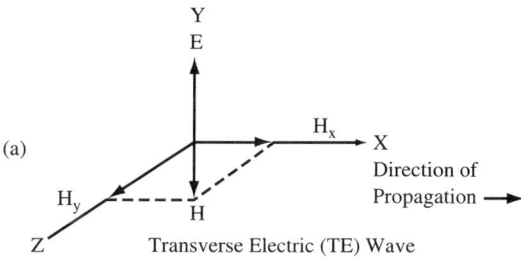

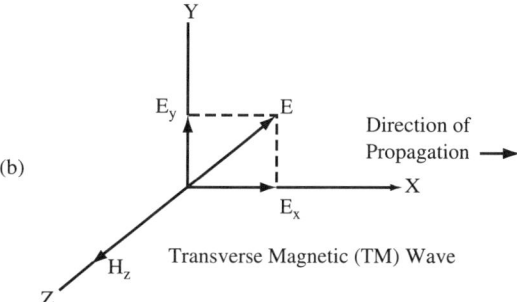

Figure 4.3 Higher-order modes of wave propagation.

The higher-order modes are classified as either TE wave or TM waves. The TE wave is a transverse electric wave for which the *magnetic* field is in the direction of wave propagation and the electric field is tranverse to it. For the transverse magnetic wave, TM, the component of the *electric* field is in the direction of propagation and the magnetic field is transverse to it. These fields are shown in Figure 4.3.

The higher-order modes are specified by subscripts m and n. They are written as TE_{mn} and TM_{mn}. The subscripts relate to the number of half wavelengths that can exist around the circumference of the circle. The lowest order of these higher-order modes is TE_{11} and TM_{01} respectively. These are shown in Figure 4.4. Many other higher-order modes can exist.

The operating frequency for coaxial cable is chosen to be below these lowest-order modes. This ensures propagation via the TEM mode, which is the dominant mode for the coaxial line. The higher-order modes are not preferred because of inefficient transfer of

Figure 4.4 Typical TE and TM modes for coaxial lines.

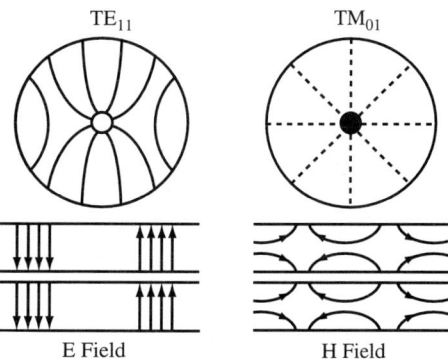

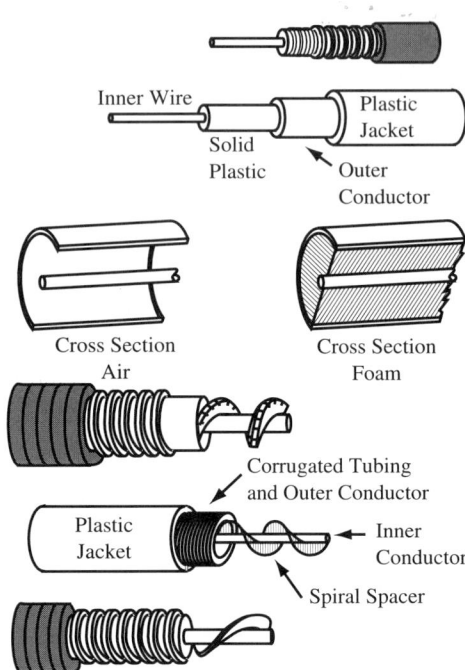

Figure 4.5   Types of coaxial cable.

energy down the line. Problems occur when higher frequencies are required, since the shorter wavelengths can generate the higher-order modes. To prevent this, the diameter of the coaxial line must be reduced. Reducing the diameter decreases the power handling capability of the line and increases attenuation due to skin effect. When diameters are reduced, power is limited because of voltage flashover (arcing). Flashover can occur between the more closely spaced conductors. Ultimately, the reduced diameter exemplifies the major limitation with coaxial lines, namely, lower power-handling capability. However, coaxial lines can be used for short lengths at higher frequencies with negligible losses.

Attenuation increases with frequency and is slightly greater for foam-filled coaxial cable than for an air dielectric. To minimize attenuation caused by the foam dielectric, some designs use discs to separate the inner and outer conductors, while others use a spiral membrane. When air is used as the dielectric, attention must be given to keeping the air dry. Sometimes air is replaced with an inert gas such as nitrogen under pressure. If nitrogen is used, the flashover problem is not as severe, because nitrogen is less subject to arcing than air is. In any case, compromises are made concerning uses and the type of dielectric. Coaxial cable is available in flexible, semirigid, and rigid configurations according to application. Various types of coaxial cable are shown in Figure 4.5.

4.4   RECTANGULAR WAVEGUIDE

A waveguide is a hollow metal tube employed for propagating electromagnetic waves. Unlike coaxial cable, waveguide utilizes no inner conductor. As the name implies, it relies on the tube to guide the wave from its source to its destination.

There are three basic types of waveguide: rigid rectangular, circular, and elliptical. Each type of waveguide has a range of unique installation hardware. Waveguide installation is often colloquially referred to as "plumbing," since many installations use bends, joints, and fixtures reminiscent of water pipe installation.

Waveguide is made of copper, brass, or aluminum. Rigid waveguide comes in sections from several inches to 10 feet. Circular waveguide is usually shipped in 20-foot. sections, while elliptical waveguide, which is flexible, is available in 400-foot rolls.

At extremely high frequencies and for short lengths, the waveguide inner walls are often coated with silver or gold. As the energy reflects off the sidewalls it tries to establish "eddy currents" in the side walls and thereby attenuates the propagating signal. This coating reduces losses in the side walls. (More on this in section 5.10.)

Table 4.1 shows several EIA designations for rectangular waveguide in terms of minimum operating frequency and inner guide dimensions. For example, "WR-90" refers to waveguide (W), rectangular (R), and 90 hundredths of an inch in the wide dimension.

Table 4.2 is a cross-reference guide for various designations. These include EIA (Electronics Industries Association), MIL (military) or JAN (Joint Army Navy), IEC (International Electrotechnical Commission), and Great Britain.

Waveguides only propagate signals with a frequency above a given cut-off, which is determined by the width of the waveguide. A half-wave length of the propagated signal must "fit" in the wide dimension of the waveguide. As can be seen from the tables, frequencies

Table 4.1 EIA Rectangular Waveguide Designations

EIA Designation	Minimum Frequency (GHz)	Size 1/100 in.	Inner Dimensions A (in.)	Inner Dimensions B (in.)
WR-2300	0.256	2300	23.000	11.500
WR-2100	0.281	2100	21.000	10.500
WR-1800	0.328	1800	18.000	9.000
WR-1500	0.328	1500	15.000	7.500
WR-1150	0.513	1150	11.500	5.750
WR-975	0.605	975	9.750	4.875
WR-770	0.766	770	7.700	3.850
WR-650	0.908	650	6.500	3.250
WR-510	1.158	510	5.100	2.550
WR-430	1.375	430	4.300	2.150
WR-340	2.737	340	3.400	1.700
WR-284	2.080	284	2.840	1.340
WR-229	2.579	229	2.290	1.145
WR-187	3.155	187	1.872	0.872
WR-159	3.714	159	1.590	0.795
WR-137	4.285	137	1.372	0.622
WR-112	5.260	112	1.122	0.497
WR-90	6.560	90	0.900	0.450
WR-75	7.873	75	0.750	0.375
WR-62	9.490	62	0.622	0.311
WR-51	11.578	51	0.510	0.255
WR-42	14.080	42	0.420	0.170
WR-34	17.368	34	0.340	0.170
WR-28	21.200	28	0.280	0.140
WR-22	26.350	22	0.224	0.112
WR-19	31.410	19	0.188	0.094

Table 4.2   Rectangular Waveguide Designation Cross-Reference Guide

Frequency Band (GHz)	EIA	MIL or JAN	IEC	Great Britain
0.75–1.12	WR-975	RG-204/U	R-9	4
0.96–1.45	WR-770	RG-205/U	R-12	5
1.12–1.70	WR-650	RG-69/U	R-14	6
1.45–2.20	WR-510	RG-103/U	R-18	7
1.70–2.60	WR-430	RG-104/U	R-22	8
2.20–3.30	WR-340	RG-112/U	R-26	9a
2.60–3.95	WR-284	RG-48/U	R-32	10
3.30–4.90	WR-229	RG-340/U	R-40	11a
3.95–5.85	WR-187	RG-49/U	R-48	12
4.90–7.05	WR-159	RG-343/U	R-58	13
5.85–8.20	WR-137	RG-50/U	R-70	14
7.05–10.00	WR-112	RG-51/U	R-84	15
8.20–12.40	WR-90	RG-52/U	R-100	16
10.00–15.00	WR-75	RG-346/U	R-120	17
12.40–18.00	WR-62	RG-91/U	R-140	18
15.00–22.00	WR-51	none	R-180	19
18.00–26.50	WR-42	RG-53/U	R-220	20
22.00–33.00	WR-34	none	R-260	21
26.50–40.00	WR-28	RG-96/U	R-320	22
33.00–50.00	WR-22	RG-97/U	R-400	23
40.00–60.00	WR-19	RG-272/U	R-500	24
50.00–75.00	WR-15	RG-98/U	R-620	25

below 3 GHz cannot be transmitted in WR-229 waveguide. This is because the 229 hundredths of an inch dimension of the waveguide is smaller than one-half wavelength for frequencies below 3 GHz.

Waveguides get their unique shape by virtue of some basic microwave theory. Consider a signal generator connected as shown in Figure 4.6. Our understanding of basic electronics is that this circuit would short the generator. However, the connection is *not* a short if the width *s* is a quarter-wavelength of the signal frequency. In that case, the short *b* shunts all the current back to the generator. The total path length traveled is one-half wavelength (or 180° phase difference). The incident and returned currents cancel, which results in zero net current. Therefore, a quarter-wave ($\lambda/4$) short circuit is effectively an "insulator" to the signal.

Impedance for this short section is proportional to the trigonometric tangent of (βs), where: $\beta = 360°/\lambda$. If *s* is much smaller than λ, as is the case for low-frequency circuits,

Figure 4.6 $\lambda/4$ wave short circuit. When the width *s* is a quarter-wavelenght ($\lambda/4$) of the frequency, the short *b* shunts all the current back to the generator.

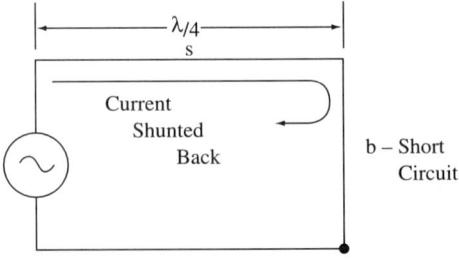

tan (βs), and therefore the impedance, is practically zero. This is consistent with our understanding of low frequency shorts. On the other hand, if s equals $\lambda/4$, then $\beta s = 90°$ and tan $90° = \infty$, i.e., the impedance of a perfect insulator.

It is worth pointing out that although $\lambda/4$ sections theoretically could be used as insulators for low-frequency circuits, they are not very practical. At 60 Hz, the insulator would be 750 miles long!

Cutoff Frequency

A rectangular waveguide may be thought of as a series of quarter-wave short (stub) circuits. As more short circuits are added, a rectangular waveguide is formed. This is shown in Figure 4.7.

The width of the rectangular waveguide is simply $2s$ which equals a. This dimension sets the lower limit of the frequency that can still propagate inside the guide. This limit on wavelength is given by $\lambda_c = 2a$. Any signal whose wavelength is larger than this cutoff wavelength is not allowed.

$$\lambda_c = 2a \qquad \textbf{(Eq. 4.3)}$$

EXAMPLE 4.3:

Consider the WR-90 rectangular waveguide. Its inner dimension for a equals 0.9 inches, or 2.286 cm. The cutoff wavelength λ_c can be calculated from equation 4.3 as:

$$\lambda_c = 2(2.286) \text{ cm}$$
$$= 4.572 \text{ cm}$$

This cutoff wavelength corresponds to a cutoff frequency as shown in equation 4.4.

$$f_c = c/\lambda_c \qquad \textbf{(Eq. 4.4)}$$

EXAMPLE 4.4:

$$f_c = (3.0 \times 10^{10})/4.572$$
$$= 6.56 \times 10^9 \text{ Hz}$$
$$= 6.56 \text{ GHz}$$

Figure 4.7 Rectangular waveguide as a series of $\lambda/4$ short-circuit stubs.

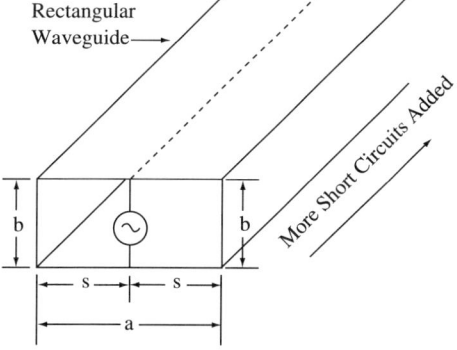

Any signal with a frequency less than 6.56 GHz cannot propagate in the WR-90 waveguide. Frequencies higher than f_c can propagate through the guide, since their half-wave dimensions are smaller than λ_c. Hence, a rectangular waveguide can be considered a *high-pass filter.*

Propagation

Propagation inside the waveguide follows a back-and-forth or zigzag reflective path. This can be explained by assuming a small antenna is placed in one end of the waveguide and is excited at some high RF frequency. Both positive and negative half-cycles are radiated as shown in Figure 4.8.

The wave front produced is like that of an expanding circle. The part that travels in the direction of arrow C goes straight down the waveguide and is quickly attenuated. However, the part of the wave front that travels in the direction of arrow A is reflected from the side wall. The side wall is a short circuit and causes the wave front to be reflected 180° out-of-phase. Meanwhile, the wave front that travels in direction B is reflected in opposite phase as well. The radiation fields are contained within the waveguide as illustrated in Figure 4.9a.

Looking through the top wall of the waveguide, as shown in Figure 4.9b, the light solid and broken lines represent the wave front going in direction A of Figure 4.8. The heavy lines and dashes represent the wave front going in direction B. Note that all parts of the wave front A are traveling upward at an angle across the guide. Wave front B is traveling at the same angle but downward. (This effect can also be seen by observing a water wave along the shoreline.)

When the wave travels in this fashion in a waveguide, propagation is possible. It resembles light propagating within a fiber optic cable. In understanding what happens, note that positive wave fronts (represented by solid lines) occur simultaneously throughout the center of the guide. These fronts add and cause a maximum voltage to occur at the center. This is shown as a maximum E field at the center in Figure 4.9c. The negative wave front adds in the same manner as the positive wave front. When the negative wave front meets the positive wave front at the side walls, the two cancel, making the net voltage zero. At a given frequency, only a single wave with angles A and B satisfies the boundary conditions that produce zero voltage at the walls. This verifies the E-field condition shown in Figure 4.9c.

Figure 4.8   How radiation fields fit into a waveguide, using an antenna to launch the signal. Energy in directions A and B reflects, while energy in direction C attenuates.

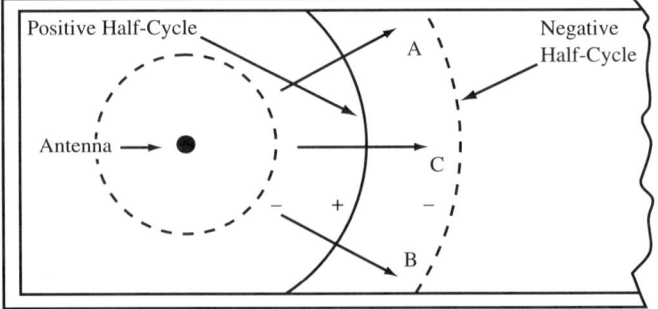

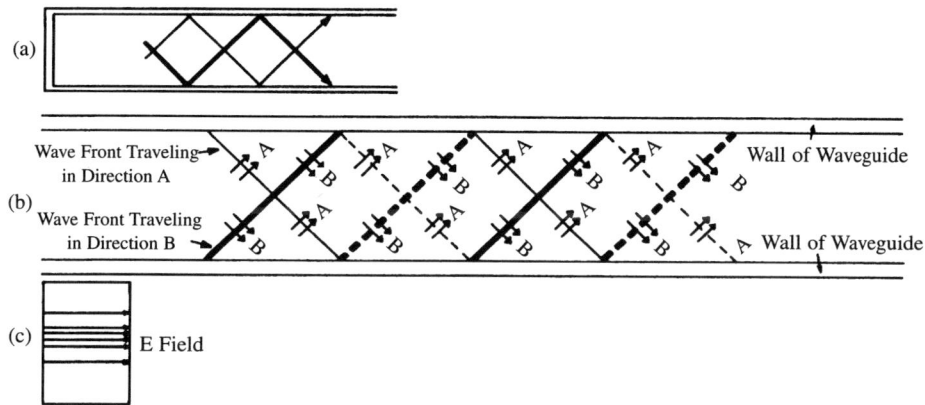

Figure 4.9 Paths of wave fronts in waveguide.

The sine of the angle at which a wave front crosses a waveguide is a function of the wavelength and the cross-sectional dimension of the guide. The angle is calculated as shown in equation 4.5.

$$\theta = \sin^{-1} \lambda/\lambda_c$$ **(Eq. 4.5)**

where
θ = crossing angle, degrees
λ = wavelength of the signal
λ_c = cutoff wavelength of the guide

At some intermediate frequency the reflection is as shown in Figure 4.10b. But as the frequency increases, the angle of incidence becomes less, and the signal travels farther before it reaches the other side wall (see Figure 4.10a). At lower frequencies (Figure 4.10c) the wave front crosses the guide at more nearly right angles to the walls. At some frequency the angle is 90°. This frequency is the cutoff frequency. At this point, energy does not propagate but is dissipated by the resistance of the walls. Quantitatively, some examples will show the calculations for various crossing angles as follows.

EXAMPLE 4.5:

Using equation 4.5, equation 1.4, and λ_c from example 4.3, find the crossing angles for a WR-90 waveguide when the frequency of the signal is 6.56 GHz, 10 GHz, and 50 GHz respectively.

$$\theta = \sin^{-1} (4.572 \text{ cm}/4.572 \text{ cm})$$
$$= \sin^{-1} 1.0$$
$$= 90° \text{ (for 6.56 GHz)}$$
$$\theta = \sin^{-1} (3.0 \text{ cm}/4.572 \text{ cm})$$
$$= \sin^{-1} 0.656$$
$$= 41° \text{ (for 10 GHz)}$$
$$\theta = \sin^{-1} (0.6 \text{ cm}/4.572 \text{cm})$$
$$= \sin^{-1} 0.131$$
$$= 7.5° \text{ (for 50 GHz)}$$

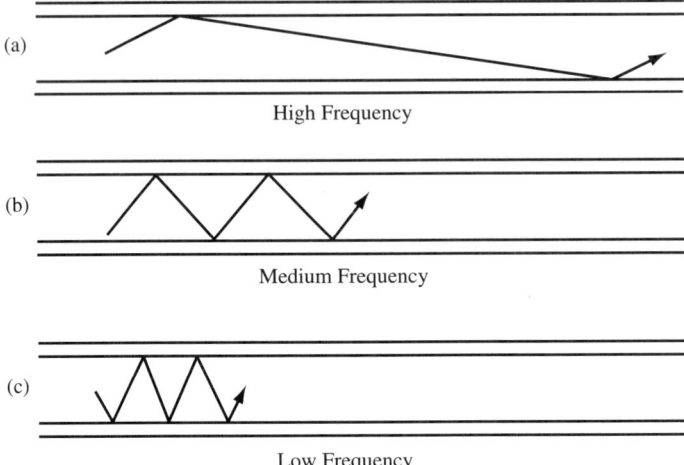

(a)

High Frequency

(b)

Medium Frequency

(c)

Low Frequency

Figure 4.10   Relative angles of reflection in a waveguide for various frequencies.

Waveguide Velocities

The velocity of propagation of a wave inside the waveguide is less than in free space. This lower velocity is due to the way the field travels. As shown in Figure 4.10c, the path of a wave front at a relatively low microwave frequency is along the zigzag arrow. But because of the long path, the wave front actually travels very slowly along the waveguide. In Figure 4.10a the frequency is higher, and the wave front or the group of waves actually travels a given distance in less time than those in Figure 4.10c.

The axial velocity of a wave front or a group of waves is called the *group velocity*. The relationship of the group velocity to diagonal velocity causes an unusual phenomenon. The velocity of propagation appears to be greater than the speed of light. As can be see in Figure 4.11, during a given time, a wave front moves from point 1 to point 2, or a distance L, at the velocity of light V_C. Due to the diagonal movement (direction of arrow), the wave front has actually moved down the guide only the distance G, so, in effect, forward movement is at a lower velocity. This is called group velocity V_G.

If an instrument were used to detect the two positions at the wall, they would be the distance P apart. This is greater than the distance L or G. The movement of the contact point between the wave and the wall is at a greater velocity. Since the phase of the signal

Figure 4.11   Relation of phase, group, and wave front velocity.

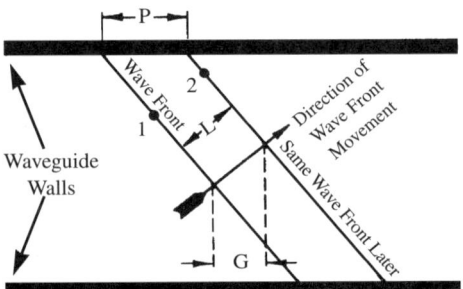

has changed over distance *P*, this velocity is called the *phase velocity* V_P. The mathematical relationship among the three velocities is stated by equation 4.6.

$$V_C = \sqrt{V_P V_G}$$ (Eq. 4.6)

where
V_C = velocity of light, c
V_P = phase velocity
V_G = group velocity

This equation indicates that it is possible for the phase velocity to be greater than the velocity of light (or apparently so). As the frequency decreases, the crossing angle increases toward 90°. In this condition the phase velocity increases. For measuring standing waves in a waveguide, it is the phase velocity that determines the distance between V_{max} and V_{min}. For this reason, *the wavelength measured in the guide is greater than the wavelength in free space.*

From a practical standpoint, the different velocities are related in the following manner: When an RF signal being propagated is sine-wave modulated, the modulation envelope moves forward through the guide at the group velocity, while the individual cycles of RF energy move forward through the modulation envelope at the phase velocity.

Since the standing wave measuring equipment is affected by each RF cycle, the wavelength is governed by the rapid movement of the changes in RF voltage. Since intelligence is conveyed by the modulation, the transfer of intelligence through the waveguide is slower than the speed of light, as is the case in other types of transmission lines.

Propagation Modes

TEM modes are the principal modes of propagation for all low-frequency transmission lines (those that have two wire conductors). TEM modes are not allowed in waveguides, since a waveguide is considered a single-wire conductor. The boundary conditions for propagation are not met by the TEM mode. The two modes that are allowed are the same two higher-order modes that can exist in coaxial lines. These are the TE and TM modes as shown in Figure 4.12.

Figure 4.12 Waveguide modes.

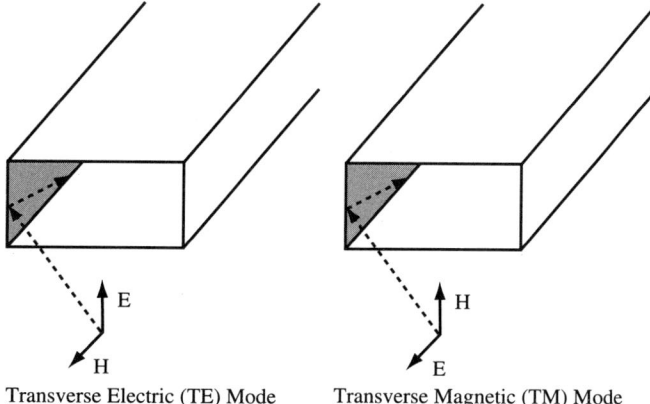

Transverse Electric (TE) Mode   Transverse Magnetic (TM) Mode

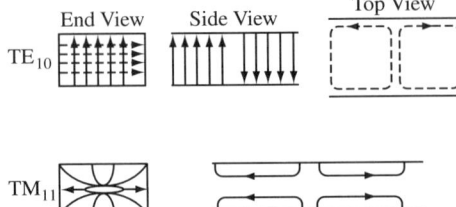

Figure 4.13 Dominant modes in rectangular waveguide.

As with coaxial lines, the lowest frequency for which it is possible to propagate through a specific waveguide is referred to as the principal, or dominant, wave mode. For the principal mode, a one-half wavelength exists across the waveguide width, permitting a zero E field to occur at the side walls. At this principal mode frequency, the waveguide acts resistive. Greater multiples of one-half wavelengths would then result in higher-order modes of propagation.

Higher-order modes are designated with subscripts *m* and *n* (just as for coaxial lines). The lowercase *m* indicates the number of half-wave variations of the electric field or magnetic field along the wide dimension of the waveguide. The lowercase *n* indicates the number of half-wave variations of the electric or magnetic field along the narrow dimension. The dominant TE mode is TE_{10}, while the dominant TM mode is TM_{11}. These are shown in Figure 4.13. The dominant mode for a waveguide has the following characteristics:

1. It has the longest operating wavelength.
2. It has the greatest energy transfer efficiency.
3. It has the simplest field configuration.
4. It is the easiest mode to induce (or extract) in a waveguide.

Guide Wavelength

Since it is possible to have the phase velocity greater than the speed of light, it follows that for frequency to remain constant, the wavelength of the signal within the guide must increase. This is shown by inspection of equation 4.7. The guide wavelength increases as V_P increases, thus keeping the frequency constant.

$$f = V_P / \lambda_g \qquad \text{(Eq. 4.7)}$$

where

f = frequency, Hz

V_P = phase velocity

λ_g = guide wavelength, m

The guide wavelength is a function of the geometries of the guide, and its calculation is shown in equation 4.8.

$$\lambda_g = \lambda_0 / \sqrt{1 - (\lambda_0/\lambda_c)^2} \qquad \text{(Eq. 4.8)}$$

where

λ_g = guide wavelength, m

λ_0 = free space wavelength, m

λ_c = cutoff wavelength, m

EXAMPLE 4.6:

Calculate the λ_g of a WR-90 rectangular waveguide at 10.6 GHz.

$$\lambda_g = 0.0283/\sqrt{[1 - (0.0283/0.04572)^2]}$$
$$= 0.0283/0.785$$
$$= 0.036 \text{ m} = 3.6 \text{ cm} = 36 \text{ mm}$$

Wave Impedance Within a Waveguide

As was shown in chapter 1, the characteristic impedance of free space (or air) is 377 Ω. Any TEM wave traversing this medium is subject to a wave impedance of 377 Ω.

A similar situation exists for EM waves (not TEM waves) in air-filled guide structures where, because of wave orientation and guide geometry, the wave impedance (Z_{TE}) is never quite equal to 377 Ω. Moreover, the wave impedance is frequency dependent since it is a function of both signal wavelength and cutoff wavelength.

For rectangular waveguides propagating TE modes, the wave impedance is given by equation 4.9.

$$Z_{TE} = 377/\sqrt{1 - (\lambda/\lambda_c)^2} \qquad \textbf{(Eq. 4.9)}$$

where
Z_{TE} = wave impedance

λ = wavelength of the signal

λ_c = cutoff wavelength of the guide

Similarly, for guides propagating TM modes, the impedance is found by equation 4.10.

$$Z_{TM} = 377\sqrt{1 - (\lambda/\lambda_c)^2} \qquad \textbf{(Eq. 4.10)}$$

Note that in the TE mode (the more common mode), as long as the signal frequency is kept higher than the cutoff frequency, the signal wavelength is smaller than the cutoff wavelength. In this case the radicand of equation 4.9 approaches unity, and Z_{TE} approaches 377 Ω, but is always higher than 377 Ω. When the signal frequency is equal to or lower than the cutoff frequency, the radicand of the equation approaches zero, and Z_{TE} approaches infinity. The signal is now considered to be evanescent (having vanished). This also demonstrates that signal frequencies below the cutoff frequency of the guide cannot propagate within the guide.

4.5   RIDGED WAVEGUIDE

Ridged waveguide has properties that increase the bandwidth of operation. The cutoff frequency is lowered while the higher-modes have their cutoff frequencies elevated. Thus, ridged waveguide is a broadband device. The addition of the ridge lessens the power-handling capability of the guide. The impedance is lower than that of rectangular waveguide.

The impedance of the guide varies with the addition of the ridge. One major application is to use ridged waveguide as an impedance matching device since its impedance can easily be changed by tapering its cross section gradually. Examples of ridged waveguide are shown in Figure 4.14.

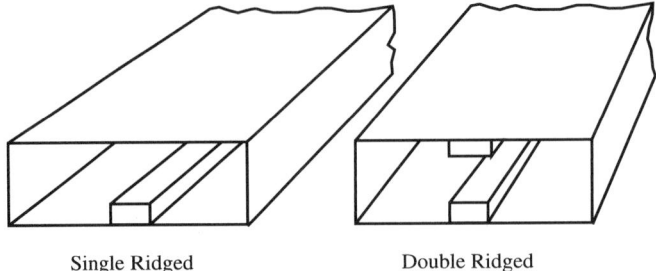

Single Ridged   Double Ridged

Figure 4.14   Ridged waveguide.

4.6 CIRCULAR WAVEGUIDE

Circular waveguide is another common form. It has characteristics very similar to coaxial lines, and in fact, it has the same dominant or principal modes as the higher-order modes of the coaxial line. These are TE_{11} and TM_{01}. The major difference is the m and n integers refer to the number of full wavelengths around the circumference of the inner dimension and the number of one-half wavelengths across the inner diameter, respectively. The dominant modes are shown in Figure 4.15.

Circular guide is larger than rectangular guide for the same frequency band, and this can be considered a disadvantage. However, its circular shape lends itself to rotary usage for radar antennas. It has higher power-handling capability than a comparably sized rectangular guide. A circular guide can support both vertically and horizontally polarized signals simultaneously. Circular guide is most often used in vertical runs between the source and antenna.

Cutoff wavelength is more difficult to calculate than it was for rectangular guide. This is because we must use the Bessel functions for the common modes used. The cutoff wavelength is calculated using equation 4.11. Bessel functions for many common modes are listed in Table 4.3.

$$\lambda_c = 2\pi r / B_{(m,n)} \qquad \textbf{(Eq. 4.11)}$$

where

λ_c = cutoff wavelength, m

r = radius of guide, m

B = Bessel roots for given mode

Figure 4.15   Dominant circular waveguide modes.

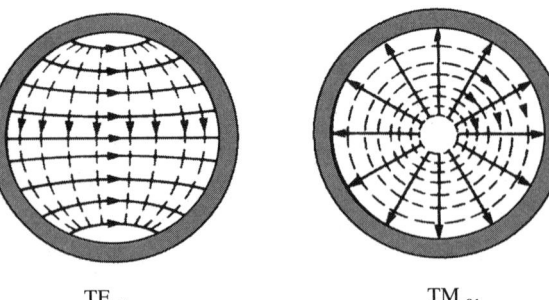

TE_{11} TM_{01}

Table 4.3 Bessel Function
Solutions (Partial Listing)

Mode	$B_{(m,n)}$
TE_{01}	3.83
TE_{11} (Dominant)	1.84
TE_{21}	3.05
TE_{02}	7.02
TE_{12}	5.33
TE_{22}	6.71
TM_{01} (Symmetrical)	2.40
TM_{11}	3.83
TM_{21}	5.14
TM_{02}	5.52
TM_{12}	7.02
TM_{22}	8.42

EXAMPLE 4.7:

Determine the cutoff wavelength for a circular guide when $r = 0.04$ m for a mode of TM_{02}.

$$\lambda_c = 2\pi(0.04)/5.52$$
$$= 0.0455 \text{ m}$$

or
$$= 4.55 \text{ cm}$$

Only one dimension can be changed in a circular guide, the diameter (or radius). Therefore, it is not possible to suppress higher-order modes by the choice of just physical dimensions. Further, the cutoff frequencies of the higher-order modes are relatively close to that of the dominant mode, so the use of circular guide is usually avoided except in the cases stated earlier.

4.7   ELLIPTICAL WAVEGUIDE

Elliptical waveguide is one of the most commonly used feeders for microwave systems operating in the range of 3.4 to 23.6 GHz. The amount of "plumbing" is reduced, since this waveguide comes in 400-foot rolls and is flexible. The waveguide is oval and usually made of corrugated copper or aluminum. The corrugated copper walls give elliptical waveguide excellent crush strength and good flexibility. It is available in two grades, standard and premium. The difference is the amount of energy loss the microwave signal experiences due to reflections as it propagates. The most affected parameter is the VSWR. It is lower in premium waveguide. Premium assemblies are recommended for long-haul or high-channel density systems. Standard assemblies are recommended for short- and medium-haul radio relay systems with low- and medium-channel densities and with medium-haul color television microwave relay systems.

The dominant mode of operation for elliptical guide is TE_{11}, which is comparable to rectangular guide TE_{10}. Operating in the frequency band where only the dominant mode can exist eliminates signal distortion due to mode conversion and minimizes VSWR.

Since the connections to the radio terminal and antenna are usually designed for rectangular waveguide, it is necessary to use a transition (or connector) to connect the elliptical waveguide. These connectors may incorporate a tuning mechanism that, by the use of

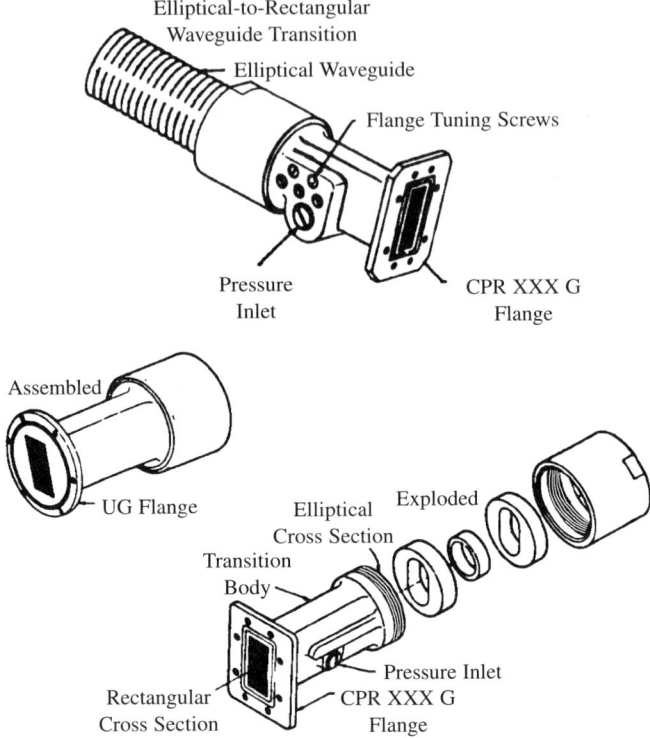

Figure 4.16   Elliptical-to-rectangular transition.

adjustable screw probes, adjusts the E field to compensate for mismatches. A typical tunable elliptical-rectangular transition is illustrated in Figure 4.16. The connector is usually provided with a pressure inlet through which dry air or nitrogen may be supplied, under pressure, to the waveguide.

4.8   WAVEGUIDE DISCONTINUITIES

Whenever there is an abrupt change in the dimensions of the waveguide, the E and H fields make a transition across the region. Higher-order modes propagate at the point of these abrupt changes or discontinuities. This effectively changes the impedance of the guide at this point. The higher-order modes rapidly attenuate outside this region because the guide is physically made for a dominant mode.

These impedance changes can be used to match impedances along the guide. Shunt susceptances can be made in a variety of ways and attached to the guide. A capacitive iris, an example of a shunt susceptance, is a thin piece of metal, as shown in Figure 4.17a. This metal diaphragm causes the E and H fields to change in order to satisfy the boundary conditions. This generates higher-order modes and changes the impedance at this point.

An inductive iris is shown in Figure 4.17b. Another type of discontinuity is shown in Figure 4.17c. The length of the protruding post determines whether it acts capacitively or inductively; the smaller the length, the more capacitive it acts. Similar results can be accomplished with a tuning screw, as shown in Figure 4.17d.

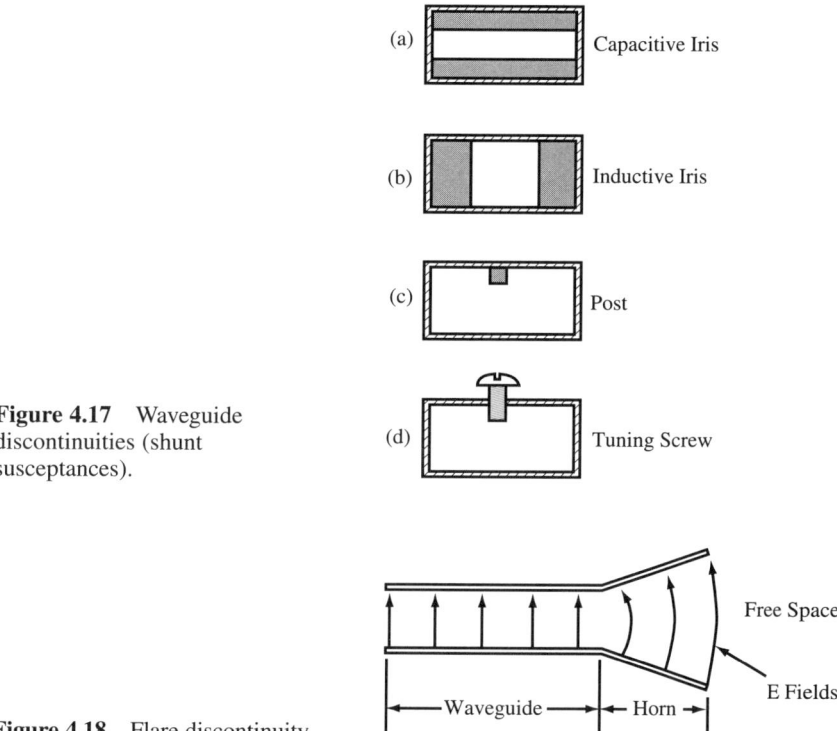

(a) Capacitive Iris

(b) Inductive Iris

(c) Post

(d) Tuning Screw

Figure 4.17 Waveguide discontinuities (shunt susceptances).

Figure 4.18 Flare discontinuity.

Free Space

E Fields

Waveguide — Horn

Another common discontinuity is to flare one end of a section of rectangular waveguide. This is shown in Figure 4.18. This effectively converts the guide into a horn antenna that radiates the propagating energy into free space. The E field has to change as the flare occurs, causing an impedance change at that point. This impedance change causes a match between the horn end and the free space impedance of 377 Ω.

4.9 METHODS OF EXCITING WAVEGUIDES

Energy must get into a waveguide somehow. We need to effectively radiate or launch energy into the dominate mode of propagation. This can be accomplished through four basic methods.

Capacitive coupling is one of the easiest methods. It involves simply putting a section of the center conductor of a coaxial line into one end of a rectangular guide. This is usually done $\lambda/4$ away from a closed end as shown in Figure 4.19a. This method effectively launches a TE_{10} mode of propagation, but it can be used for TM modes as well depending on placement. Mode jumping is restricted because of the placement of the probe. The same method and placement can be used to get power back out of the waveguide. Commercial coaxial-to-waveguide adapters are available to perform this task.

Figure 4.19b shows *inductive coupling* via a loop in one end of the guide. The loop can be in the end or at the side wall. Either TM (preferred) or TE modes can be launched this way.

Radiation can occur through slots or holes in the walls of the waveguide. *Slot couplers,* shown in Figure 4.19c, are efficient and make good couplers at high frequencies.

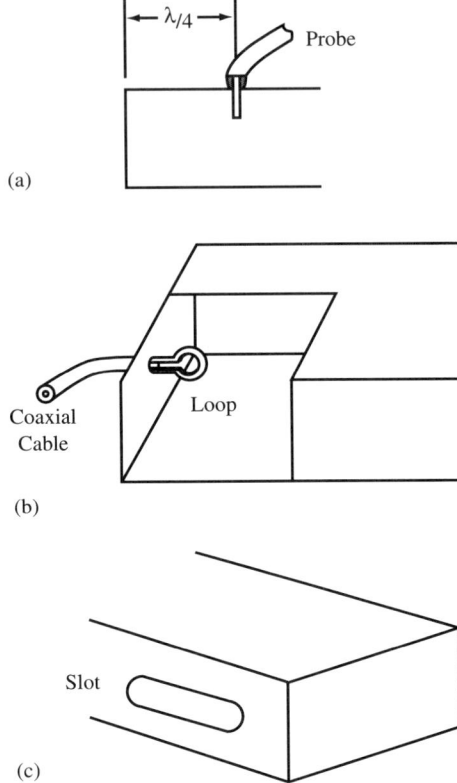

Figure 4.19 Waveguide coupling methods.

(a)

(b)

(c)

Hole coupling is used in directional couplers (chapter 5) in the walls between the main and auxiliary arms of the guide.

The fourth method of coupling was shown in Figure 4.18. This is the *horn antenna* that can effectively couple energy into or out of the guide.

4.10   STRIPLINE AND MICROSTRIP

Stripline is a general class of miniature transmission lines. Stripline can take the form of a flattened coaxial line, referred to as embedded stripline or, in some references, as triplate. When one conductor plane is removed and the center conductor is on top of the substrate, the line is known as a microstrip. Examples of these are shown in Figure 4.20.

Stripline has the advantage of being fabricated with a wide variety of dielectric materials and substrates using printed circuit techniques. It is smaller than waveguide and coaxial lines and has a greater bandwidth than either. Operating frequencies start at the L band (1.0 GHz) and can go into the millimeter band (above 40 GHz).

Stripline is a two-wire structure, so propagation is done via the TEM mode. Most other two-wire characteristics are also true for stripline. Radiation loss is minimized due to the construction nature of the embedded stripline, but dielectric and skin effect losses are common.

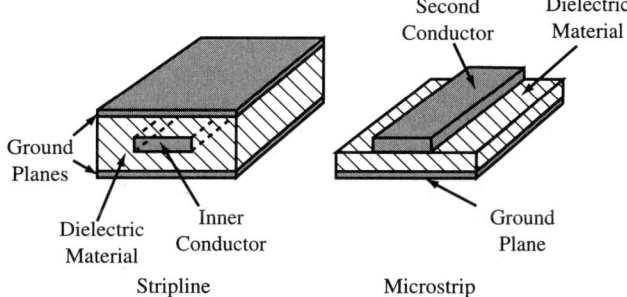

Figure 4.20 Miniature transmission lines.

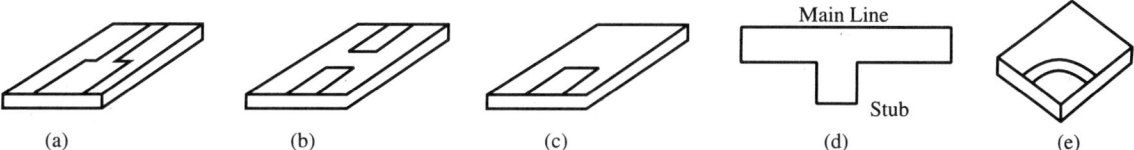

Figure 4.21 Microstrip discontinuity patterns.

    The microstrip line is one of the most commonly used transmission lines. With the advent of high-frequency and low-power MMIC and HMIC technologies, the microstrip can have its associated passive and active components fabricated directly atop or embedded within the printed circuit material. Both monolithic ICs and hybrids can be used. The copper traces become the conducting paths of the transmission line. By changing the trace patterns, discontinuities can be formed to make equivalent LC parameters or to use in impedance matching. A step in width is shown in Figure 4.21a and shows that impedance mismatches occur with width changes. This step change can be used to match an existing mismatch. Figure 4.21b shows a gap, which is equivalent to a series capacitor. An open end is shown in Figure 4.21c. This can be used as a radiator. The stub shown in Figure 4.21d is used for impedance matching. Because abrupt right-angle corners can cause capacitance effects, this shunt capacitance effect can be minimized with smooth radius corners, as shown in Figure 4.21e. Design for these circuits is done using modern CAD/CAM and simulation software, with testing done via high-frequency spectrum and network analyzers.

    Attenuation losses for the microstrip are ohmic (skin effect), dielectric, or radiative. Because a conductor plane has been removed from the top, effective shielding is lost. This means that microstrip has greater losses due to radiation. Most losses are frequency dependent. The relative small size of microstrip circuits means that most losses are tolerable. Newer substrates and dielectrics can also help overcome losses.

4.11   SUMMARY

1. The four general categories of transmission lines, are (1) two-wire, (2), coaxial, (3) waveguide, and (4) stripline.

2. Two-wire lines exhibit radiation loss, dielectric loss, and skin effect loss.

3. The coaxial line effectively replaces the two-wire. It has limited power-handling capability, as well as dielectricl loss and skin effect loss.

4. Coaxial lines are very common in microwave radio.

5. Higher-order modes can propagate in coaxial lines or waveguides due to the geometry of the line and the frequency of operation.

6. Attenuation increases with frequency for most transmission lines.

7. The three basic types of waveguide are rigid rectangular, circular, and elliptical. Each can be made of copper, brass, or aluminum.

8. Rigid rectangular waveguide and circular waveguide are most often used in higher-power radar applications.

9. Circular waveguide is used mostly for vertical runs between source and antenna.

10. Elliptical waveguide is one of the most commonly used guides in microwave radio and television relay systems.

11. Waveguide discontinuities can be used for impedance matching purposes.

12. The common methods for exciting a signal into a waveguide are (1) probe (capacitive coupling), (2) loop (inductive coupling), (3) slots or holes, and (4) a horn.

13. Stripline is a form of a "flattened" coaxial line.

14. Microstrip is a form of stripline and is commonly used from 1.0 GHz to more than 40 GHz. It is one of the most popular of the microwave transmission lines.

Key Equations:

$$Z_O = (276/\sqrt{\epsilon_r})(\log_{10} 2D/d) \qquad \textbf{(Eq. 4.1)}$$

$$Z_O = (138/\sqrt{\epsilon_r})(\log_{10} D/d) \qquad \textbf{(Eq. 4.2)}$$

$$\lambda_c = 2a \qquad \textbf{(Eq. 4.3)}$$

$$f_c = c/\lambda_c \qquad \textbf{(Eq. 4.4)}$$

$$\theta = \sin^{-1} \lambda/\lambda_c \qquad \textbf{(Eq. 4.5)}$$

$$V_C = \sqrt{V_P V_G} \qquad \textbf{(Eq. 4.6)}$$

$$f = V_P/\lambda_g \qquad \textbf{(Eq. 4.7)}$$

$$\lambda_g = \lambda_o/\sqrt{1 - (\lambda_o/\lambda_c)^2} \qquad \textbf{(Eq. 4.8)}$$

$$Z_{TE} = 377/\sqrt{1 - (\lambda/\lambda_c)^2} \qquad \textbf{(Eq. 4.9)}$$

$$Z_{TM} = 377\sqrt{1 - (\lambda/\lambda_c)^2} \qquad \textbf{(Eq. 4.10)}$$

$$\lambda_c = 2\pi r/B_{(m,n)} \qquad \textbf{(Eq. 4.11)}$$

PROBLEMS

1. Calculate the characteristic impedance of a two-wire line where the diameter of the conductors is 1.6 mm and the spacing between centers is 0.85 cm. Assume that air is the dielectric.

2. Calculate Z_O for a two-wire line when each conductor has a radius of 2.5 mm and the spacing between centers is 1.4 cm. Assume that air is the dielectric.

3. Repeat problem 2 using a new medium having a dielectric constant of 2.25.

4. The outer conductor of a coaxial cable has a diameter of 2.25 cm and the inner conductor has a diameter of 1.75 mm. Using polyethylene (ϵ_r = 2.26) as the insulator, determine the value of Z_O.

5. Determine the cutoff wavelength for a WR-75 waveguide in inches. (Use Table 4.1 as a guide.)

6. Convert your answer to problem 5 into mm.

7. Determine the cutoff frequency for the waveguide of problem 5.

8. Determine the cutoff frequency for a waveguide with an *a* dimension of 18 mm.

9. Determine the crossing angle the wave front makes in a WR-51 waveguide when the signal frequency is 12 GHz.

10. Repeat problem 9 using a frequency of 30 GHz.

11. Calculate λ_g for a WR-51 waveguide at 14 GHz.

12. Determine the cutoff wavelength of a 3.815 cm diameter circular waveguide operated in the TE_{11} mode.

QUESTIONS

1. Which of the transmission lines are best for low-power, high-frequency work?

2. Describe how the coaxial line is constructed.

3. Describe the difference between unbalanced and balanced line.

4. Explain the nature of energy loss in the dielectric.

5. What is the dominant mode of propagation in a coaxial line?

6. What are the higher-order modes in coaxial line propagation? Are they preferred? Why or why not?

7. The rectangular waveguide can be modeled as an extension of the _____-wave stub.

8. The dominant modes in the waveguide are

_____.

9. List three different velocities pertaining to a waveguide.

10. Why are short lengths of waveguide coated inside with gold or silver?

11. What does "WR-75" refer to?

12. What is meant by the cutoff wavelength? the cutoff frequency?

13. The waveguide may be considered to be a _____ filter.

14. Describe the propagation path in a rectangular waveguide.

15. Describe the uses of the ridged waveguide.

16. Describe the advantages and disadvantages of circular waveguide.

17. Describe several forms of waveguide discontinuities and their possible uses.

18. How can energy be excited into or out of a waveguide?

19. Describe the characteristics of the stripline.

20. Describe the characteristics of the microstrip.

5   PASSIVE MICROWAVE COMPONENTS

OBJECTIVES

1. To identify many common passive microwave components.
2. To describe the operating characteristics of common passive microwave components.
3. To describe applications where passive microwave components are used.

5.1   INTRODUCTION

Microwave components can be identified as either passive or active devices that are commonly used in the laboratory or on test benches. However, many components are used in the field as well to meet physical and electrical system requirements. Passive components include antennas (chapter 8), attenuators, connectors, joints, couplers, filters, and nonreciprocal devices. Thermistors and diode detectors are passive components, but they are used in circuitry with active components. Active devices are covered in chapters 6 and 7.

Passive microwave components can terminate the wave, split the signal into two or more paths, control the direction of the signal, switch power, reduce power (attenuate), sample power, mix signals, detect signals, and shift the phase of the signal.

5.2   CONNECTORS

Connectors are used primarily with coaxial lines or as adapters to waveguide or stripline assemblies. As adapters, their purpose is to "launch" the microwave signal into the waveguide or stripline. Connectors come in the familiar "male" and "female" connector pairs. Connectors must meet minimum losses, minimum reflection coefficients, and the like. The standard characteristic impedance is 50 Ω. Connector fittings and threads are standardized for compatibility in interconnecting. Connectors come in standard and subminiature sizes. Table 5.1 lists some of the common connector types.

One of the most common connectors is the bayonet-type BNC (Bayonet Navy Connector). The BNC is used in frequency ranges up to 4 GHz. The TNC (Threaded Navy Connector) has an improved thread design and can be used to 12 GHz. Miniature connectors such as the SMC (Subminiature C) are useful to 7 GHz and are recommended in applications where interconnections are few. The SMA (Subminiature A) is useful to 24 GHz. It

Table 5.1 Common Connector
Types

Connector Type	Maximum Frequency (GHz)
APC-2.4	50
APC-3.5	34
APC-7	18
1.85 mm	65
BNC	4
TNC	12
SMA	24
SMC	7
Type K	40
Type N	18

is one of the least expensive microwave connectors. APC is a registered trademark of the Amphenol Corporation, which makes precision connectors that come in 7 mm, 3.5 mm, and 2.4 mm sizes. The 7 mm is the preferred connector for the most demanding applications; it works to 18 GHz. The 3.5 mm can be mated with an SMA connector and is a more rugged choice. It can work to 34 GHz. The 2.4 mm works to 50 GHz and is a more rugged connector than the SMA. The 1.85 mm was developed by Hewlett-Packard and works to 65 GHz. It was offered as public domain in 1988. Many experts consider this connector to be the smallest possible coaxial connector for common usage, making waveguide the choice above 65 GHz. Type K is a registered trademark of the Wiltron Corporation; these connectors are used to 40 GHz. The type N series (the original Navy connector, circa World War II) can be used to 12 GHz. By eliminating the slots in the outer conductor, type N connectors can work to 18 GHz. Assorted coaxial connectors and their operating frequencies and communications applications are shown in Figure 5.1.

Connector bodies are made of corrosion resistant steel, beryllium copper, phosphor bronze, or brass. Connector finishes include gold plate, nontarnish silver plate, silver plate, and passivated treatments for resistance to corrosion. Many connectors are made in a rugged form as well.

5.3   DIRECTIONAL COUPLERS

A directional coupler allows a designated (sample) portion of the microwave power traveling in the main line to be coupled to the secondary (auxiliary) arm in a preferred direction of flow. The operation of a two-hole coupler can be shown with the aid of the diagram in Figure 5.2. When a signal is fed into the main waveguide at port 1, a fraction of that signal couples into the secondary waveguide at the first hole. This signal splits into two directions, 2 and 5. Meanwhile, in the main waveguide the signal couples into the secondary waveguide again at the second hole. This hole is one-quarter wavelength ($\lambda_g/4$) from the first hole. Energy from the second hole splits into two signals, 3 and 4. Signals 4 and 5 have traveled the same distance from port 1, so they are in phase and add to become the coupled output at port 3. However, as you can see, signal 3 has traveled one-half wavelength farther than signal 2. This means that 2 and 3 have opposite phase and therefore cancel. Any signal not fully canceled is absorbed by the resistive termination.

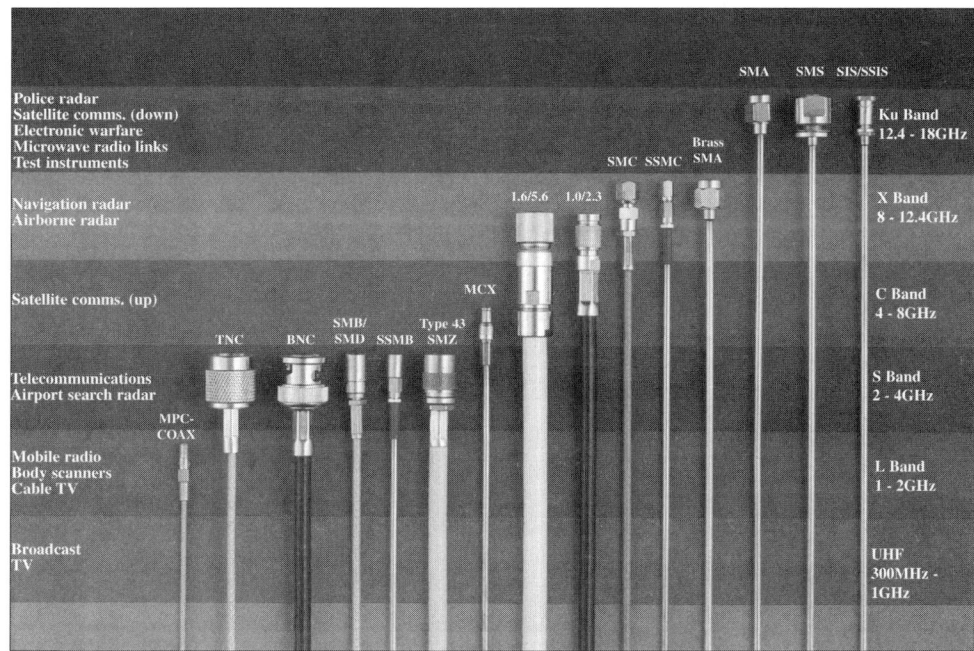

Figure 5.1   Assorted coaxial connectors. (Photo courtesy of ITT Cannon RF Products.)

Directional couplers are usually used in various types of bench tests and measurements. A 20 dB directional coupler allows 1% of the main-line power to be coupled to the secondary arm. Common coupling factors are 10 dB, 20 dB, and 30 dB. The coupling factor *(CF)* can be found using equation 5.1. The coupling factor is given in positive dB values, thus the (–) sign in the equation.

$$CF_{(dB)} = -10 \log_{10} (P_3/P_1)$$ **(Eq. 5.1)**

where $CF_{(dB)}$ = coupling factor

P_3 = coupled port power

P_1 = input port power

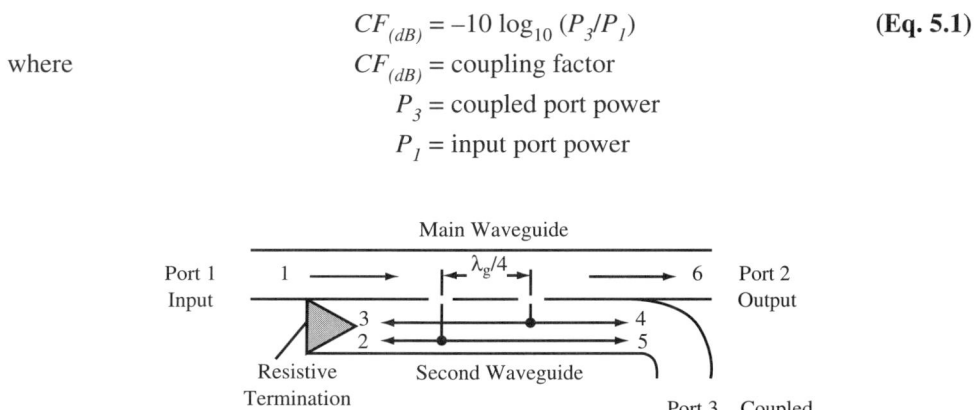

Figure 5.2   Two-hole directional coupler. Signals 4 and 5 are in phase because they have traveled the same distance from port 1. Signal 3 has traveled one-half wavelength farther from port 1 than signal 2 has, so signals 2 and 3 have opposite phase and cancel.

EXAMPLE 5.1:

Determine the coupling factor if the input power P_1 is 228 mW and the coupled power P_3 is 228 µW.

$$CF_{(dB)} = -10 \log_{10} (228 \text{ µW}/228 \text{ mW})$$
$$= -10 \log_{10} .001$$
$$= 30 \text{ dB}$$

Note that the power out of the main arm is simply the difference between the input power and the secondary arm power.

In the example just given, the main arm output power is 228 mW – 228 µW or 227.772 mW.

When the power in and power coupled out of a coupler are specified in dBm, the coupling factor in decibels is the difference of the two values.

$$CF_{(dB)} = \text{dBm}_1 - \text{dBm}_2 \qquad \textbf{(Eq. 5.2)}$$

where

$$\text{dBm}_1 = \text{input power}$$
$$\text{dBm}_2 = \text{coupled power}$$

EXAMPLE 5.2:

What is the coupling factor of a coupler, in decibels, if the coupled power at the secondary arm is –40 dBm and the input power is –20 dBm?

$$CF_{(dB)} = -20 - (-40)$$
$$= 20 \text{ dB}$$

Coupling is done via a hole or holes between the arms of the coupler. A Bethe hole coupler employs a single hole in the wide dimension of the guide. This type improves the degree of signal directivity into the secondary arm.

The holes need not be circular. A common hole form takes the shape of a cross. Although more difficult to make, the discontinuity is less than that produced by a circular hole.

In real devices, some power appears at the resistive termination and a measure of the quality of the coupler is in its directivity. The more energy delivered to the coupled output at port 3 (the forward direction) instead of the termination (the reverse direction) yields a higher quality factor. Directivity can be calculated using equation 5.3. A typical value is around 30 dB. This indicates that the ratio of forward to reverse power is 1,000 to 1. A 40 dB value indicates a ratio of forward to reverse power of 10,000 to 1.

$$D_{(dB)} = 10 \log_{10} \frac{\text{Output Power (forward)}}{\text{Output Power (reverse)}} \qquad \textbf{(Eq. 5.3)}$$

EXAMPLE 5.3:

If the forward power at the secondary arm output is 100 mW and the reverse power is determined to be 10 µW, find the directivity in decibels.

$$D_{(dB)} = 10 \log_{10} (100 \text{ mW}/10 \mu\text{W})$$
$$= 10 \log_{10} 10,000$$
$$= 40 \text{ dB}$$

The directional coupler shown in Figure 5.3 (along with symbol) is from Hewlett-Packard's HP X752A series and have three ports. Some directional couplers have four ports. In the case where just three are used, the fourth port is terminated with a matched termination in order to reduce reflections.

5.4   ATTENUATORS

Attenuators are components that can reduce the microwave power by absorbing energy and dissipating heat. There are two types of attenuators: fixed and variable. Terminators (terminations) are a special kind of attenuator.

Attenuation can be accomplished in two ways. The first method uses graphitized sand located in one end. When a microwave signal encounters the sand, the power loss is generated as heat. This method is most useful for terminations. The second method uses a resistive rod or vane placed at the center of the electric field. This is the center of the waveguide propagating the principal TE_{10} mode. Microwave energy induces current into the vane, which results in an ohmic power loss. The vane method is useful for variable attenuators, since the vane can rotate or slide between the edge and center of the guide.

Fixed attenuators reduce the input signal power by a fixed amount, such as 3 dB, 6 dB, 10 dB. Variable attenuators can have a range from 0 to 60 dB, which can vary continuously or in steps. On variable attenuators the vane is manipulated by a knob. Scales on some attenuators can be calibrated into dB directly, while others use a micrometer adjustment in millimeters. In this case, a conversion table is needed between millimeters and dB values. A fixed and a variable attenuator are shown in Figure 5.4 along with their symbols.

Reflections can occur in the attenuator, so one factor to look for on a data sheet is the value of VSWR associated with the attenuator. Another factor to consider is the insertion loss. Insertion loss is the difference in power levels before and after a component is inserted in the line. Even with the vane moved to one side, some attenuation, usually around 0.5 dB, will occur due to an insertion loss.

Terminators totally absorb the incident energy and convert it to heat. Therefore, one end is closed because no output power is expected. A terminator is shown in Figure 5.5 along with its symbol. A terminator using a resistive rod or vane usually has the internal geometry tapered to minimize reflections. Terminators are extremely useful in bench tests to act as a matched load for various tests. They are also known as a "dummy" load.

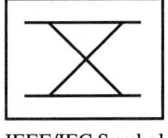

IEEE/IEC Symbol

Figure 5.3   Three-port waveguide directional couplers. (Photo courtesy of Hewlett-Packard Co.)

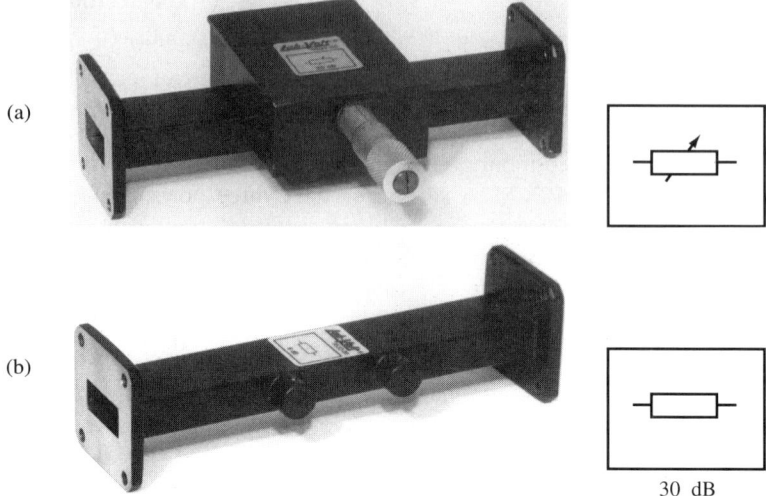

Figure 5.4   (a) Variable attenuator with symbol. (b) Fixed attenuator with symbol. (Photos courtesy of Lab-Volt Systems.)

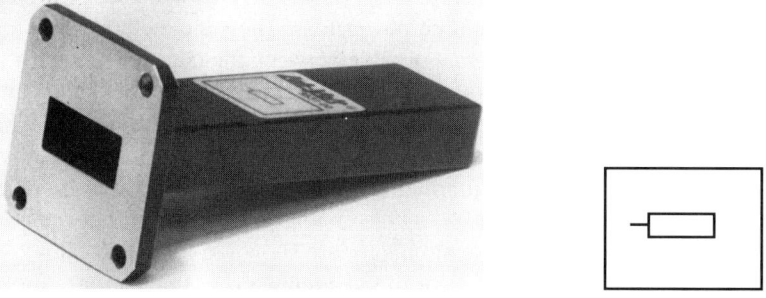

Figure 5.5   Termination with symbol. (Photo courtesy of Lab-Volt Systems.)

Attenuators are used to provide protection, reduce power, and extend the dynamic range of the test equipment. In choosing an attenuator, you must select an attenuator for the proper frequency range, since accuracy depends on employing it in the proper range.

5.5   ISOLATORS

Isolators are devices that pass electromagnetic energy in one direction only. Ferrites are at the heart of how isolators and circulators work. Attenuation and phase shifting are produced by using ferrites. The changes that occur depend on the direction the signal travels through the device. Components that possess these attributes are known as *non-reciprocal* devices.

Ferrites are ceramic materials containing compounds of iron with zinc, manganese, cobalt, aluminum, or nickel oxide in various combinations. Ferrites behave as iron alloys at low frequencies, but at high frequencies their high electrical resistance prevents eddy currents, and they exhibit resonance within the iron atoms themselves.

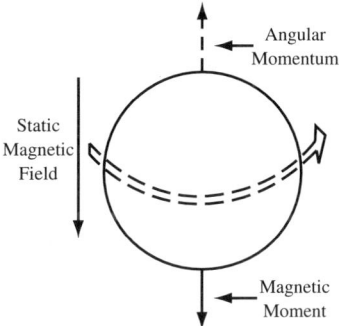

Figure 5.6 Spinning electrons
in DC magnetic field.

A fundamental property of atoms is that both electrons and protons spin on their axes. As an electron spins, it creates a magnetic moment, or field, along its spin axis. If a spinning electron is placed in a static (DC) magnetic field, the electron's magnetic moment becomes aligned with the static field. This action is similar to the way a permanent bar magnet can be used to align the magnetic domains within a piece of nonmagnetized material. The magnetic moment and its alignment with a DC magnetic field are shown in Figure 5.6.

Because of their spinning motion, electrons behave like very small gyroscopes. When a force is applied to the axis of an electron that would cause it to tilt, the electron behaves like a gyroscope. It *precesses*, or wobbles. Precession is defined as a movement of the axis of rotation at right angles to its original axis.

A steady magnetic field applied to the ferrite lines up the axes of spinning electrons. This field causes any random precession (wobbling) to die out quickly. If an RF signal is now applied at the ferrite's natural frequency, precession builds up. This increases the frictional damping effects of the atoms because the entire iron atom is vibrating. Energy is extracted from the RF field under these conditions and is dissipated as heat in the ferrite. Therefore, certain RF signals can be attenuated by ferrites under the right conditions. The natural precession frequency of an electron in a DC field is in the low GHz range, depending on the field strength of the static magnetic field.

In an isolator, ferrites allow microwave energy to travel in one direction but absorb (and attenuate) energy that is traveling in the opposite direction. When the electron's resonant frequency and the signal frequency are the same, precession occurs in the electrons within the ferrite. The frequencies can be made to be the same by changing either the magnetic field strength of the static field or the frequency of the microwave signal. The precession that occurs is due to the direction the microwave signal is traveling through the ferrite material. The ferrites absorb energy from the microwave signal and attenuate the signal when the signal passes through in a specific direction. A wave traveling in the opposite direction does not cause precession to occur, therefore no energy is absorbed from the wave. Phase shifting also occurs with attenuation and is maximum during maximum attenuation.

One application of the isolator is to place it between the signal source and the transmission line. If any impedance mismatches occur down the line, no reflections can come back to the source and cause loading or pulling of the source oscillator. Loading can change the oscillator's frequency or cause damage to it from high levels of reflected signals.

When a higher power level needs to be absorbed, the ferrite is placed in the center of the waveguide, with the static magnetic field placed outside the guide across the ferrite material. This type of isolator is called a *resonant absorption isolator.*

5.6   CIRCULATORS

A circulator is another device that takes advantage of the properties of ferrites. It is a three-port component in which the adjacent ports are effectively connected in one direction but are isolated in the reverse direction.

The internal structure and signal flow for a circulator are shown in Figure 5.7. The three ports are labeled A, B, and C. The arrows indicate the direction of signal flow. Hence, a signal entering port A exits at port B; no signal from port A appears at port C. Any signal incident at port B exits at port C, and so forth.

One application of a circulator is to separate the transmitted signal from the received signal at an antenna that can both transmit and receive signals. In this application the second circulator's magnetic field is reversed via bias current in the solenoid used to generate the magnetic field for this circulator. In the "transmit" mode very little power from the transmitter enters the second circulator. Any that does is dissipated in the matched load. In the "receive" mode, the magnetic bias is (externally) reversed for the second circulator, so that the signal from the antenna is coupled to the receiver. This is illustrated in Figure 5.8.

These ferrite switches have distinct advantages over diode switches because they use less DC power, they lead to less insertion loss, their operation is more reliable, and they can work at higher RF power.

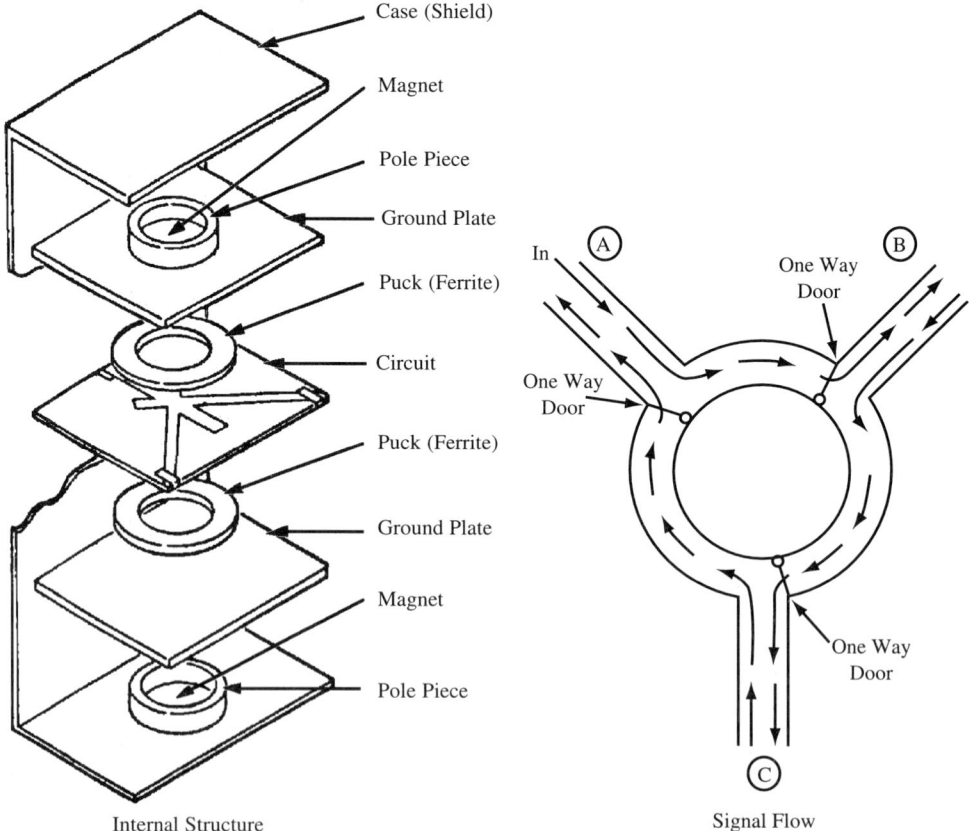

Figure 5.7   Three-port circulator.

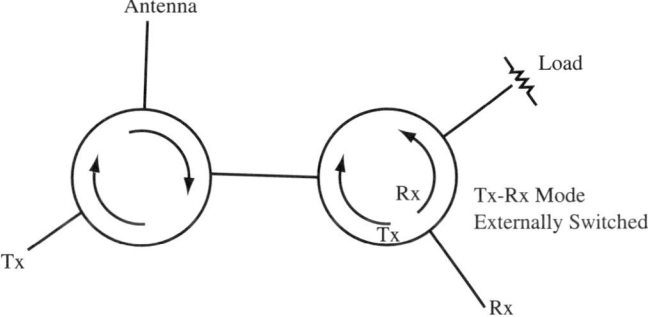

Figure 5.8   Ferrite switch using two circulators.

In choosing a circulator, the user should make sure the device will operate at the needed frequency band. Circulators operate efficiently only over a narrow bandwidth.

5.7   SLOTTED LINE

The slotted line is a handy component on a microwave laboratory bench. Figure 5.9 shows a waveguide slotted line and its symbol. The slotted line is made from a section of waveguide with a longitudinally oriented slot. A carriage attached to the guide has a mount for a crystal detector (diode). The carriage moves along the slot. The slot has a scale along the edge that is calibrated in millimeters in order to make precision measurements. The crystal detector has a probe connected to it. The probe's depth of penetration into the guide determines the level of an induced signal. The diode detects this signal and provides a DC output that is fed via a coaxial cable into a VSWR meter. This makes the slotted line very useful in measurements such as standing wave ratio, guide wavelength, and those required in the shifted-minima technique.

The slotted line is limited at the highest frequency of operation by scale graduations. There are also radiation losses through the slot.

5.8   SLIDE-SCREW TUNER

The slide-screw tuner (also known as a stub-tuner) is a close cousin to the slotted line. The difference is that the slide-screw tuner does not have a crystal detector mount. Instead,

Figure 5.9   Waveguide slotted line and symbol. (Photo courtesy of Lab-Volt Systems.)

Figure 5.10   Slide-screw tuner and symbol. (Photo courtesy of Lab-Volt Systems.)

there is just a probe mount, whose depth into the guide can be controlled by a vernier. The probe creates a discontinuity within the guide that can be used for impedance matching purposes. A slide-screw tuner and symbol are shown in Figure 5.10.

The probe acts capacitively when its depth of penetration is less than one-quarter wavelength. It acts inductively when the penetration is greater than one-quarter wavelength.

The probe is usually placed at a distance from the load where an inductive susceptance is present. There are two reasons for this: First, it is generally easier to adjust the susceptance of the slide-screw tuner in its capacitive region rather than its inductive region, and second, since the probe penetration is shallower in the capacitive region, less microwave power is lost because of probe penetration.

5.9   MIXERS

A mixer is a device with two input signals that are *heterodyned,* or mixed, into several frequencies. The output frequencies are the familiar sum and difference frequencies, as well as the original frequencies and their harmonics. At microwave frequencies the device is usually a microwave diode. The diode is usually a point-contact or a Schottky barrier type. A diode is a nonlinear device that has a "square-law region" of operation. The mixing action of the two input signals is done within this region.

Mixers are used in receiver/transmitters, in modulation, and in frequency translation. The most notable output frequency is the IF value (the difference frequency). Figure 5.11 shows a typical receiver mixer stage.

Mixing is usually done with one diode (single-diode mixer) or two or more diodes (balanced mixer). Single-diode mixers are low in cost but have higher conversion losses and higher noise levels than their two-diode counterpart. This is because the two-diode balanced mixer conveniently cancels noise through its push-pull circuit arrangement. Figure 5.12 shows a more common double-balanced (four diodes) ring mixer.

Figure 5.11   Receiver mixer stage.

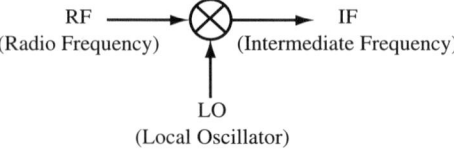

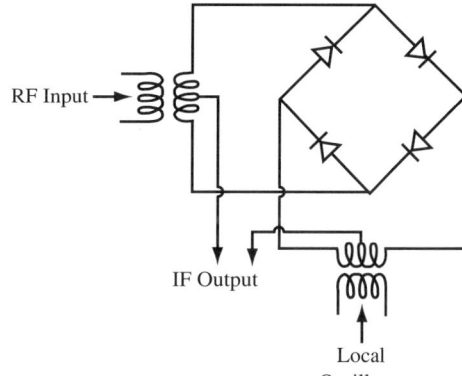

Figure 5.12   Double-balanced
ring mixer.

RF Input →

IF Output

Local
Oscillator

5.10   CAVITY RESONATORS

Radio frequency (RF) energy can be stored in a resonant circuit consisting of an inductance and a capacitance. The stored energy is maximum when the RF signal has the same frequency as the resonant frequency of the circuit. The electrical energy is stored in the capacitance, and the magnetic energy is stored in the inductance. The stored energy oscillates between the capacitance and the inductance at the resonant frequency. A similar circuit configuration built in a microwave circuit is called a *microwave resonator*. In the high-frequency range of microwaves, a microwave resonator is usually a *cavity* in which the electromagnetic waves are enclosed. The cavity is just a waveguide with walls on all sides. The development of the cavity is shown in Figure 5.13.

Figure 5.13a shows the conventional low-frequency resonant circuit. At higher frequencies, inductance (L) is reduced to a half-turn coil, while the capacitance (C) consists only of the stray capacitance across the coil (Figure 5.13b). By adding additional loops (Figure 5.13c) we can increase the current-handling ability of the resonant circuit, which also reduces its resistance. This effectively raises the Q-value of the cavity. By adding the loops in parallel, the effective changes in L or C offset one another and keep the resonant frequency the same.

If the loops are made in one-quarter wavelength sections (Figure 5.13d), they act like parallel $\lambda/4$ lines that are resonant at this dimension. When more loops are added (Figure 5.13e), the assembly eventually becomes a closed resonant box. The cavity doesn't have to be circular; rectangular shapes are also possible.

Fields within the cavity resemble those of a waveguide. The E field (voltage) is across the cavity, while the H field (current) follows the surface walls. Current is affected by the skin depth or wave penetration into the surface of the walls. Since the walls are a metal surface with good conductivity (and in high-Q cavities are even plated with gold or silver), for all practical purposes the current is confined to a very thin layer (the skin) of the conductor surface. Under these conditions, I^2R losses are negligible because the current is confined to the surface of the metal only. Skin depth at 1.0 GHz into a silver surface is only 0.002 mm, while at 60 Hz it is 8.27 mm. Similar values exist for copper and gold.

The Q-value of low-frequency circuits can be as high as a few hundred. With the resonant cavity, Q-values can reach several thousand, and thus narrow bandwidth and selectivity are available.

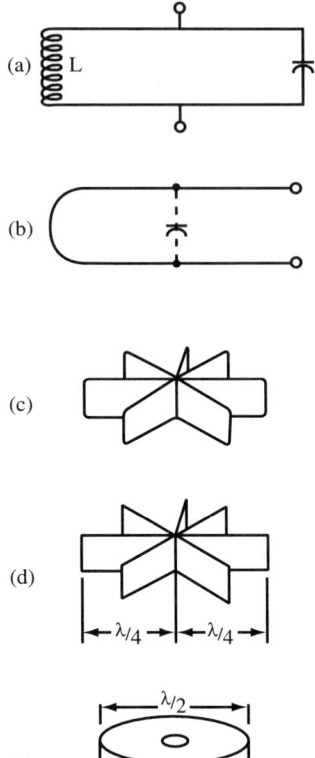

Figure 5.13 Development of cavity from λ/4 sections. (a) Conventional low-frequency circuit. (b) Coil-capacitor combination reduced for highest resonant frequency. (c) Half-turn loop in parallel. (d) Quarter wave sections. (e) Closed metal container.

Energy can be excited into or removed from the cavity the same way it was done in waveguides (through probes, loops, or slots). The resonant frequency of the cavity can vary by changing its dimensions. Adjustable discs, plugs, plungers, or screws can vary the volume or capacity of the cavity.

Resonant cavities are used as oscillators, filters, mixing chambers (mixers), impedance matching devices between waveguide sections, or as "ringing circuits" or echo boxes for radar applications. They are also used in wave meters.

When a resonant cavity is used as an oscillator, the flywheel effect is started due to "shock excitation" by noise pulses when DC is turned on between the cathode and anode. Because harmonics of the resonant frequency are also possible, the harmonics compete in an attempt to sustain the oscillations in the cavity. The output is not a stable sine wave but is in the form of pulses.

To overcome this problem, the cavity is not symmetrical in shape. Instead, the practical cavity has an irregular geometry that ensures the oscillation frequency is not harmonically related. One such shape, shown in Figure 5.14, is called a *reentrant cavity* because one of its surfaces reenters the cavity itself. The klystron oscillator (discussed in chapter 6) uses this type of cavity. The cavity has a plunger that can be adjusted through a vernier to vary the cavity size and therefore the frequency of resonance.

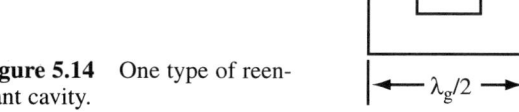

Figure 5.14 One type of reentrant cavity.

5.11 FILTERS

A filter is a frequency-sensitive network of reactive components. They were first taught in a basic AC course. They include low-pass, high-pass, and band-pass filters. Their attenuation-vs.-frequency responses are shown in Figure 5.15.

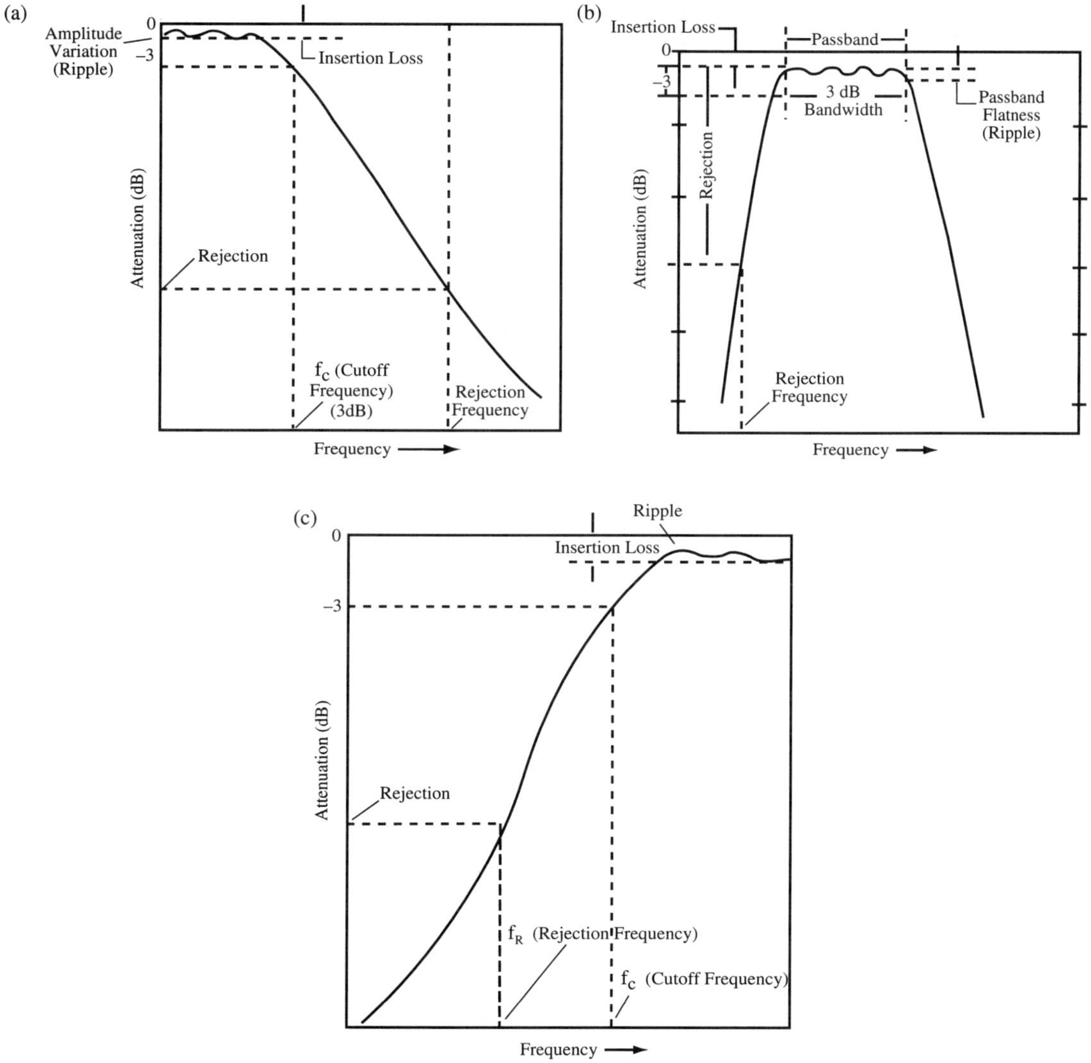

(a)

Amplitude
Variation
(Ripple)

Insertion Loss

Attenuation (dB)

Rejection

f_c (Cutoff
Frequency)
(3dB)

Rejection
Frequency

Frequency

(b)

Insertion Loss

Passband

3 dB
Bandwidth

Passband
Flatness
(Ripple)

Rejection

Attenuation (dB)

Rejection
Frequency

Frequency

(c)

Ripple

Insertion Loss

−3

Attenuation (dB)

Rejection

f_R (Rejection Frequency)

f_c (Cutoff Frequency)

Frequency

Figure 5.15 Filter responses. (a) Low-pass filter. (b) Band-pass filter. (c) High-pass filter.

The low-pass filter passes all frequencies that are lower than a specified cutoff frequency. The high-pass filter passes all frequencies above a specified cutoff frequency. The band-pass filter passes a specific band of frequencies. The cutoff frequency is the familiar –3 dB point on the response curves.

Microwave filters have responses different from their low-frequency counterparts due to the Q-value of the filter stage. The basic element of a microwave filter is a cavity that possesses a very high Q-value. Ferrites are often used in conjunction with a cavity. Multicavities are often used at microwave frequencies. There are cascaded individual cavities with tuning screws that adjust the individual cavities' operation frequencies and thus shape the filter response.

YIG Filters

Another common microwave filter is the YIG (yttrium iron garnet) filter. It is a highly polished sphere of ferrite material that can conform to use in stripline assemblies or within a cavity or section of waveguide. It is a tunable band-pass filter with a narrow bandwidth. It has a natural resonant frequency that can be controlled over a fairly wide range by the strength of an external electromagnetic field.

Figure 5.16 shows a simplified illustration of a single-stage YIG band-pass filter. The RF input and output circuits consist of two semicircular coupling loops located orthogonally (90°) around a tiny YIG sphere. The assembly is positioned between two pole pieces of an electromagnet and is contained within a cavity or a section of waveguide. Incoming RF energy must be at a proper frequency to activate the electron precession of the YIG sphere in order to transfer energy magnetically from input to output loops. Off-frequency values are attenuated. The bandwidth is controlled by the ratio of the diameter of the sphere to the diameter of the coupling loop. The YIG may also be used as a resonator or an oscillator.

SAW Filters

Surface acoustic wave (SAW) components have been in production for many years. The performance advantages and design flexibility of SAW technology have been featured in numerous publications for more than a decade.

Figure 5.16   YIG filter.

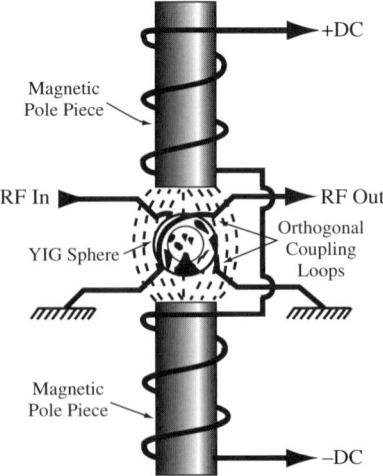

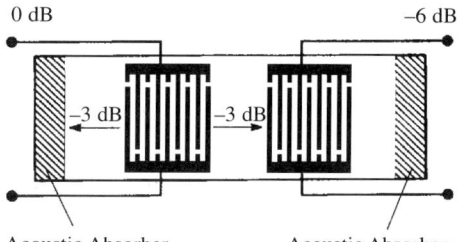

Figure 5.17   Simple transversal
SAW filter configuration.

         Acoustic Absorber   Acoustic Absorber

    Many advantages of SAW devices are derived from their physical structure. They are inherently very rugged and reliable. Because their operating frequencies and responses are set by photolithographic processes, they do not require complicated tuning operations nor do they become detuned in the field. They are used as filters, delay lines, resonators, and resonator filters with differences primarily in their physical construction.

    Surface acoustic waves were quantitatively described by Lord Rayleigh in 1885. They are mechanical (acoustic) rather than electromagnetic. Much of an earthquake's destructive force is carried by this type of wave. Surface waves achieved little recognition for their application in RF until less than two decades ago when SAW devices began to be developed for spread spectrum use in military radar equipment.

    In applications of the surface acoustic wave phenomenon to electronic devices, piezoelectric materials are required to convert the incoming electromagnetic signal to an acoustic one, and vice versa.

    In its simplest form, a transversal SAW filter consists of two transducers with interdigital arrays of thin metal electrodes deposited on a highly polished piezoelectric substrate, such as quartz or lithium niobate (Figure 5.17). The electrodes in these arrays alternate polarities so that when an RF signal voltage of the proper frequency is applied across them, the surface of the crystal expands and contracts. This generates the Rayleigh wave, or surface wave, as it is more commonly called. These interdigital electrodes are generally spaced at one-half or one-quarter wavelength of the operating center frequency. Since the surface wave or acoustic velocity is 10^{-5} slower than the speed of light, an acoustic wavelength is much smaller than its electromagnetic counterpart. For example, a continuous wave signal at 100 MHz with a free space wavelength of three meters has a corresponding acoustic wavelength of about 30 microns. This results in the SAW's unique ability to incorporate an incredible amount of signal processing or delay in a very small volume.

    As a result of this relationship, physical limitations exist at higher frequencies, at which the electrodes become too narrow to fabricate with standard photolithographic techniques, and at lower frequencies, at which the devices become impractically large. Hence, at this time, SAW devices are most typically used from 10 MHz to about 3 GHz.

    The basic SAW transducer is a bidirectional radiator. That is, half of the power is directed toward the output transducer while the other half is radiated toward the end of the crystal and is lost. By reciprocity, only half of the intercepted acoustic energy at the output is reconverted to electrical energy; hence, an inherent 6 dB loss is associated with this structure, as noted in Figure 5.17. Numerous second-order effects, such as coupling efficiency, resistive losses, and impedance mismatch, raise the insertion loss of practical filters to 15 to 30 dB.

    Newer, low-loss structures have been developed. The most useful of these is the single-phase unidirectional transducer or SPUDT. These devices are generally as straightforward to fabricate as ordinary bidirectional transducers. In addition, the required impedance

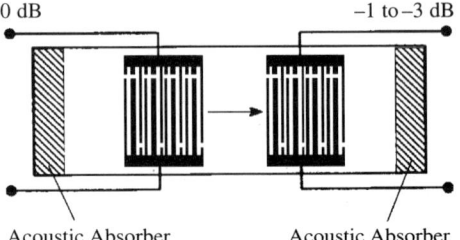

0 dB   −1 to−3 dB

Figure 5.18 Single-phase
unidirectional transducer
(SPUDT) configuration.

Acoustic Absorber   Acoustic Absorber

matching network is greatly simplified, usually consisting of an LC network on each port. Most SPUDT structures contain acoustic reflectors within the interdigital pattern, illustrated in Figure 5.18, as wider electrodes. These internal reflectors serve to redirect most of the acoustic energy that is normally lost in a conventional bidirectional device. The practical insertion loss is in the range of 5 to 12 dB.

While a variety of materials are used to form SAW filters, four types of materials are most prominent. ST-cut quartz is often employed in narrowband intermediate-frequency (IF) filters for its thermal stability, while zinc oxide thin films are used in low-cost television IF filters. The wideband filters used in cellular telephone applications employ lithium niobate and lithium tantalite. The high coupling coefficient of these materials results in low insertion loss.

5.12   T-SECTIONS

T-sections, or *tees*, are sections of waveguides with one or more side ports. They are used to split a wave so that coupling can occur into two other waveguides.

The *shunt tee*, also known as H-plane tee, is shown in Figure 5.19. A signal entering port C is split evenly and exits at port A and port B as two signals in phase with each other. Another feature of this tee is to allow two in-phase signals to enter ports A and B and exit at port C with a power level equal to the sum of the two entering signals.

A *series tee*, also known as an E-plane tee, is shown in Figure 5.20. A signal entering port D splits evenly and exits at ports A and B, but the two exiting signals are 180° out of phase.

A shunt tee and a series tee can be combined to form the *magic (hybrid) tee* as shown in Figure 5.21. Energy traveling into port D splits evenly between A and B. The signal at A is 180° out of phase with the signal at B. Energy entering port C splits evenly between A and B with both signals in phase. Energy into port C cannot exit port D and vice versa.

Figure 5.19 Waveguide shunt
tee. A signal entering at C is
split into two signals that exit at
A and B, respectively, in phase
with each other.

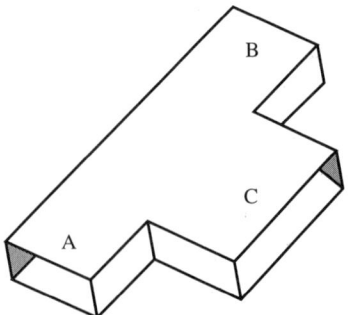

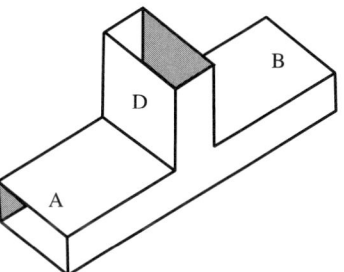

Figure 5.20 Waveguide series tee. A signal entering at D splits into two signals that exit at A and B, respectively, 180° out of phase.

The magic tee is used when it is necessary to couple two signals in or out of phase into the same waveguide. It can also be used for the opposite condition, when it is necessary to split a wave into either two in-phase or out-of-phase signals.

If one arm is not required for an application it can be terminated with a matched load. One application for this setup is as a mixer stage for a receiver (Figure 5.22). Here an antenna signal at port D mixes (combines) with the local oscillator signal that is supplied to port C. The combined output appears at ports A and B. One of these ports is connected to the receiver input (port A), and the other (port B) is terminated with a matched load to prevent reflections. This means that the input signal suffers a 3 dB loss in the tee.

5.13   FLANGES/JOINTS

It is virtually impossible to install a rigid waveguide system that is one continuous piece since most rigid rectangular waveguide sections are about 10 feet long.

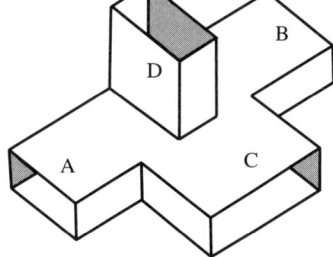

Figure 5.21 Waveguide magic (hybrid) tee. Energy entering at D splits into two out-of-phase signals exiting at A and B. Energy entering at C splits into two in-phase signals exiting at A and B. Energy entering at D cannot exit at C, and vice versa.

Figure 5.22 Receiver front end using magic (hybrid) tee.

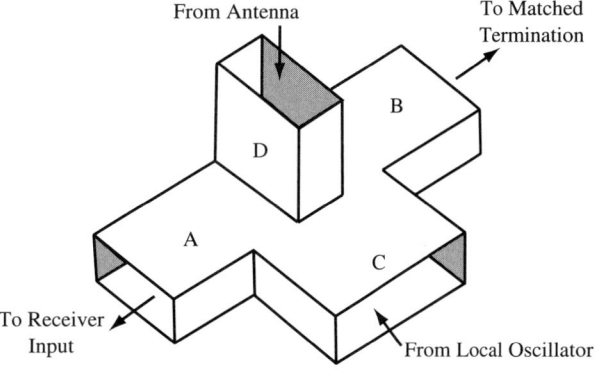

From Antenna

To Matched Termination

To Receiver Input

From Local Oscillator

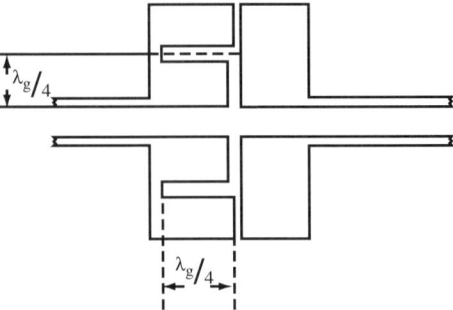

Figure 5.23 Choke joint.

Flanges are used to connect individual sections of rigid waveguide. The flange contains holes that can be butted against and bolted to a similar flange on the adjoining section of waveguide. Flanges of this type are simply called "butt" or "plain" flanges. Flanges are available with rectangular, square, and circular external shapes for rectangular waveguide, while circular waveguide flanges are nearly always circular. The size of the flange is determined by the particular waveguide used and its operating frequency ranges.

As the operating frequency gets higher, the quality of the flange gets more critical. Discontinuities can occur when the smooth surface of the flange or the alignment of the joined sections is not perfect. At these higher frequencies, any discontinuity causes a more pronounced reflection, and this can degrade the electrical performance of the waveguide.

A superior solution to the butt joint is the choke joint. Figure 5.23 illustrates that the right-hand flange is flat, while the other has a slot one quarter-wavelength ($\lambda_g/4$) deep cut into its surface. The slot is positioned at a distance $\lambda_g/4$ from the point where the flanges are joined. Since two quarter-wavelengths equal a half-wavelength section, this represents a short circuit at the place where the walls are joined. Thus, an *electrical* short exists at the junction of the two waveguides even though a physical gap occurs. A gasket may be placed in the gap to seal the interior of the waveguide from any moisture penetration. Moisture within a waveguide introduces a discontinuity, creates standing waves, alters the guide's impedance, increases the attenuation, and may cause corrosion. It is also possible to use a seal or gasket so the guide can be pressurized with an inert gas.

Another adaptation of the choke joint is for use in rotating joints. A rotating radar antenna needs a rotating joint to adapt from the rectangular waveguide coming from the transmitter to the rectangular waveguide going into the antenna. This is shown in Figure 5.24.

Flanges are classified as C (choke) flanges or P (plain or butt) flanges. A C flange contains a gasket or O-ring for pressure-tight connections. A P flange has a smooth face. A representative sample of cross-references for various flanges is given in Table 5.2.

Figure 5.24 Rotary joint.

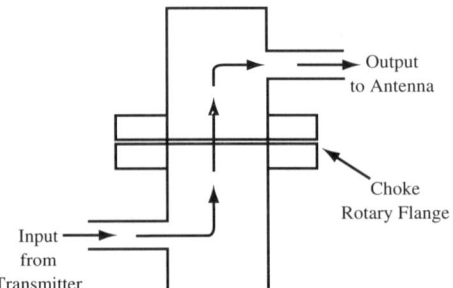

Table 5.2 Sample Cross-Reference of Flanges

Frequency Band (GHz)	IEC		MIL or JAN		Great Britain		EIA	
	Plain	**Choke**	**Plain**	**Choke**	**Plain**	**Choke**	**Plain**	**Choke**
2.60–3.95	PDR 32	CAR 32	UG 53/U	UG 54/U	083-0010	083-0009	CPR 284	—
3.95–5.85	PDR 48	CAR 48	UG 149/U	UG 148/U	083-0041	—	CMR 187	—
5.85–8.20	PDR 70	CAR 70	UG 344/U	UG 343/U	083-0038	083-0037	CMR 137	—
8.20–12.40	PDR 100	CBR 100	UG 39/U	UG 40/U	083-0004	083-0003	CMR 90	—
12.40–18.00	PBR 140	—	UG 419/U	UG 420/U	083-0030	083-0029	—	WR-62
18.00–26.50	PBR 220	CBR 220	UG 595/U	UG 596/U	011-9666	011-9667	—	WR-42
26.50–40.00	PBR 320	CBR 320	UG 599/U	UG 600/U	083-0018	083-0019	—	WR-28
40.00–60.00	PBR 500	—	—	—	083-0026	083-0027	—	WR-19
50.00–75.00	PBR 620	—	UG 385/U	UG 386/U	083-1612	083-1613	—	WR-15

PDR is rectangular pressurized; PBR is square pressurized; CAR is choke circular. CBR is choke square.

In accordance with MIL specifications, the letters UG stand for flanges. The IEC (International Electromechanical Commission) standard for flanges is compatible with the EIA (Electronics Industry Association) or MIL items in most cases.

Sometimes the designation CPR is used. This indicates a contact pressure rectangular flange intended for pressurized applications. A suffix indicates whether it is grooved to accept a gasket (G) or plain (P). A CMR (contact miniature rectangular) is a plain flange not intended for pressurization.

There is also a plethora of adapters, transitions, bends, and twists available (see Figure 5.25) to accommodate the "plumbing" of the waveguide assembly. An adapter is used to connect a choke flange to a plain flange. Transitions are used to connect adjacent

Figure 5.25 Waveguide "plumbing" adapters, bends, transitions, and twists. (Photo courtesy of Andrew Corporation.)

size waveguide or rectangular waveguide to circular or elliptical waveguide, or to a coaxial cable. Bends are used to make a 90° turn, while twists are used for making a transition from an E dimension to an H dimension.

5.14   SUMMARY

1. Passive microwave components include connectors, couplers, attenuators, nonreciprocal devices, measuring devices, resonators, filters, and joints.
2. Connectors come in the familiar "male" and "female" pairs. They are useful over a variety of ranges according to physical characteristics.
3. Directional couplers allow for the signal to be split into various ports. Coupling is done via holes between adjacent arms.
4. Attenuators are components that can reduce the microwave signal power. They come in fixed or variable types.
5. Isolators are a form of a nonreciprocal device. These devices employ ferrites that allow signal travel through the device in one direction only. Attenuation is done as an electron wobbles or *precesses* while spinning and thereby absorbs energy from the signal.
6. Circulators also use ferrites. They couple the microwave signal into a connecting port in one direction only.
7. The slotted line contains a crystal detector and is used in taking measurements such as VSWR, guide wavelength, and those required in the shifted-minima technique.
8. The slide-screw (stub) tuner is used as an impedance matching device.
9. Mixers mix or *heterodyne* incoming signals. They are made from a point-contact or Schottky barrier diode.
10. Cavity resonators employ distributed values in order to resonate and thereby replace the lower frequency LC "tank" circuit. They exhibit a very high Q-value.
11. Filters are frequency-sensitive networks using various characteristics to influence their output functions.
12. The YIG is a common microwave filter that employs the ferrite characteristics.
13. The SAW device is a transducer that uses piezoelectric materials to convert an incoming electromagnetic signal to an acoustic one and vice versa. This reduces the signal wavelength and allows for greater signal processing or delay in a small volume. SAW devices are used as filters, delay lines, resonators, and resonator filters.
14. T-sections are sections of waveguide used to split a wave so that coupling can occur into two other guides.
15. Flanges and joints are available in order to couple sections of waveguide together or make necessary bends or transitions. Rotary applications usually use choke joints.

Key Equations:

$$CF_{(dB)} = -10 \log_{10} (P_3/P_1) \qquad \textbf{(Eq. 5.1)}$$

$$CF_{(dB)} = dBm_1 - dBm_2 \qquad \textbf{(Eq. 5.2)}$$

$$D_{(dB)} = 10 \log_{10} \frac{\text{Output Power (forward)}}{\text{Output Power (reverse)}} \qquad \textbf{(Eq. 5.3)}$$

PROBLEMS

1. Determine the coupling factor (*CF*) for a directional coupler when the power at the input is 476 mW and the output of the secondary arm has a power of 476 μW.

2. What is the value of a coupler, in dB, if the output power at the secondary arm is –40 dBm and the input power is –20 dBm?

3. Determine the output power of the secondary arm to a 6 dB coupler when the input power is 64 mW.

4. Determine the main arm output power in problem 3.

5. Determine the main arm output power in problem 1.

6. What is the input power to a 10 dB coupler if the secondary arm output is 150 mW?

7. Determine the directivity when the forward power in the secondary arm is 450 mW and the reverse power is 0.71 μW.

QUESTIONS

1. Describe the uses for passive microwave devices.

2. Describe the types of connectors commonly used in microwave applications.

3. What does *coupling factor* indicate?

4. What is the function of a directional coupler?

5. What does *directivity* indicate?

6. How is coupling accomplished in a directional coupler?

7. How are attenuators made?

8. What are the uses for attenuators in microwave applications?

9. Describe the differences between variable and fixed attenuators.

10. Why do terminators employ a tapered rod?

11. What are ferrites made of?

12. Describe how ferrites work.

13. Define the term *precession*.

14. What is an application of an isolator?

15. Describe how a circulator works.

16. Describe how a slotted line works.

17. What is the difference between a slotted line and a slide-screw tuner?

18. Define what is meant by *mixing*.

19. Which is better, a single-diode or two-diode mixer stage? Why?

20. Describe how a cavity resonator works.

21. What is a *reentrant cavity*?

22. What is meant by *skin depth*?

23. How do microwave filters differ from their low-frequency counterparts?

24. Describe how the YIG filter works.

25. Describe how the SAW device works.

26. How are waveguide T-sections used?

27. Describe a choke joint and its advantages.

28. Why must some waveguide sections be sealed from moisture penetration?

29. What does CPR stand for?

30. What devices are available to accommodate the "plumbing" of a waveguide assembly?

6 ACTIVE MICROWAVE DEVICES (THERMIONIC)

OBJECTIVES

1. To describe methods used for generating RF power in the early radio days.
2. To describe the limitations of the early devices used for RF power generation.
3. To describe what differentiates the properties of O-type and M-type microwave tubes.
4. To describe general characteristics and applications of various microwave tubes.
5. To compare and contrast power levels, operating frequencies, bandwidth, and efficiency of various microwave tubes.

6.1 INTRODUCTION

As stated in chapter 1, the development of radar gave the impetus to find sources that could generate higher frequencies at sufficient power levels. Useful power levels originally were limited to devices in the very low frequency (VLF) bands. Various problems limited the early devices, such as the commonly used vacuum tube. But throughout the 1920s, 1930s, and 1940s, advances in technology used the inherent physical limitations of vacuum tube devices to good advantage to generate the necessary higher-frequency signals.

The student should understand that there is a considerable body of knowledge about vacuum tubes that is beyond the scope of this text. Few schools continue vacuum tube technology in their curriculum because solid-state devices have replaced tubes in all but the highest-power applications. This is true even in engineering schools, where many engineers have been graduating with limited background in vacuum tube theory. To help ameliorate this problem, one leader in microwave tube technology, Litton Electron Devices, San Carlos, California, commissioned a microwave tube tutorial. They provided this tutorial to their recent engineering graduates to help bring them up to par on microwave tube technology.

Litton has shared this tutorial with schools and colleges as well. This tutorial, in diskette form, is provided with this text, courtesy of Litton Electron Devices. It is a handy remedial reference, or an in-depth source, for just about anything you need to know about microwave tube devices and applications. Some sections of this chapter direct you to the tutorial to supplement your understanding and to see via animation how electrons travel in the various microwave tubes.

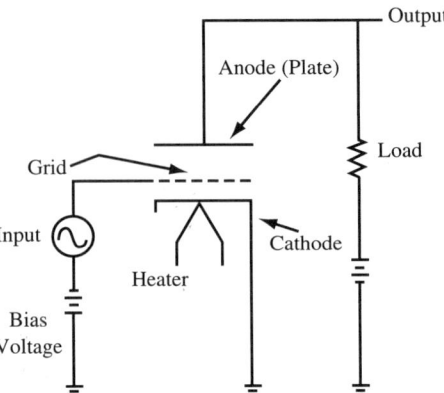

Figure 6.1   Triode vacuum tube amplifier (indirectly heated cathode).

A review of vacuum tube fundamentals is in order. The vacuum tube was and still is a viable source for high-frequency, high-power microwave applications (it is still being used at audio frequencies as well). It is considered an active device because of its ability to actively change (amplify) the power level of an incoming signal. The term *thermionic* denotes that one element is necessarily heated in order to "boil" electrons off the surface of a metal element. Therefore, this effect is known as the *thermionic emission* of electrons.

The first practical vacuum tube was credited to an Englishman, Sir John Fleming. Thomas Edison had noted what became called the "Edison effect" some twenty years earlier than Fleming's work. The Edison effect is the fact that a positively charged electrode inside an evacuated glass light bulb draws a current. Fleming used that fact in 1905 to build the Fleming "valve," which would now be called a two-element diode vacuum tube, to rectify radio signals. It consisted of a "hot" (thermionic) cathode that emitted electrons and an anode that collected the electrons. In 1907 Lee DeForest inserted a third element into the Fleming valve to form a *triode* amplifier. This third element was a grid that controlled the flow of electrons.

A schematic of the triode tube, with bias supplies and load, is shown in Figure 6.1. The three elements of the conventional triode vacuum tube are the cathode, the anode (plate), and the control grid. In some models the cathode is a filament, while in others it is a hollow tube with an internal heater. Whether directly or indirectly heated, the intent is to raise the surface temperature of the cathode to the point at which electrons boil off the surface. The electrons that boil off the surface form what is known as a *space cloud* or *space charge*. When the anode is positively charged, it attracts the negatively charged electrons to form an anode current.

The grid is a porous element placed between the cathode and the anode. It is biased negatively so that it will not accept electrons that would otherwise end up at the anode. The negative bias, if high enough, can shut off the electron flow to the anode. It is normally kept at such a level that when an incoming signal is applied, the signal voltage rides (superimposes) on the bias voltage, thus causing an increasing and decreasing bias level. This in turn causes the anode current to vary proportionately. This varying current is translated into a varying voltage by an anode load resistor.

6.2   EARLY VACUUM TUBE DEVICES

Early vacuum tube devices were severely limited in operating frequency. The primary problems were lead inductance, interelectrode capacitance, and transit time effects. Interelectrode capacitance is a function of electrode area and spacing. Making the electrodes smaller helped

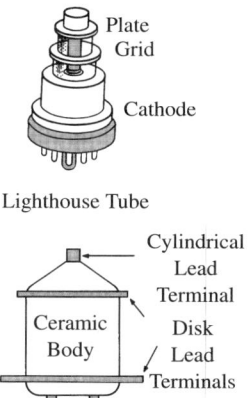

Figure 6.2   Early vacuum tubes.

lower the capacitance but severely limited operating power. Early forms of these devices were called the "acorn" and "lighthouse" tubes. The newer version of the acorn tube is the "planar" tube, which has a ceramic body. This and the lighthouse tube are shown in Figure 6.2. These tubes are UHF band devices.

The solution to the frequency problem in vacuum tube devices was found in working with the inherent transit time limitation and turning it to good advantage. One of the earliest attempts was the *Barkhausen-Kurz oscillator* (BKO).

The BKO circuit (Figure 6.3a) used the triode device but reversed the bias polarities. In the BKO circuit, the anode is negative and the grid is positive. Figure 6.3b shows how the circuit works. Electrons from the cathode are accelerated by the positive charge on the grid. Most of the electrons pass through the grid, since it is porous, and head toward the anode. But due to the negative potential on the anode, the electrons are slowed and repelled back. As the electrons attempt to move toward the negative cathode a similar action takes place. The result is that electrons travel in a circular path around the grid structure. The operating frequency is set by the rate of rotation. This process resolved the transit time limitation.

Output power from the BKO is taken at the grid. This fact limits the BKO power capability, since the grid is a small structure that runs "white hot."

Later developments used magnetic fields to control the electron flow and replaced the grid with a resonant cavity. These devices became the klystron and magnetron. They are described in the sections that follow.

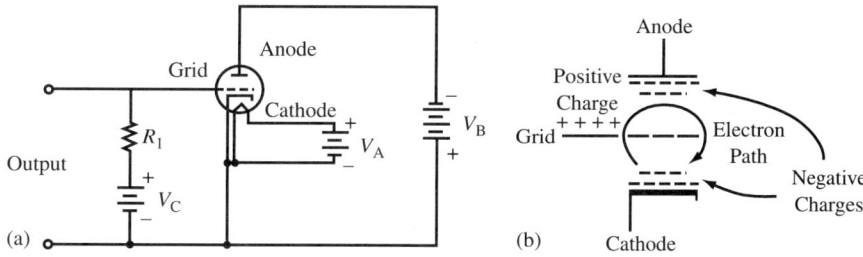

Figure 6.3   BKO vacuum tube.

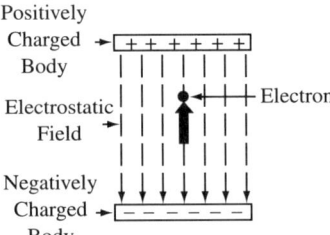

Figure 6.4 Moving electron gaining velocity and energy.

6.3   VELOCITY MODULATION

An electron has mass and thus contains kinetic energy when in motion. This energy can be determined with the following formula:

$$\text{Energy} = \tfrac{1}{2}MV^2 \qquad \textbf{(Eq. 6.1)}$$

where

$$\text{Energy} = \text{joules or watts/second}$$
$$M = \text{mass of electron } (9.1 \times 10^{-31} \text{ kg})$$
$$V = \text{velocity of electron (m/s)}$$

Thus, the kinetic energy of an electron is directly related to its velocity. That is, the greater the velocity, the higher the electron's energy level. This basic relationship of electron energy level to electron velocity is the key principle leading to energy transfer and amplification in microwave tubes.

An electron can be accelerated or decelerated by an electrostatic field. Figure 6.4 shows an electron moving in an electrostatic field. The direction of travel of the electron (shown by the heavy arrow) is against the electrostatic lines of force. The electron sees a positive potential on top and a negative potential on the bottom. The electron, being negatively charged, is naturally attracted to the positive potential, and it increases in velocity. From equation 6.1, if velocity increases, so must the energy of the electron. Where does this additional energy acquired by the electron come from? The obvious answer is that it comes from the electrostatic field. Thus an electron traveling in the opposite direction to electrostatic lines absorbs energy from the electrostatic field and, therefore, accelerates.

The opposite condition also holds true. An electron traveling in the same direction as the electrostatic lines gives up energy to the electrostatic field and, therefore, decelerates. This is shown in Figure 6.5. The negative potential has a repelling force on the electron. This lowers its velocity, causing it to give up energy to the electrostatic field.

A velocity-modulated tube is one in which operation depends on a change in the velocity of electrons passing through it. By means of this change in electron velocity, the

Figure 6.5 Moving electron losing velocity and energy.

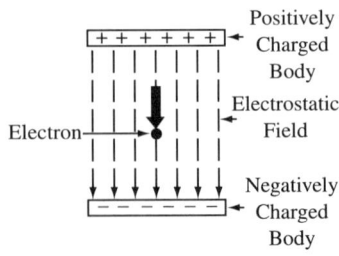

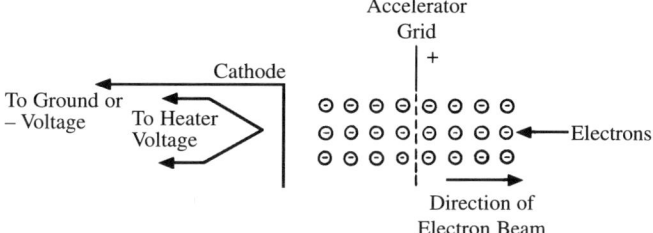

Figure 6.6   Electron gun.

tube produces bunches of electrons separated by space in which there are few electrons. *Velocity modulation* is defined as the variation of the velocity of a beam of electrons by alternately accelerating and decelerating the electrons with a period (time) comparable to the transit time in the space concerned. This usually is done by a voltage applied between two grids through which the beam must pass.

The first requirement in obtaining velocity modulation is to produce a stream of electrons, all of which are traveling at the same speed. The electron stream is developed with an electron gun, as shown in Figure 6.6. Electrons emitted by the cathode are attracted toward the positive accelerator grid. All but a few of the electrons pass through the grid wires and form a beam.

The electron beam then passes through a pair of closely spaced grids, called the *buncher grids*, each of which is connected to one side of a tuned circuit, as illustrated in Figure 6.7. The parallel resonant circuit represents a doughnut-shaped cavity resonator. The grids are the perforated center of the cavity. This configuration is sometimes called a "gridded drift tube" with resonant cavity. The grids are at the same DC potential as the accelerator grid. The alternating voltage that exists across the cavity causes the velocity of the electrons leaving the buncher grids to differ. This difference depends on the magnitude and direction of the electrostatic field within the cavity as the electrons pass through the grids.

The manner in which the buncher produces groups of electrons can be better understood by considering the motion of individual electrons. For example, the velocity of an electron passing through the buncher grid when the RF cavity voltage is zero is not affected. Electrons decelerate if they pass through the buncher when the cavity has grid #1

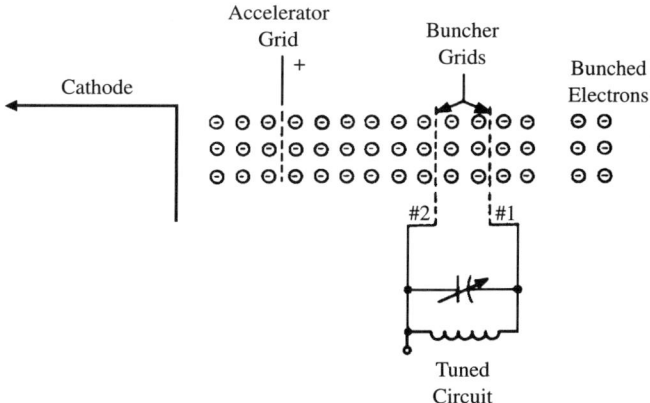

Figure 6.7   Electron gun and buncher grids.

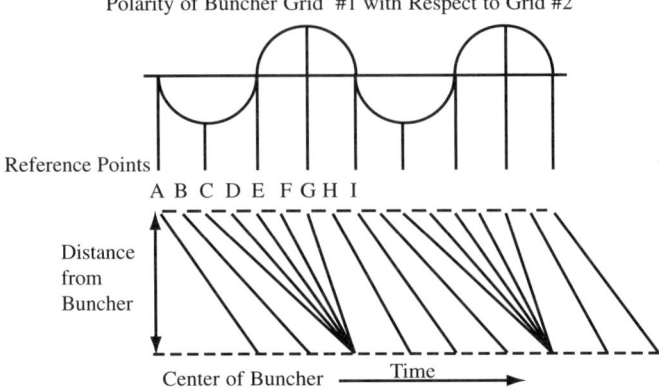

Figure 6.8   Electron bunching.

negative with respect to grid #2. As the electrons slow down, they give up energy to the field set up by the grids. This occurs from point B through point D in Figure 6.8.

Those electrons that pass through the buncher after its voltage passes zero in the opposite direction (grid #1 positive with respect to grid #2) are accelerated. This occurs from point F through point H in Figure 6.8.

The velocity modulation of the beam is a means to an end. No useful power has been produced because practically no power is needed in the process. That is, the energy imparted to accelerated electrons is offset by the energy received from the decelerated ones. However, a new beam distribution is formed if these velocity-modulated electrons are allowed to drift in a field-free area.

Three electron groups leave the buncher grid: those electrons not changing velocity, those accelerated, and those decelerated. These three electron groups then drift in a field-free area beyond the buncher grids. Due to new velocity relationships among these groups, some electrons fall back and are overtaken by another group. This forms a new electron bunch.

As the electrons continue to drift, they "overbunch" or "debunch" until at some later time they re-form the original beam distribution with other electrons in front of and behind them. If permitted, this bunching and debunching process continues indefinitely until the beam reaches a collecting element or is acted upon by other means.

By the formation of electron bunches via velocity modulation, a periodic variation in the density of the electron beam takes place. In other words, velocity modulation causes the formation of current density modulation. This periodic variation in the beam density takes place at the same rate as the RF gap voltage of the cavity. In terms of frequency, the beam density changes occur at the same frequency as the cavity field.

Velocity modulation transforms the DC beam into a current-modulated beam. The next step is to extract useful RF energy from the beam.

The current-modulated (bunched) electron beam in Figure 6.9a is shown in various stages of formation and decay. To extract useful RF energy from this beam, a second resonant cavity must be placed at a point of maximum bunching. This is shown in Figure 6.9b. The electron bunches induce an RF voltage in the grid gap causing the second cavity to oscillate. By proper placement of this second cavity, the oscillating gap voltage is of the proper polarity to decelerate the electron bunches and accelerate the small number of electrons between

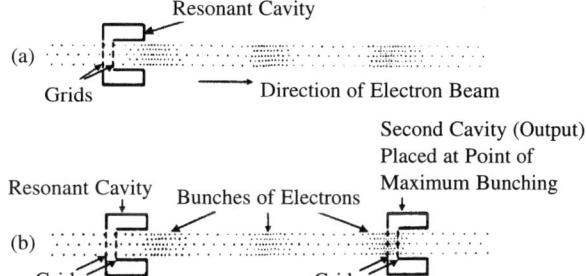

Figure 6.9 Extracting energy from bunched electrons.

the bunches. With many more electrons *in* the bunches than *between* the bunches, a net transfer of energy to the output cavity occurs. In this manner, RF power is extracted from a drifting beam of electrons.

6.4 KLYSTRON OSCILLATORS

The *reflex klystron* is an electron tube (Figure 6.10) that uses the velocity modulation of an electron beam to achieve oscillation in the microwave frequency region. The klystron is an O-type (from the French *ordinaire* tube), parallel-field microwave tube. The term *O-type* means the electron beam travels in a straight (linear) path and is under the influence of a parallel magnetic field. A parallel magnetic field tends to prevent the electron beam from spreading.

There are three basic regions of a reflex klystron: (1) the electron gun; (2) the RF structure, also referred to as the resonator or resonant cavity circuit, where the interaction takes place that causes electron bunching; and (3) the drift space. The repeller, an electrode with a more negative potential than the cathode, is placed at a point at which the electrons are repelled back. After the electrons turn around, they may arrive back at the interaction space when bunches are being formed. If their travel time in the drift space, coming back into the RF structure, is such that they are in phase (electrons are being bunched), then they provide the necessary positive feedback to sustain oscillation. Oscillation can be obtained

Figure 6.10 Reflex klystron.

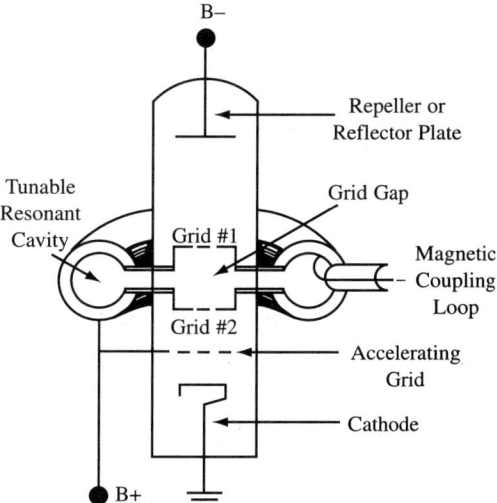

as the repeller voltage is varied. The transit time from the interaction space into the drift space and back to the interaction space is varied with repeller voltage. If bunches returning from the drift space are not in phase with bunches being formed, oscillations do not occur.

Figure 6.10 shows the schematic diagram of a reflex klystron. The diagram shows the repeller, a cathode, a resonant cavity that is tunable, and a magnetic coupling loop for the output. The electrons move toward the gap of the resonant cavity accelerated by the positive voltage on the accelerating grid and then are affected by an instantaneous voltage appearing across the gap. This instantaneous voltage oscillates at the natural resonant frequency of the cavity. The cavity is of the reentrant type that ensures the oscillation frequency is not harmonically related and is a stable single-frequency sine wave. The cavity is a resonant circuit, and when power is applied and an electron beam starts, the shock starts oscillations in the cavity.

As electrons pass through the gap they are either accelerated or decelerated. The distance these electrons travel in the space separating the grid and the repeller depends on their velocity. Those moving at slower speeds move only a short distance from the grid gap before they are overcome by the repeller voltage. As soon as an electron reaches the point at which its velocity is overcome by the field set up by the negative repeller, it stops, reverses its direction, and thus returns toward the grid gap. Those electrons traveling at higher speeds travel farther into the space beyond the grid gap before reversing their direction. If the repeller voltage is correct, the electrons form a bunch about the constant-speed electrons and return to the grid gap at the instant the RF field is at its maximum decelerating point (grid #1 maximum positive with respect to grid #2). With the grid field providing maximum deceleration, the returning electrons give up maximum energy to the grid gap field in the form of induced currents that are in phase with cavity current. Thus, the returning electrons supply regenerative feedback that sustains cavity oscillations.

For the electron bunch to arrive back at the grid gap at the exact instant, the unaffected electrons (those at constant speed, labeled "B" in Figure 6.11) must remain in the reflecting field space for a time equal to 3/4 cycle of the grid gap field. This period of time is determined by the magnitude of the repeller's negative voltage. Electrons at C are decelerated by the negative grid voltage and arrive at the gap at the same time as the constant speed electrons. However, as shown in Figure 6.11, for the circuit to oscillate, it is not necessary for the constant speed electrons to remain in the reflecting space for 3/4 cycle. Instead, they can remain in the reflecting space for any number of cycles as long as they return to the grid gap when its field is decelerating. Figure 6.11 shows the electron bunching in the reflecting space for 3/4 cycle and also for 1 3/4 cycles. Electrons at A are accelerated by the positive grid voltage and arrive at the gap after 1 3/4 cycles. Although not shown, it is possible for the constant speed electrons to remain in the reflecting space for any number of cycles plus 3/4 cycles. This difference in transit time is referred to as a *mode*. Mode 1 is the mode of operation obtained by the repeller voltage with the shortest transit time. The usual number of modes available is four. The available power from each mode and the bandwidth are different.

Power limitations are due to debunching of the electrons. They simply spread or reduce their density before they reach the cavity on their return trip. The farther they have to travel, the more debunching can occur. Thus, mode 1 (with the shortest transit time) has more output power because its electron bunch is more concentrated and induces more power into the cavity.

The resonant frequency of the cavity can be changed by adjusting the cavity's dimensions. This can be done capacitively by using a tuning strut to vary the distance between the grids. The cavity can also be inductively tuned by paddles and plugs, as explained in the section on cavity resonators (chapter 5).

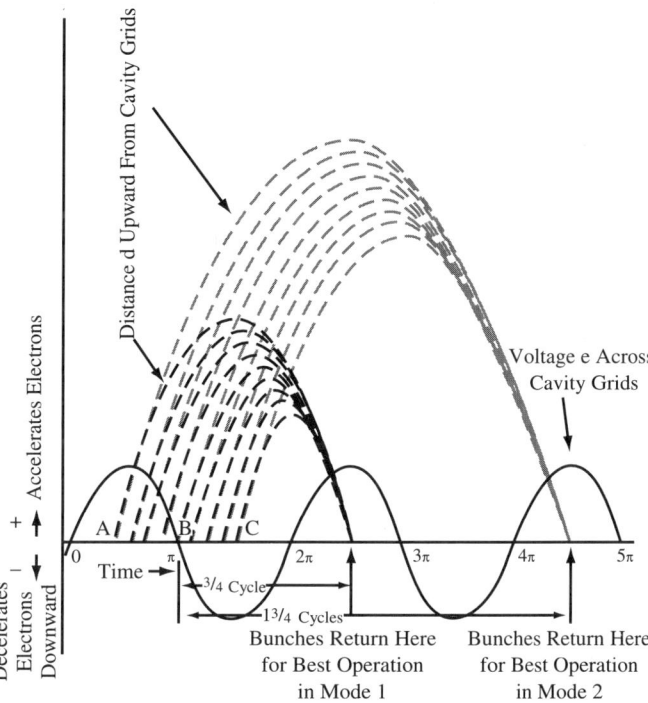

Figure 6.11   Bunching action of reflex klystron.

Electronic tuning is done by changing the repeller voltage. This method adjusts the bandwidth rather than altering the resonant (center) frequency. Mode 1 has a higher output power but a narrower bandwidth than the other modes. Modes 2 through 4 have increasingly lower output power and wider bandwidths.

Another form of a klystron oscillator employs two cavities. A collector takes the place of a repeller, and a coupling hole between the two cavities allows a regenerative feedback that sustains the oscillations. Oscillations start from noise as the electron beam is directed toward the first cavity. The cavity oscillates, providing an alternating voltage that begins velocity modulation and electron bunching. Bunched electrons give up their energy to the output cavity. Modes of oscillation occur with changing beam voltages.

The two-cavity oscillator has the lowest spectral noise characteristics of any oscillator and is ideal for Doppler radar applications.

To see a pictorial representation of how this oscillator works, access the diskette that accompanies this book. Under Table of Contents select Microwave Tube Types (#4) and then the subject Klystron Oscillators (letter C).

6.5   KLYSTRON AMPLIFIERS

Klystron amplifiers differ from reflex klystrons by virtue of having multiple cavities and an input and output port. Two-cavity klystrons are rarely used as amplifiers because they have insufficient power. Most modern power klystrons have at least three cavities, which allow for greater bandwidth and higher-output power. Power klystrons are used extensively for high-power communications links, radar, television/broadcasting, particle accelerators, and industrial processing.

A three-cavity klystron is illustrated in Figure 6.12. The heater and cathode assemblies, brought out at the base of the tube, are insulated from the entrance to the drift tube by a glass insulating sleeve. The input cavity is located close to the drift tube entrance, the middle cavity is located farther along the drift tube, and the output cavity is at the other end of the drift tube. The collector is beyond the output cavity. The entire drift tube assembly, the three cavities, and the collector are at ground potential. This eliminates hazards and complications in tuning the cavities and in liquid cooling of the body and collector. With the drift tube assembly at ground potential, the cathode is pulsed with a negative voltage to accelerate electrons from the cathode toward the drift tube entrance.

The output of any klystron (regardless of the number of cavities employed) is developed by velocity modulation of the electron beam. Electron bunching then occurs, and the electron beam is acted upon by the RF fields developed across the input and middle cavities.

Only a small degree of density modulation (bunching) occurs within the electron beam in the interval of travel from the input cavity to the middle cavity. The amount is very small compared to the degree of bunching required at the output cavity. The amount of bunching is sufficient, however, to excite the middle cavity and, because of the high Q-value of the cavity, to maintain a large oscillating voltage across the input gap. This voltage is responsible for most of the velocity modulation and the subsequent current density modulation produced within the klystron.

Figure 6.12   Three-cavity klystron used for radar applications.

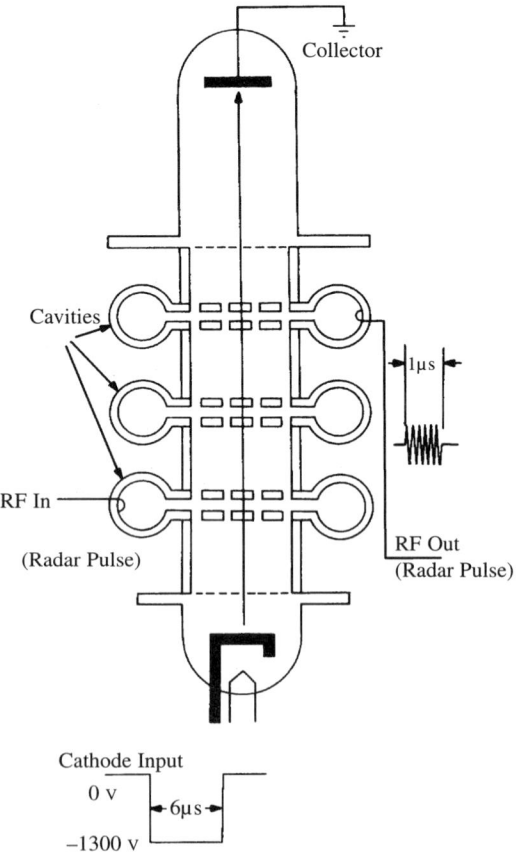

Table 6.1   Klystron Amplifier Litton L-5892

RF Performance	Electrical Parameters	Physical Characteristics
Frequency Range: 2800–3200 MHz	Cathode Voltage: –117 kV	Dimensions: $43 \times 9.0 \times 16$ inches
Minimum Power Out: 3 MW	Heater Voltage: 7.5 V	Output Connector: WR-284
Gain (saturated): 35 dB	Solenoid Voltage: 250 V	Weight: 140 lbs
Bandwidth (–3 dB): 400 MHz	Cathode Current: 80A	Cooling: Water/Glycol
Duty Cycle: 0.2%	Heater Current: 33.5 A	
Modulation: Cathode	Solenoid Current: 10.5 A	

The large voltage across the middle cavity gap produces much more velocity modulation than that in the input cavity. Greater bunching now occurs. When the electron bunches cross the output gap as the gap voltage is maximum negative, maximum energy transfers from the electrons to the output cavity, and hence maximum power gain is achieved. Wider bandwidth can be achieved by detuning each cavity slightly; however, output power is degraded in so doing. Table 6.1 shows the typical RF performance characteristics for an S-band, high-power klystron produced by Litton Electron Devices.

6.6   TRAVELING WAVE TUBES

Helix Traveling Wave Tube

The *traveling wave tube* (TWT) is an extremely useful device at microwave frequencies. It utilizes velocity modulation of an electron beam to achieve microwave signal amplification. In contrast to the klystron amplifier, the *helix TWT* device does not use resonant cavities. It employs a direct interaction between a nonresonant transmission line and an electron beam to form electron bunches.

The TWT is also an O-type device in that it utilizes a magnetic field that is parallel to the electric field rather than at right angles to it. The TWT offers high gain (30–60 dB), linear amplification characteristics, versatile modulation, efficiency up to 50%, and wide bandwidth. Bandwidths of one octave are common. (A bandwidth of one octave is one in which the upper frequency is twice the lower frequency.) The schematic section of a TWT is illustrated in Figure 6.13.

The long, thin glass tube houses a wire helix having many turns (approximately fifty per inch). There is an input port at one end of the helix and an output port at the other end. At the far right end of the tube is a positive potential collecting electrode. The whole assembly is typically around twenty inches long.

The electron gun produces a pencil-shaped beam of electrons having uniform thickness. For the electron beam to impart energy to the RF traveling wave that is applied to the input connector, the speed of the traveling wave and the speed of the electron beam must be about the same. The function of the helix is to slow down the traveling wave to slightly less than the speed of the electrons.

As has already been demonstrated, a traveling wave in space travels at the speed of light. The helix contains shunt capacitance between turns and series inductance within the turns so that it corresponds to a *delay line*. As the slowed wave travels the length of the tube (via the helix), the electron beam moves through the center of the tube in the same direction as the traveling wave.

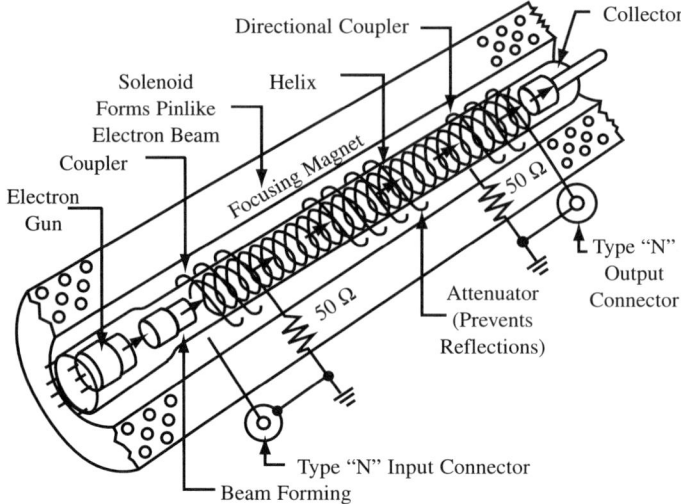

Figure 6.13   Schematic of traveling wave tube (TWT) amplifier.

The traveling wave applied to the helix has both electric and magnetic field components. The magnetic component is not useful and tends to scatter the electrons in the pencil beam. To counteract this effect, the tube is contained in a focusing magnet that establishes a magnetic focusing field around the tube. This magnet may either be a periodic permanent magnet (PPM) or an electromagnet (solenoid).

The electric field component of the traveling wave is used for interaction with the electron beam. The field's polarity accelerates and decelerates the electrons of the beam. This causes the beam to alternately be made more or less dense as it travels the length of the tube. The newly formed bunches add a small amount of voltage to the signal on the helix. The slightly amplified signal then produces a denser stream of electrons, which in turn adds a greater voltage to the signal, and so on.

Each time the beam is slowed down, it gives up some of its kinetic energy to the traveling wave. Thus, the traveling wave is increased in strength (amplified). The greater the bunching action, the greater the amplification.

If the amplified wave is reflected at the output end of the tube, it may set up unwanted oscillations. Attenuators, shown in Figure 6.13, are placed near the center of the helix to prevent reflected waves from returning to the tube input.

All waves are reduced to nearly zero at the center. The electron bunches travel through the attenuator unaffected, then emerge from the attenuator to induce a new signal on the helix. The new signal is an exact replica of the original signal. The field of the newly induced signal interacts with the bunched electrons to begin the amplification process over again.

To summarize, amplification increases as the greater velocity of the electron beam pulls the electron bunches nearer in phase with the electric field of the wave on the helix. At the point of the desired amplification, the amplified signal is coupled out of the helix. Note that the amplified signal is a new signal, whose energy is wholly supplied by the bunched electron beam.

The bandwidth of a helix circuit can be increased by adding loading segments inside the barrel of the tube to make the circuit less dispersing, making ultrawideband tube performance

Table 6.2   Traveling Wave Tube Litton L-5785

RF Performance	Electrical Parameters	Physical Characteristics
Frequency Range: 16.0–17.0 GHz Minimum Power Out: 3.0 kW Gain: 60 dB Duty Cycle: 5%	Cathode Voltage: –11.0 kV Collector Voltage: 8.6 kV Grid Volts (On): 180 V Grid Volts (Off): –200 V Heater Voltage: 6.3 V Modulation Element: Sh. Grid Cathode Current: 1.7 A Helix Current (Max): 0.6 A Grid Current (On): 10 mA Grid Current (Off): 0.1 mA Heater Current: 2.0 A Focus: PPM	Dimensions: 13.6×3.0×2.0 inches Output Connector: UG-541/U Weight: 6.0 lbs Cooling: Conduction

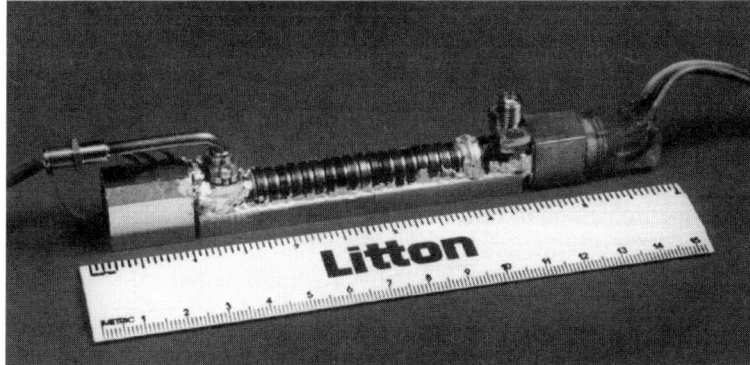

Figure 6.14   Helix mini-traveling-wave tube. (Photo courtesy of Litton Electron Devices.)

possible. Multioctave bandwidth tubes are now available for wideband electronic warfare applications. Table 6.2 shows the typical RF performance characteristics for a Litton Ku-band helix TWT.

Figure 6.14 shows a picture of a helix mini-traveling-wave tube, manufactured by Litton Electron Devices. Its operating frequency range is from 6 to 18 GHz at 100 watts of output power. This device comes in a 6-inch package with SMA RF connectors used for the input and output connections.

Coupled-cavity TWT

Coupled-cavity traveling wave tubes (CCTWT) are devices that use a series of coupled cavities along the tube rather than a helix. The cavities are overcoupled to have band-pass filter characteristics. RF waves are slowed down by the reentry path, with the phase velocity close to the beam velocity. Interaction between the RF field and the beam occurs. This beam energy enters the cavity, which acts like a drift tube, which in turn interacts with the beam, causing bunching of the electrons. Bunching increases as the wave travels down the tube, and interaction continues along the length of tube as with the helix TWT.

Table 6.3   Coupled-cavity TWT Litton L-5633-50

RF Performance	Electrical Parameters	Physical Characteristics
Frequency Range: 12.0–16.0 GHz Minimum Power Out: 10 kW Modulation: Grid Gain (saturated): 39 dB Duty Cycle: 2%	Cathode Voltage: –32.1 kV Cathode Current: 3.7 A Grid Volts (On): 450 V Heater Voltage: –10.0 V Collector Volts: 21.1 kV Body Current: 590 mA Grid Bias: –350 V Heater Current: 3.0 A Focus: PPM	Length: 24 inches RF Connectors: WR-62 & WR-62 Weight: 22 lbs Cooling: Coolanol or FC-77

Coupled-cavity traveling wave tubes have higher power outputs and operate at higher frequencies than is possible with the helix TWT with its use of a delay line. Table 6.3 shows the typical RF performance characteristics of a Litton Ku-band CCTWT. (See the tutorial diskette, Table of Contents, Microwave Tube Types (#4), Coupled-Cavity TWTs (H) for more information on this tube.)

6.7   MAGNETRON

The *magnetron* oscillator (Figure 6.15) is a diode that has a nearly constant magnetic field superimosed on it. It has lines of force that are parallel to the cathode axis and perpendicular to the electric field of the cathode and anode. This right angle magnetic direction means this tube is considered an *M-type* (from magnetron) rather than the O-type used in klystrons and TWTs. Another difference is that the electron beam does not travel a linear (straight) path, but rather travels in a circular path. The magnetron can be pulse modulated or operated as a continuous wave (CW) oscillator. It continues to be the basic building block of a radar set. (You will also find one in your microwave oven in the kitchen, operating at a frequency of 2.45 GHz.) The basic parts consist of an electron gun and the RF microwave circuit. Output power ranges from watts to multimegawatts, with tunability up to 5% and efficiency typically 40% to 60%.

The magnetron anode is constructed from a solid block of copper (Figure 6.16). It is cylindrical in shape, with its external surface connected to cooling fins. The resonant cavities

Figure 6.15   Magnetron.

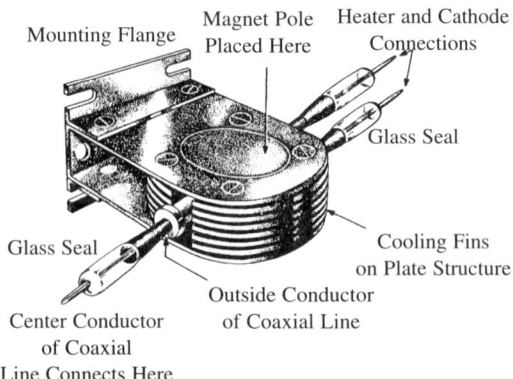

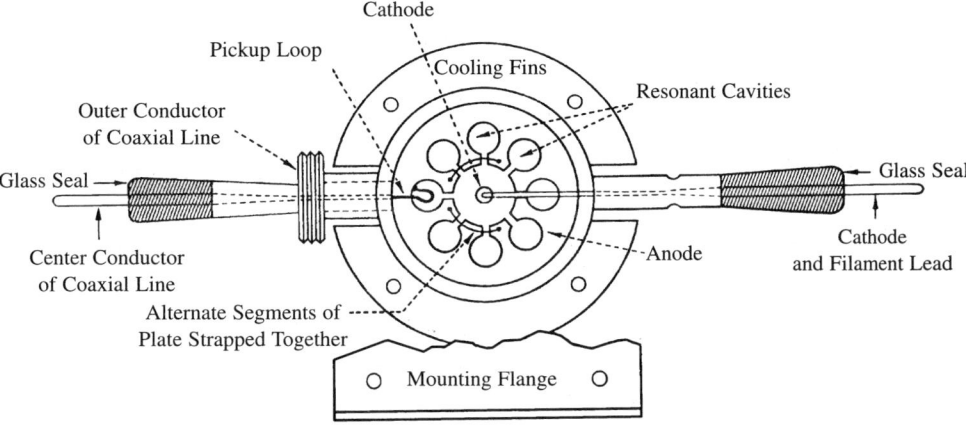

Figure 6.16   Cutaway view of magnetron.

are drilled into the anode block and open into the space provided for the cathode (electron gun assembly). The size of the cavity is the primary factor that determines the operating frequency. The number of cavities (always an even number and usually eight) is determined by the physical size of the anode. The greater the power requirement, the larger the anode that is needed (but not the cavity size).

Anode cavities may be in various shapes. One of the more common shapes is the cylindrical cavity that is connected to the central space by a slot. This is called the hole and slot that is illustrated in Figure 6.16. Three other anode configurations are illustrated in Figure 6.17.

Another anode shape is shown in the coaxial magnetron (Figure 6.18). This shape offers the most stable and reliable device since the design allows larger anodes and cathodes. It employs a high-Q stabilizing cavity that can be easily tuned.

The operation of the magnetron uses the effects of an electric field and a magnetic field on a moving electron. A moving electron generates its own magnetic field. The strength of this field is directly proportional to the velocity of the electron. An electron can be made to deflect through the interaction of its own magnetic field and that of a stationary magnetic field.

In a magnetron, the stationary magnetic field is developed by a permanent magnet whose field is perpendicular to an electric field in which the electrons move. The electrons in such a

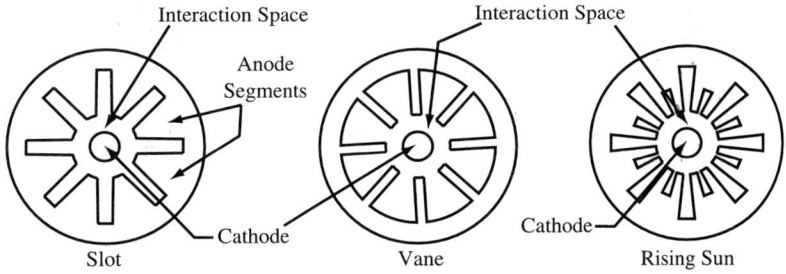

Figure 6.17   Other anode configurations.

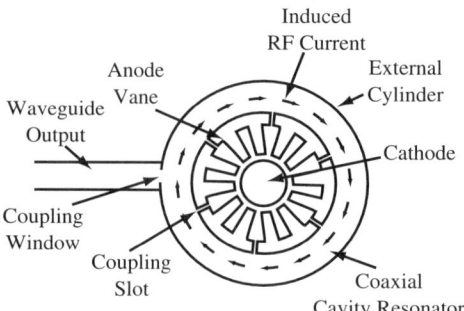

Figure 6.18 Coaxial magnetron.

field do not follow a straight path. Instead, they move in a series of arcs and loops. This path is called a *cycloidal* path. The number of cycloids is a function of the strength of the permanent magnetic field and the strength of the electric field. As the electron is accelerated by the electric field, the magnetic field around it increases, causing the electron path to curve even more. When the force on the electron from its magnetic field and the permanent magnetic field become greater than that exerted by the steady electric field, the electron curves the other way (toward a negatively charged body). In following this path, the electron is moving against the electric field and it slows down. As the electron slows, its magnetic field decreases and the force becomes less than that exerted by the electric field, so the electron's path is once again reversed. It starts to accelerate back toward a positively charged body. The familiar electron bunching is now occurring.

This curved path continues as RF electrostatic fields build up at the anodes and continues to influence the electron paths, depending upon the phase of the anode voltage. Some electrons are attracted toward the anode, while others are turned back to the cathode. The returning electrons cause secondary emission from the cathode and increase the cathode temperature. Heater power is then reduced to achieve proper cathode temperature for good performance and the longest life. Each active electron eventually lands on an anode, where it gives up energy to the microwave cavities.

The RF field in the cavities increases as the electrons give up their energy to the cavity. The design parameters are chosen so that adjacent vanes are 180° out of phase, providing the proper electrostatic fields to influence the electrons. As the cavities oscillate, the electrostatic fields change, causing the active electrons to orbit the cathode until they impact on one of the vanes. This produces "spokes" of electron current from cathode to anode (vanes) that produces high oscillation current in the cavities. The resulting high power is coupled out of the circuit (from any one cavity) via a loop or waveguide.

The pi mode employs a phase shift between cavities to be some multiple of 360° divided by the number of cavities and to be self-consistent from one cavity to the next. With eight cavities, and four complete 360° cycles, this interprets to 180° (π) between cavities, hence the name "pi" mode. To assure stability in the pi mode, in which adjacent vanes are 180° out of phase and every other anode is in phase, the hole-and-slot and the vane (trapezoidal) cavities employ metal straps on each side of the cavity to electrically connect alternate anodes. The geometry of the rising sun cavity does not require strapping and is more suited to millimeter frequencies where the size of the cavity makes strapping difficult.

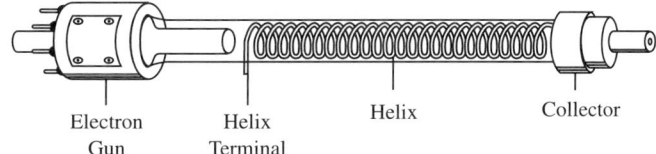

Figure 6.19   Backward-wave oscillator (BWO) tube.

6.8   OTHER MICROWAVE TUBE DEVICES

Backward-wave Oscillator

The *backward-wave oscillator* (BWO), shown in Figure 6.19, is a common O-type device with characteristics similar to the TWT. Its output frequency is voltage-tuned over bandwidths as high as 5:1. The output frequency is determined by a frequency-selective feedback rather than by resonant cavities.

The BWO consists of an electron gun, a helix structure, and a collector. Physically, the BWO resembles the TWT; however, for comparable frequencies it is larger in diameter and somewhat shorter in length.

As the BWO is turned on, oscillation evolves from shot noise generated by the electron beam. Voltages are induced on the helix, and these voltages produce electron bunching. The electron bunches move toward the collector at a velocity controlled by the accelerating potentials. As the electron bunches pass the spaces between helix turns, their electric fields appear outside the helix. At some frequency these electric fields are in step (resonant) with the electron bunches along the helix and a backward-moving wave is generated. The backward wave further bunches the beam, the beam in turn amplifies the backward wave, and so on, until a maximum bunch density is reached. At this state the backward wave has maximum amplitude for existing operating conditions.

The RF output of the BWO is a result of the interaction between the electron beam and the electric fields accompanying a microwave signal present on the helix. The term "backward-wave oscillator" is quite appropriate owing to the fact that the RF energy moves and builds up in a direction opposite to that of the electron beam and is coupled out of the tube near the electron gun via the helix.

Klystrode

The *klystrode* is a hybrid tube, combining the attributes of a klystron and a triode tube in an attempt to achieve a device that has higher beam efficiency. The klystrode's efficiency improvement comes from density modulating the beam much like a power grid tube that draws cathode current during only a part of an RF cycle.

The modulated beam is coupled to a klystron-type output cavity and produces high output power similar to a klystron amplifier, although at lower gain. The klystrode does not lend itself to operating frequencies higher than 1 GHz, so most applications are in the UHF frequency band, especially UHF-TV. The klystrode name is the designation given to the tube by Varian Associates, an early developer of the tube. Similar tubes are built by European manufacturers and are called Inductive Output Tubes (IOTs).

Twystron

The *twystron* is a hybrid tube, a combination of a klystron and a TWT. The combination of klystron driver cavities and a coupled-cavity output provides a flat output power characteristic over a wide frequency range.

By tuning the klystron driver cavities to provide more gain at the band edges, the power output versus frequency characteristics can be flatter over a wider bandwidth than can be achieved with a TWT.

Cross-field Amplifiers

The *cross-field amplifier* (CFA) is an M-type tube, but it uses a slow wave structure similar to a TWT. It offers power levels from kilowatts to multimegawatts, a bandwidth typically 5% to 20%, and an efficiency that is typically more than 50%. Lower voltage operation and small size are also important features. Most CFAs are distributed beam amplifiers using magnetronlike emitters.

Distributed emission CFAs can be made in either a linear format or a circular format. Figure 6.20 shows a circular format CFA. The input and output ports can be made close enough to allow the electrons to reenter the interaction area at the circuit input, enhancing efficiency. The direction the wave takes can either be forward or backward in relation to the electron beam.

A DC electric field exists between the cathode and anode. The electron velocity is made to approximate the velocity of the wave (via the slow wave structure). The velocity can be adjusted by the ratio of the DC electric field to magnetic fields. Interacting with the RF wave on the circuit, electrons either gain or lose energy, depending on the direction of the RF field they are in and moving along with. Electron "spokes" result, eventually striking the anode and giving up energy, producing amplification of the RF wave.

Gyrotrons

The *gyrotron* (Figure 6.21) family of devices generates or amplifies microwave and millimeter-wave signals, providing very high RF power for radar, scientific, and industrial equipment. Power output levels are from kilowatts to multimegawatts, with bandwidths for oscillators typically 0.1% and for an amplifier typically 0.1% to 10%. Efficiency is typically from 30% to 50%. The gyrotron provides extremely high power levels for microwave signals above 20 GHz, which is not possible with present state-of-the-art linear beam tube technology.

Figure 6.20   Cross-field amplifier (circular format).

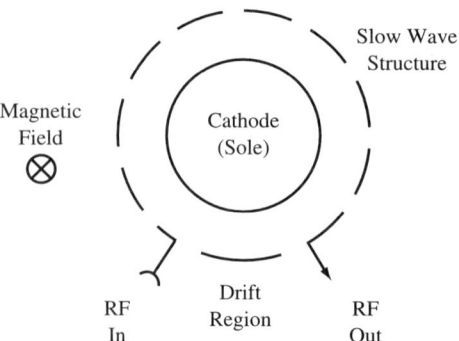

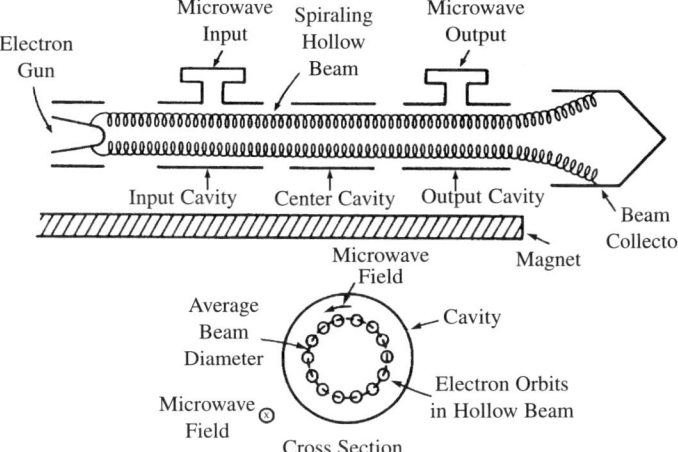

Figure 6.21   Gyrotron amplifier.

Unlike linear beam tubes, such as klystrons, TWTs, and the like, in which electron bunches are formed as the electron beam travels through the tube, gyrotron devices are based on a completely different type of interaction between beam and circuit. The gyrotron is based on the fact that an electron in a magnetic field has a rotational frequency, called the *cyclotron frequency*. This frequency is a function of electron charge, electron mass, and the DC magnetic field.

Electrons travel through the tube in individual helical paths as determined by the cyclotron frequency and the beam voltage. When affected by an electric field of an RF circuit, the electron's angular velocity in its orbit can either be accelerated or decelerated, changing the mass of the electron and thereby changing the cyclotron frequency. Orbital bunching occurs transverse to the direction of the beam, because angular bunching is increased for some electrons and decreased for others. As the bunches rotate in near synchronism with the alternating fields of the RF wave, energy is given up to the wave, providing a means of making either a gyrotron oscillator or an amplifier.

The operating frequency is determined by the cavity resonance of the RF circuit and the value of the DC magnetic field. In general, the frequency is approximately linearly related to the DC magnetic field.

For more detailed information on some of the tubes in this section, see the tutorial diskette. Select Table of Contents, Microwave Tube Types (#4).

6.9   SUMMARY

1. Thermionic emission occurs when a cathode is heated to the point of "boiling" electrons from its surface.
2. The first vacuum tubes were the diode and triode.
3. The triode is an amplifier that was developed by DeForest.
4. Early vacuum tube devices were limited by lead inductance, interelectrode capacitance, and transit time.
5. The BKO was one of the first attempts to use the transit time to good advantage.

6. Velocity modulation is defined as the variation of the velocity of a beam of electrons by alternately accelerating and decelerating the electrons with a period comparable to the transit time in the space concerned.

7. Velocity modulation causes the density modulation of the electrons, which in turn become bunched.

8. An O-type tube uses a magnetic field that is parallel to the electric field, while an M-type has these fields at right angles. Further, the O-type employs a straight (linear) current path, while the M-type employs a circular current path.

9. A reflex klystron is a single-cavity device in which a negatively charged repeller electrode is used in place of the collector. The repeller turns back the electron beam, causing velocity and density modulation. This action causes electron bunching, and the bunches shock excite the cavity into oscillation and impart energy into the cavity to sustain the oscillations. It is an O-type tube.

10. A klystron amplifier has three or more cavities and no repeller. The interaction process is similar to that of the reflex klystron, but multiple cavities allow higher power to be acquired in the output cavity.

11. The traveling wave tube (TWT) is an O-type microwave amplifier. A slow wave structure (helix) serves as a delay line to reduce the RF wave velocity to the electron beam velocity. This synchronism allows velocity and density modulation of the beam. The RF wave voltage builds up along the length of the line, causing amplification.

12. The coupled-cavity TWT uses cavities rather than a helix. It can supply higher power at higher frequencies than the helix TWT.

13. The magnetron is an M-type tube, in which the electron beam travels in a circular path, shock exciting a series of resonant cavities. "Spokes" of electron current from cathode to anode produce high oscillation current in the cavities. Output power is extracted via a loop or waveguide.

14. Other microwave tube devices include the backward-wave oscillator (BWO), which has characteristics similar to that of a TWT; the klystrode, a hybrid consisting of a triode-klystron combination; and the twystron, another hybrid consisting of a klystron-TWT combination. Cross-field amplifiers (CFA) are M-type tubes with a slow wave structure. CFAs can have wave directions in either the forward or backward direction.

15. The gyrotron is an extremely high-frequency, high-power device for frequencies above 20 GHz. The gyrotron does not use linear beam characteristics. It is based on the fact that an electron in a magnetic field has a rotational frequency, called the cyclotron frequency. The angular velocity on an electron in orbit is thus accelerated or decelerated. This angular bunching is used to impart energy to a cavity to make either an oscillator or an amplifier.

Key Equation:

$$\text{Energy} = \tfrac{1}{2}MV^2 \qquad\qquad \textbf{(Eq. 6.1)}$$

QUESTIONS

1. Describe the fundamental limitations of various tubes at microwave frequencies.

2. Describe an early tube device that was used to overcome transit time problems.

3. List the three elements of a triode tube.

4. Describe how a triode tube amplifies. What is meant by the term *thermionic emission*?

5. What are the characteristics of the BKO tube? What is its drawback?

6. Define *velocity modulation.*

7. Explain how velocity modulation causes density modulation.

8. What is *electron bunching*?

9. Describe the basic operation of a reflex klystron.

10. What are the power-vs.-bandwidth characteristics of the different modes of a reflex klystron?

11. What is the difference between a reflex klystron and a two-cavity klystron oscillator?

12. What frequency adjustments are available in the reflex klystron?

13. How does the klystron amplifier differ from the klystron oscillator?

14. What are the advantages of the TWT over the klystron?

15. Describe the action of the helix in a TWT.

16. What is the purpose of the periodic permanent magnet or solenoid of the TWT?

17. What is the purpose of the attenuator in a TWT?

18. How does a CCTWT differ from a helix TWT?

19. Describe the operating characteristics of the magnetron.

20. What is the *pi mode* of operation in a magnetron?

21. How does a BWO differ from a TWT?

22. What are the operating characteristics of the klystrode?

23. Describe the operation of a CFA.

24. Describe the operation of a gyrotron.

25. What is the *cyclotron frequency*?

7 ACTIVE MICROWAVE DEVICES (SOLID-STATE)

OBJECTIVES

1. To describe basic characteristics of high-frequency solid-state devices.
2. To compare the physical and atomic differences among semiconductor materials.
3. To describe amplifier characteristics of solid-state amplifier devices.
4. To describe the basic "doping" characteristics of semiconductor devices.
5. To define the term "mobility" as it relates to semiconductor devices.
6. To differentiate between passive and active devices.

7.1 INTRODUCTION

It's assumed the student already has a background in solid-state devices from previous course work (particularly for low-frequency circuits). By way of review, some basic characteristics of this study and an introduction to microwave solid-state devices is in order.

Recall that all elements can be classified under three categories: insulator, semiconductor, and conductor. Insulators have a maximum of eight electrons in their valence (outermost) band. A tremendous amount of energy is required to remove them from this orbit. At the other extreme, the valence electrons of an ideal conductor require virtually no energy to be removed. Most good conductors have one or two valence electrons. Semiconductors have four valence electrons, which require a moderate amount of external energy in order to be removed.

If valence electrons gain energy, they move up and out of the valence band into what is known as the conduction band and become free electrons. Here they randomly drift, unless acted on by external forces. The area between the conduction band and the valence band is known as the *forbidden gap*. Charge carriers cannot exist in this region. Figure 7.1 shows the familiar band-gap approach to explain the characteristics of these elements.

To improve the conductivity of semiconductors and to have a means of control over their current carriers, we "dope" their atoms with impurities. When pentavalent impurities (ones having five valence electrons) are added to a pure semiconductor, there is an excess electron that is not covalent bonded with an atom. This forms an *N-type* semiconductor

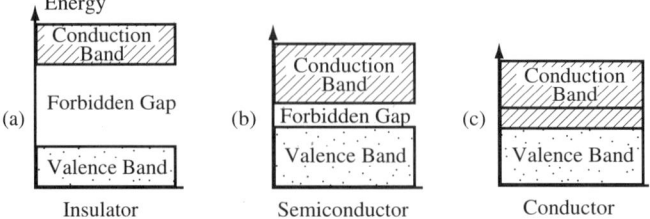

Figure 7.1   The band-gap explanation of conductivity.

(donor material). Likewise, trivalent impurities (ones having three valence electrons) form a *P-type* semiconductor (acceptor material) when added to a pure semiconductor. The P-type has a deficiency of valence electrons. This deficiency is called a *hole* and is equivalent to a positive charge. This vacancy can be filled by a nearby orbiting electron, creating a new vacancy in the nearby atom. Thus, the hole "moves" to the nearby atom. This motion is equivalent to the flow of positive charge.

A conduction electron is negatively charged and relatively light. A hole is positively charged and heavy. Since the actual mass of an electron is being shared via covalent bonding, the effective mass of an electron (2.4×10^{-31} kg) is somewhat less than the actual mass.

In an N-type semiconductor, the electrons are the dominant constituents of electric current and are called the *majority carriers*. Holes still exist, but they are relatively immobile and therefore are called *minority carriers*. On the other hand, in a P-type semiconductor, the holes are the majority carriers and the electrons are the minority carriers.

A semiconductor can be lightly or heavily doped. Symbolically, a heavily doped N-type semiconductor is N^+ and a heavily doped P-type semiconductor is P^+. The superscript + is not representative of the polarity of electric charge.

Typical high-frequency semiconductors include silicon (Si), germanium (Ge), gallium arsenide (GaAs), and indium phosphide (InP). The latter two are compound semiconductors. As is shown later, the compound semiconductors work best for high-frequency applications.

Figure 7.2a illustrates the random movement of free electrons between atoms. The electrons become free when they acquire thermal energy from ambient temperature and overcome the inherent atomic attraction. They then move into the conduction band. Under no other influence, their motion is totally random. The free electron has no average displacement in any preferred direction over a long period of observation time. The average distance that an electron travels before striking an atom is called the *mean free path*. This dimension is typically from 10^{-8} to 10^{-6} meters as compared to the typical atomic dimension of 10^{-11} meters.

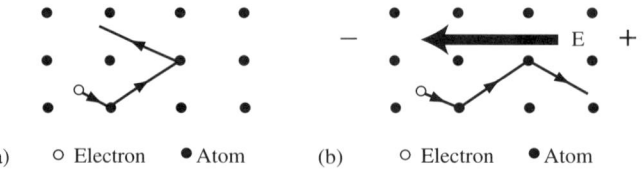

Figure 7.2   Electron movement. (a) Free electrons move randomly in between atoms. (b) The application of an electric field (E) causes a net displacement of the electrons.

The velocity of a free electron is a function of thermal energy. By approximating the electron's kinetic energy (equation 6.1, $E = 1/2MV^2$) to thermal energy ($TE = 3/2kT$, where k is Boltzman's constant and T is the absolute temperature of the ambient), the thermal velocity is found to be 1.8×10^5 m/s. Thus, its random velocity is 1.8×10^5 m/s. Therefore, the time needed to cover a distance equal to the separation of two atoms, with an average mean free path of 10^{-7} m, is approximately 5.6×10^{-12} s, i.e., a few picoseconds. (Remember, the average velocity in any preferred direction is still zero over a long period of time).

Under the influence of an external electric field, the electron flow is now in a preferred direction. The electrons tend to "drift" in a direction parallel to the electric field with a drift velocity, V_d. This is illustrated in Figure 7.2b. The drift velocity obviously depends on the strength of the electric field (E) and the effective mass of the electron. The drift velocity is usually less than the thermal velocity. Various samples of semiconductor can have different values of this velocity due to their inner-atomic structure and the amount of impurities present.

In a given sample, the ratio of the electron drift velocity to the applied electric field is called the *mobility*.

$$\mu = V_d/E \qquad \textbf{(Eq. 7.1)}$$

where
$$\mu = \text{electron mobility}$$
$$V_d = \text{electron drift velocity}$$
$$E = \text{electric field strength}$$

It is common to convert this mobility in m^2/V-s to cm^2/V-s.

The mobility, therefore, is related to the mass of an electron, the mean free path, and the ambient temperature. The concept of mobility also applies to holes. A hole is an evacuated electron site and can be regarded as a positively charged and more massive particle.

For comparison purposes, Table 7.1 gives approximate values of mobility for electrons and holes at room temperature (300°K) for several semiconductors.

What all this means is that higher-mobility materials make good high-frequency devices. At the present time silicon and gallium arsenide technologies are competing with each other for use in high-frequency applications.

Silicon is considered a "mature" technology, with good mechanical strength and thermal conductivity. High-quality oxides are easily formed on silicon, allowing for the fabrication of field-effect transistors (FETs), which are the basis for complementary metal-oxide semiconductor (CMOS) technology. Silicon wafers are currently produced in eight-inch diameters.

Table 7.1   Electron and Hole Mobility at Room Temperature (300°K)

Semiconductor Material	Electron Mobility (cm²/V-s)	Hole Mobility (cm²/V-s)
silicon (Si)	1500	600
germanium (Ge)	3900	1900
gallium arsenide (GaAs)	8500	400
gallium antimonide (GaSb)	4000	1400
indium phosphide (InP)	4600	150
indium arsenide (InAs)	33,000	460
indium antimonide (InSb)	78,000	750

Silicon has its disadvantages as well. Its intrinsic resistivity is on the order of 10 Ω/cm. This low resistance causes exceedingly high dielectric losses, making it almost impossible to transmit microwave signals on silicon. Silicon is fundamentally a slower material than germanium (Ge) or gallium arsenide (GaAs), both of which have higher electron mobilities than silicon. Germanium also has higher hole mobility, but it has low intrinsic resistivity that is not appropriate for high-frequency applications. This leaves GaAs as Si's chief competitor for high-speed applications. (Indium has great potential for high-frequency work, but it is currently not being commercially used.)

GaAs has its own disadvantages. Its low hole mobility severely limits its ability to support complementary circuitry. High-quality oxides cannot be formed on GaAs, a short-coming that has been partially circumvented by using modulation-doped FET (MODFET) structures.

To help alleviate the inherent low resistivity of silicon, high-resistivity silicon (HRS), silicon-on-insulator (SOI), and highly doped silicon wafers are used to fabricate silicon-based microwave circuits. Work on SiGe compounds is being done as well. This compound holds great promise.

Even with increased speed silicon devices, eventually silicon reaches its frequency limit, and an inherently faster material, such as GaAs, must take over for very high frequencies. It is not a matter of whether silicon will dominate microwave applications, but where it is that silicon is no longer a practical choice.

7.2   PASSIVE MICROWAVE DIODES

Recall that the conventional PN junction diode has a barrier formed at the junction and in silicon approximates 0.6 to 0.7 V. The barrier prevents recombination of carriers in the junction. Thus, a depletion region is formed at the junction.

When the carriers are forward biased by a potential greater than the barrier voltage, they move to the center, recombine, and a current flows. This current is the result of bipolar (both electron and hole) carriers. The diode conducts heavily when forward biased.

When the carriers are reverse biased, they are "held back" and no conduction occurs. The width of the depletion region depends on the amount of reverse bias. In practice, a small amount of reverse current flows, due to thermally generated electron-hole pairs. The forward and reverse currents are illustrated in Figure 7.3.

Note that "breakdown" occurs at some level of reverse bias. This is where a sudden increase in current occurs due to avalanche breakdown in the device. (Avalanche breakdown is a type of secondary electron emission.) The constant voltage beyond the knee of the VI curve is extremely useful in the cases of the Zener diode and the microwave IMPATT diode.

Conventional (passive) diodes have many applications. These include rectification, limiting, switching, and detection.

PIN Diode

One of the common passive diodes used at microwave frequencies is the PIN diode. It is almost a pure resistance at RF frequencies. The control current through a PIN diode can vary its resistance value from approximately 10 kΩ to less than 1 Ω. Most diodes exhibit this characteristic to some degree, but the PIN diode is optimized in design to achieve a relatively wide resistance range, good linearity, low distortion, and low current drive. The

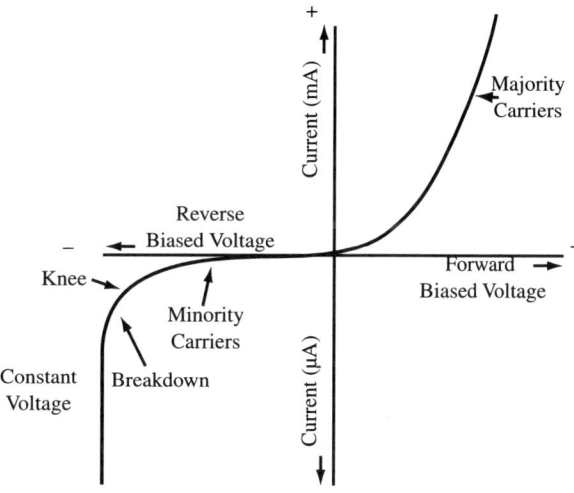

Figure 7.3 Diode characteristic curve.

characteristics of the PIN diode make it suitable for use in switches, attenuators, modulators, limiters, phase shifters, and other signal control circuits.

The PIN diode structure (Figure 7.4) consists of an I (intrinsic) layer of very high-resistivity material sandwiched between a P region of highly doped P$^+$ (positively charged) material and an N region of N$^+$ (negatively charged) material, hence the acronym PIN. With reverse or zero bias, the I layer is depleted of charges and the PIN exhibits very high resistance. When forward bias is applied across the PIN diode, positive charge from the P region and negative charge from the N region are injected into the I layer, thus increasing its conductivity and lowering its resistance.

The high off resistance and low on resistance make the PIN diode attractive for switching applications. The performance of a PIN diode switch can be simply approximated by treating the PIN diode essentially as a resistor in the forward biased state and a capacitor in the reverse biased state.

A linear change of resistance with bias makes the PIN diode useful for attenuator applications. An attenuator is operated throughout its dynamic range (or resistance range in the case of a diode attenuator).

A shunt PIN diode limiter is essentially an attenuator that uses self-bias rather than externally applied bias. As the RF input increases, the rectified current generated by the PIN diode biases the diode to a low resistance state. Most of the input power is attenuated, allowing very little to be transmitted. The sensitive equipment that follows is thus protected.

For limiting to be efficient, the PIN diode must have fast switching characteristics. This can be accomplished with an auxiliary Schottky diode.

PIN diodes come in a variety of packaging formats. They are well suited for microstrip and stripline assemblies. Operating frequencies extend from the HF to Ku bands.

Schottky Barrier Diode

A Schottky barrier diode contains a metal-semiconductor barrier formed by deposition of a metal layer on a semiconductor. The resulting nonlinear diode is similar to point contact diodes and PN junction diodes. The point contact diode was an early microwave diode that

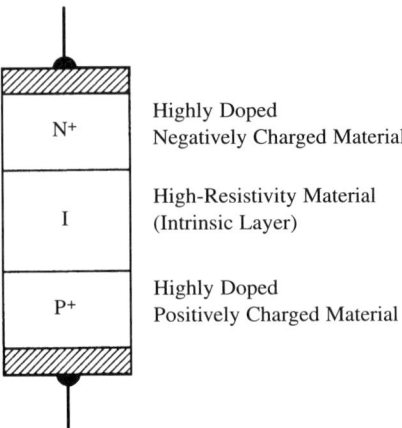

Figure 7.4   PIN diode structure.

utilized a galena crystal (a lead compound) in the form of a cat's whisker to provide contact with the semiconductor material. The Schottky diode is more rugged than the point contact diode because the contact is not subject to change under vibration. Its advantage over the PN junction is the absence of minority carriers, which limit the response speed in switching applications and the high-frequency performance in mixing and detection applications.

Several assembly geometries are used for Schottky barrier diodes. The popular offset junction Schottky diode is shown in Figure 7.5. The offset junction was developed because in conventional Schottky chips the barrier lies directly beneath the bond pad, and cracking was likely to occur when sonic energy was applied during thermosonic wire bonding. This geometry overcomes lead bonding problems, since the bond pad sits on a relatively thick oxide rather than on the barrier.

Because of their low noise levels and fast response, Schottky diodes are commonly used in mixer stages. They are also found in detector stages. Schottky diodes are zero-bias detectors. Frequencies to 40 GHz are available with silicon Schottky diodes. For higher-frequency applications GaAs is used.

7.3   ACTIVE MICROWAVE DIODES

Active microwave diodes differ from passive diodes in that they are used as signal sources to *generate* or *amplify* microwave frequencies. These include varactor, step-recovery, tunnel, Gunn, and IMPATT diodes.

Figure 7.5   Offset junction Schottky diode.

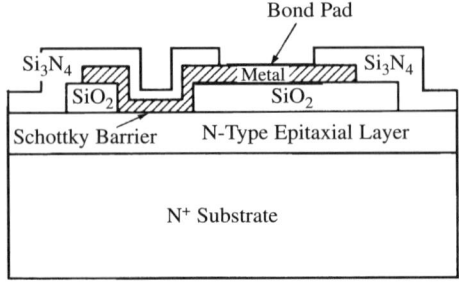

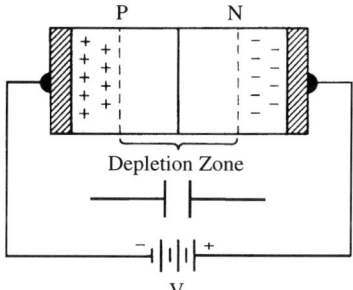

Figure 7.6   Varactor diode.

Varactor Diode

The varactor diode is a heavily doped PN diode (Figure 7.6). When the carriers are reverse biased, they are pulled away from the center. A large depletion zone forms, which acts as a dielectric insulator between the highly conductive carriers. Thus, the varactor diode acts like a *capacitor*. In fact, by varying the reverse voltage, a variable capacitance is available. As variable capacitors, varactor diodes have been used in tuned RF circuits and in VCOs (voltage controlled oscillators) for many years.

For higher-frequency microwave applications, the Si varactor has been replaced with GaAs. For microwave purposes, the varactor is used in frequency multiplier chains and parametric amplifiers. The capacitance of the varactor diode compared to reverse voltage is shown in Figure 7.7.

Step-Recovery Diodes

The step-recovery diode (SRD) is most graphically described as a *charge-controlled switch*. That is, a forward bias stores charge, a reverse bias depletes this stored charge, and when the charge is fully depleted, the SRD ceases to conduct current. The action of turning off, or ceasing current conduction, takes place so fast that the diode can be used to produce an impulse. If this is done cyclically, a train of pulses is produced. A periodic series of impulses is equivalent to a series of frequencies, all of which are multiples of the fundamental frequency. If these impulses are used to excite a resonant circuit, much of the total power in the spectrum can be concentrated into a single frequency. Thus, input power at one frequency can be converted to output power at a higher frequency.

The very sharp and narrow pulses contain harmonics of the fundamental (exciting) frequency. Comb generators utilize the SRD's production of a multitude of frequency com-

Figure 7.7   Capacitance vs. reverse voltage for a varactor diode.

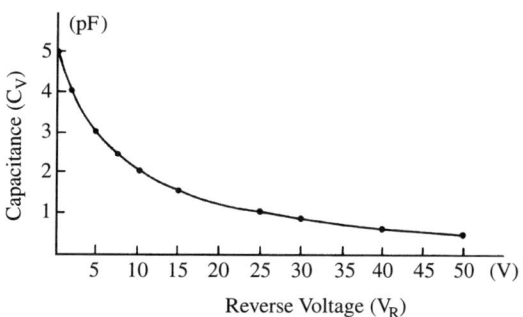

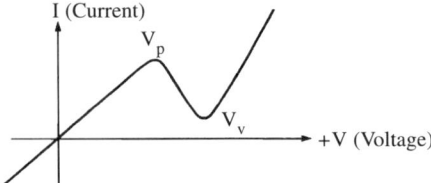

Figure 7.8 Tunnel diode characteristic curve. The region of negative resistance falls between V_p and V_v.

ponents. Comb generators are used in measurement equipment, such as spectrum analyzers, to produce locking signals.

Another circuit that picks out a single harmonic and optimizes the power output around that harmonic is called a *multiplier*. The end result of a multiplier is output power at some multiple ($2f$, $3f$, etc.) of the input frequency. Multipliers are used as local oscillators, low-power transmitters, or transmitter drivers in radar, telemetry, telecommunication, and instrumentation.

Tunnel Diode

The tunnel diode was the first semiconductor device shown to possess a *dynamic negative resistance*. Negative resistance is a phenomenon that occurs when current and voltage, which normally have a direct proportionality, vary inversely. Another way to describe negative resistance is to say that V (voltage) and I (current) are 180° out of phase. Negative resistance is a dynamic property in that it occurs only under actual circuit conditions. It is not static; it cannot be measured with an ohmmeter. Figure 7.8 shows the characteristic curve for a tunnel diode. Common symbols for the tunnel diode are shown in Figure 7.9.

The tunnel diode is a heavily doped (about 100 to 1,000 times more than the normal 10^{17} impurity atoms per cubic centimeter) PN junction made of either Si, Ge, or GaAs. This causes the junction to be extremely narrow. Note that with very small forward bias applied, there is an increase in current. At point V_p in Figure 7.8 there is a sudden decrease in current for an increasing voltage. This occurs down to point V_v, and thereafter the diode acts like a conventional diode. Between points V_p and V_v is the region of negative resistance.

With no bias voltage applied, electrons "tunnel" their way into the P material with high velocity, although they do not have sufficient energy to surmount the barrier. (This phenomenon is a function of quantum physics and difficult to understand. The electrons can randomly appear on the other side of the junction, as shown in Figure 7.10a.) As forward bias increases to point V_p (Figure 7.8), the energy levels of the electrons in the N region are the same as the vacant energy level states of the valence electrons in the P region. Thus, the maximum number of electrons has tunneled through the depletion layer into the P region.

Beyond point V_p (Figure 7.8), the energy level state of the valence electrons is further reduced, so that empty energy levels in the P region drop below the energy levels of the electrons in the N region. At this point the forbidden energy gap between the conduction

Figure 7.9 Common symbols for the tunnel diode.

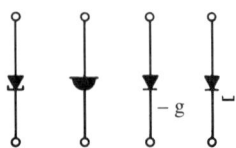

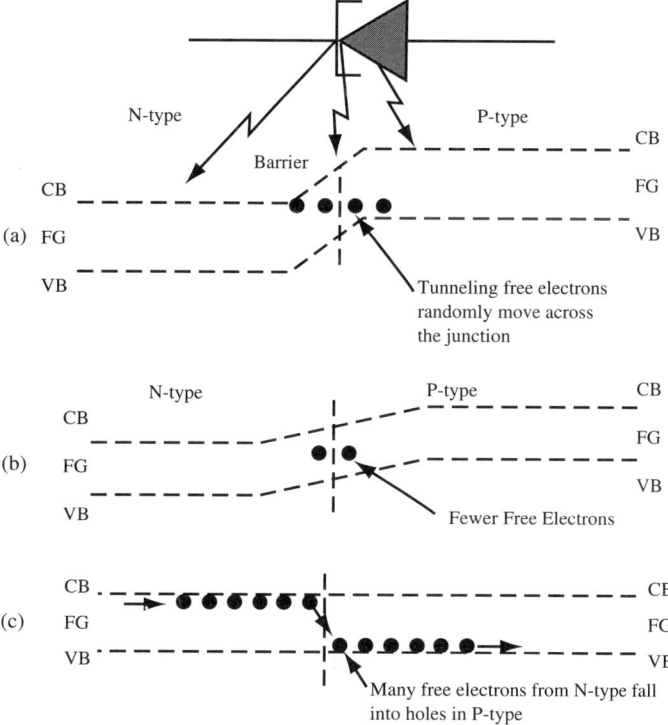

Figure 7.10   Tunnel diode current flow. (a) Energy levels when V = 0. Increase in current begins. CB = Conduction Band; FG = Forbidden Gap; VB = Valence Band. (b) Negative resistance energy level situation. Decrease in current begins. (c) Normal diode action energy level diagram. Increase in current begins.

and valence bands comes into play. The conduction band of the N region lines up with the forbidden gap, causing a decline in current (Figure 7.10b). This occurs down to point V_v (Figure 7.8). After this point, the diode acts like a conventional forward biased diode. The conduction and valence bands coincide during this time (Figure 7.10c).

What do we do with this negative resistance phenomenon? We use this device in either an oscillator or an amplifier circuit. For oscillation to occur in a resonant circuit, the negative resistance of the diode must be greater than or equal to the resistive losses of the circuit. (In terms of conductance, $-G_D \geq G_O$.)

When a tunnel diode is used as an amplifier, care must be taken to isolate the input and output. One way to accomplish this is with a circulator (Figure 7.11). Because the tunnel diode exhibits negative resistance to the system, the reflected power is greater than the incident power. Thus, amplification takes place and isolation is afforded by the circulator. This circuit is known as a *reflection amplifier*. Most other negative resistance devices can be used in place of the tunnel diode.

Gunn Diode

The Gunn diode is another device exhibiting negative resistance characteristics. These effects were first observed by John Gunn of IBM in 1963 as he was studying the properties of N-type GaAs. The Gunn diode falls under the family of transferred electron devices

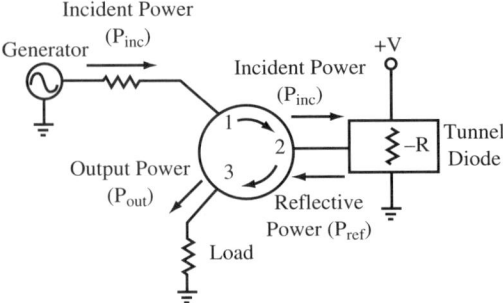

Figure 7.11   Tunnel diode amplifier with circulator.

(TEDs). Using the Gunn diode to generate microwave frequencies is a very common application for this device.

A Gunn diode is typically made of an N-type compound semiconductor such as GaAs or indium phosphide (InP). Two modes of operation are common: the nonresonant bulk (transit-time) and the resonant limited space-charge accumulation (LSA). When biased by a battery, the Gunn diode functions as a piece of bulk resistance. No junction effect or barrier occurs because the semiconductor is one piece. Unlike semiconductors such as Si, the atomic structure of the compound semiconductor causes an uneven and highly localized distribution of the external electric field strength. Thus, the electron's motion is not uniform, but bunched. Domains of electrons are said to have formed.

The bunching phenomenon is a bulk effect because the electrons are slowed down by the uneven distribution of the local electric field. Atomically, two conduction-band regions exist. A valence electron starts out at the lower conduction region and then transfers to the higher conduction region (hence the name "transferred electron devices"). Such an effect can be characterized by a decrease in electron mobility, which in turn is equivalent to negative resistance. While in the lower conduction band, the electrons exhibit high mobility and light effective mass. When in the higher conduction band, they exhibit low mobility and heavy effective mass. This can be seen in Figure 7.12.

Figure 7.12   Gunn diode—two conduction-band regions.

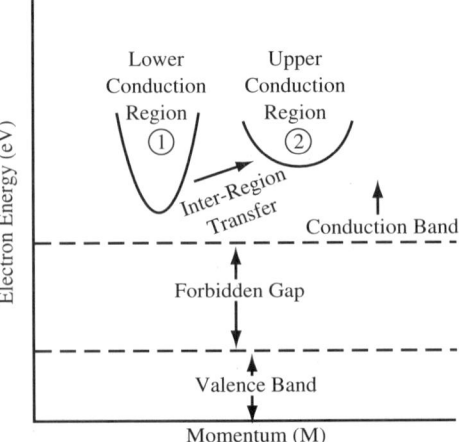

As the electric field increases, the electron velocity increases (as would be expected). Beyond a threshold electric field, the velocity suddenly slows down due to the switchover to a higher conduction band, and a negative resistance is formed.

The electron domains burst across the semiconductor at a frequency dependent on the width of the semiconductor layer, which is typically about 10 μm. The typical drift velocity is around 1×10^5 m/s, so by using the basic distance = rate × time relationship we can calculate the transit time as about 0.1 ns. The electron burst, therefore, repeats at a rate of 10 GHz. This frequency depends on the path length and the drift velocity.

The bulk (transit-time) mode is very inefficient. Power output is less than 1000 mW with efficiencies on the order of 1% to 5%. Typical DC power supply voltages are less than 10 VDC for operating frequencies from 1 to 18 GHz.

A simplified circuit for a limited space-charge (LSA) oscillator circuit is shown in Figure 7.13a. The LSA mode depends on current pulses to shock excite a resonant circuit. The resonant circuit output affords better efficiencies, higher frequency operation, and some adjustment over its oscillation frequency.

Rather than have the electrons slow down, causing domains to build as they burst across the semiconductor, the LSA mode effectively limits the "waiting" mode for the domain to build up. The slowing down of an electron to "wait" for the upcoming one

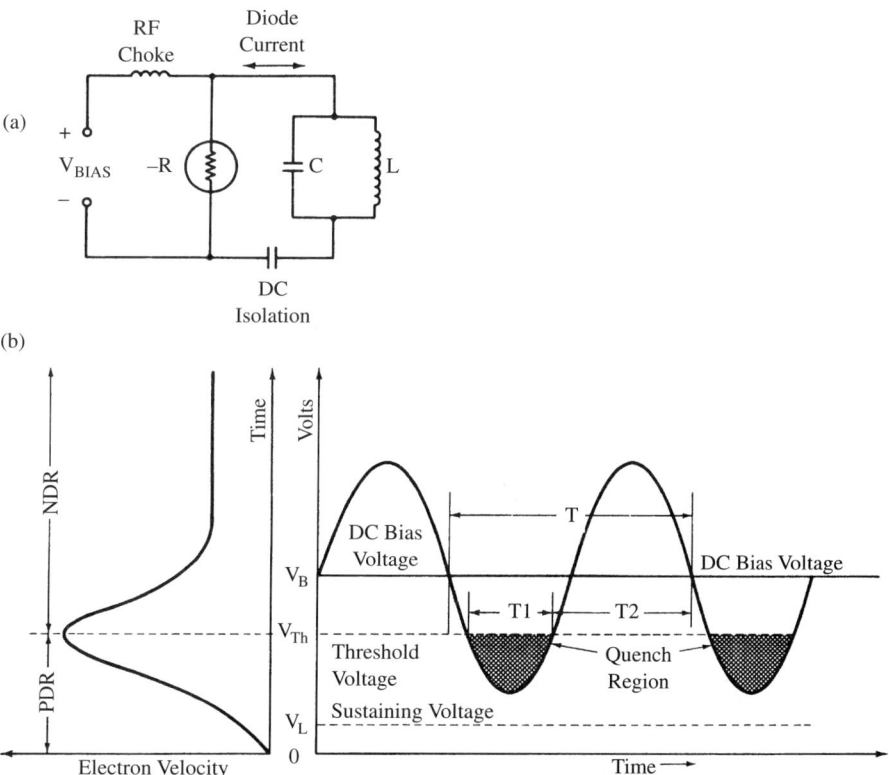

Figure 7.13 Gunn diode—LSA mode of operations. (a) LSA mode oscillator circuit. (b) LSA mode waveforms. PDR = Positive Dynamic Resistance. NDR = Negative Dynamic Resistance.

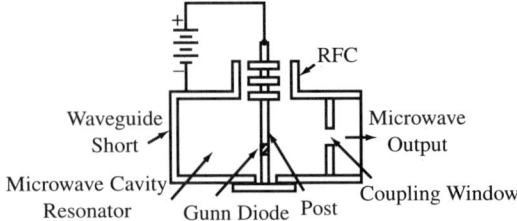

Figure 7.14 Gunn diode in cavity resonator with waveguide attached.

accumulates space-charges within the domain. The LSA eliminates this by operating directly on the negative resistance region of the diode (Figure 7.13b). This is accomplished by biasing with a voltage greater than the threshold voltage. The domain-creation phenomenon of the transit-time mode causes several initial output pulses to excite the external circuit into oscillation. This action causes a continuous RF sine wave to build up that has a frequency equal to the frequency of the resonant circuit. The RF voltage rides on the DC bias such that the total bias is greater than V_{th} on positive peaks and less on negative peaks. When going negative, the domains are quenched. If the previous domain reaches the anode while the bias is below V_{th}, the next domain is delayed until the RF cycle brings the bias back above the threshold potential. This phenomenon causes the output current pulse period to adjust automatically to the period of the external resonant circuit.

The LSA mode is typically accomplished by mounting the diode in a resonant cavity with waveguide attached. Energy is coupled out of the cavity through an iris on one end. By mechanical adjustment of the cavity dimensions, the frequency can be adjusted. This is shown in Figure 7.14.

IMPATT Diode

The IMPATT diode is the offshoot of investigations by W. T. Read of Bell Laboratories that came to fruition in 1965. Read theorized that negative resistance could be realized by an appropriate combination of delay in generating a current avalanche pulse and its transit time through the device. The IMPATT diode is therefore a current-operated device (where the Gunn diode is considered a voltage-controlled device). The combination of these two parameters produces the necessary 180° phase shift between voltage and current, causing the negative resistance phenomena. These parameters also give the device its name: *imp*act *a*valanche and *t*ransit-*t*ime (IMPATT) diode.

There are several versions of the IMPATT diode. Some are simple reverse biased PN junction diodes, some are complicated multigrade doped layers of a reverse bias PN junction, and others are reverse biased PIN diodes. The PIN diode version is shown in Figure 7.15a.

The diode must be connected to a resonant circuit. At bias turn-on, noise excites the tuned circuit into a natural oscillation frequency. This voltage adds algebraically across the diode's reverse bias voltage (Figure 7.15b). Near the peak positive half-cycle, the diode experiences impact avalanche breakdown. (Avalanche breakdown is a form of secondary emission, since charge carriers become so energetic they are able to knock additional valence electrons out of their bonds to form excess electron-hole pairs. The common Zener diode works on this principle.)

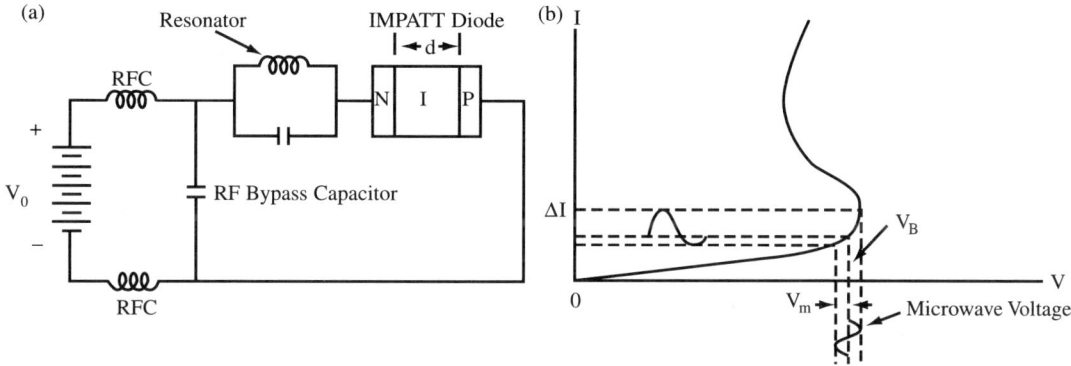

Figure 7.15   (a) IMPATT diode oscillator. (b) VI characteristics of a reverse biased PIN IMPATT diode.

When the voltage falls below this peak value, no avalanche breakdown occurs. As the electron-hole pairs are formed, the holes are immediately discharged at the cathode (P-type region), which is negatively biased. Meanwhile, the generated electrons start to drift toward the anode of the N-type material through the highly resistive I region. When the bunched electrons reach the anode they discharge there and produce a current pulse. The frequency of the current pulses is a function of the electron transit distance(d) and the drift velocity of the electron. The process then repeats.

A 90° shift occurs between the current pulse and the applied voltage in the avalanche process. A further 90° shift occurs during the transit time, for a total 180° shift. This translates to negative resistance (V and I are 180° out of phase).

When the breakdown voltage is high and the microwave signal is large, the ions created by impact avalanche ionization multiplication do not completely exit from the transit domain of the diode during the negative half-cycle of the microwave signal. The remaining ions comprise both positive ions (holes) and electrons, forming *plasma*. The plasma is trapped in the diode and participates in producing a large microwave current during the positive half-cycle. This kind of IMPATT diode is termed the TRAPATT (*tr*apped *p*lasma *a*valanche-*t*riggered *t*ransit-time) diode.

Electron injection is not necessarily at the PI junction, as in thin PIN diode structures. A Schottky barrier with a metal-semiconductor contact can create similar avalanche electron injections. When this device is used, the diode is called a BARRITT (*barr*ier *i*njection *t*ransit-*t*ime) diode.

An IMPATT oscillator has higher output power than the Gunn equivalent. However, the Gunn oscillator is relatively noise-free, while the IMPATT is noisy due to avalanche breakdown.

7.4   PARAMETRIC AMPLIFIER

When nonlinear devices are used to mix two RF signals of different frequencies, many frequencies are generated. The four common values are the input frequency (f_s), the local oscillator frequency (f_p), the sum of these $(f_p + f_s)$, and the difference of these $(f_p - f_s)$.

In a parametric amplifier, as the $f_p - f_s$ output occurs, a part of f_p is transferred to signals of frequency f_s through a nonlinear mixer impedance. In effect, the f_s signal is intensified.

Parametric amplification is considered a regenerative mixing of electromagnetic waves by means of variable impedance. The variable parameters that can be employed for this device are usually an inductance (L), a capacitance (C), or a resistance (R). For the variable-capacitance parametric amplifier, semiconductor diodes are commonly used. For the variable-inductance parametric amplifier, ferrite is used. For the variable-resistance parametric amplifiers, the electron beam is commonly used. The microwave parametric amplifier uses either variable inductance or variable capacitance and is sometimes termed the *mavar*, which stands for *m*icrowave *a*mplification by *va*riable *r*eactance.

Among the many types of parametric amplifiers, the most common employs the varactor diode to form the variable-capacitance parametric amplifier. In the parametric amplifier, the local oscillator is called the *pump oscillator* since it "pumps up" the signal level. Consequently, the frequency f_p is the *pump frequency*. The intermediate frequency $f_i = f_p - f_s$ is termed the *idler frequency*. This frequency is contained in a circuit following called the *idling circuit*. The word "pump" comes from mechanical engineering. It is analogous to the concept of a push being applied to a pendulum at the proper phase to sustain the pendulum arc. Otherwise, the pendulum would stop oscillating due to dampening effects.

A schematic diagram of the variable-capacitance diode parametric amplifier is shown in Figure 7.16. The varactor capacitance controlled by DC bias and an inductance form the idling circuit, which resonates at $f_i = f_p - f_s$.

Pump-oscillation power is coupled to the idling circuit and changes the capacitance of the diode at the pump frequency. Because f_p is a different frequency than f_i, power transfer between the two circuits is limited. Because the idling circuit is resonant at f_i, signals fed from the input circuit cause a large current with frequency f_i to flow in the resonant circuit. Therefore, much of the pump power is transferred to the idling circuit. Now, if the output circuit resonates at the same frequency as f_s, a large signal can be coupled out of the circuit. The idler frequency must be close to the signal frequency. The variable capacitance, which is changing at frequency f_p, contributes to intensifying the idling circuit current. Usually the pump frequency f_p is chosen to be twice f_s. When $f_p = 2f_s$, $f_i = 2f_s - f_s = f_s$.

By looking at Figure 7.17, we can see that the interrelation among bias voltage, pump voltage, pump frequency, variation in capacitance, and voltage across the capacitor at the signal frequency can cause an increase in voltage with the proper phase relation. From AC theory we know that voltage is a function of charge (Q) versus capacitance (C) or V = Q/C.

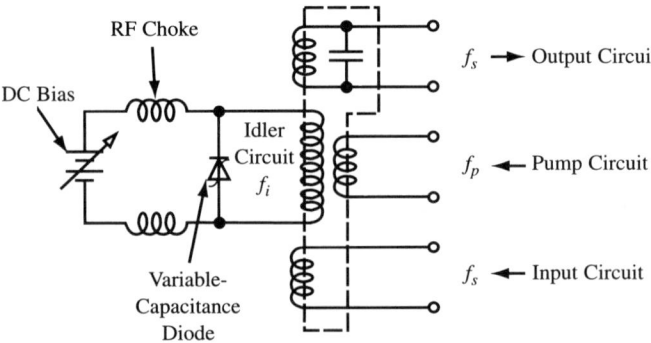

Figure 7.16   Variable-capacitance diode parametric amplifier.

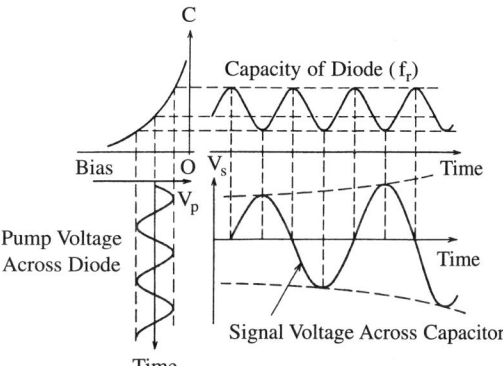

Figure 7.17 Variable-capacitance vs. signal voltage across capacitor.

Thus, it is evident that when the capacitance is at minimum, the signal voltage across the diode is at maximum.

Parametric amplifiers are usually more complex, of lighter weight, and less expensive than thermionic devices. They also generate less noise because the active device is not thermionic and there is very little shot noise. On the downside, parametric amplifiers require a pump oscillator, which may be some form of thermionic or solid-state amplifier.

7.5   MICROWAVE BIPOLAR TRANSISTORS

The basic operation of a conventional bipolar transistor is such that three doped elements forming two PN junctions constitute a bipolar transistor. As shown in Figure 7.18, the three elements are the emitter, base, and collector in either an NPN or a PNP arrangement.

The base is very thin, and it is lightly doped in comparison to the other elements. The base is the control element of the device; without base current, there is no current flow in the circuit. As a current-controlled device, its current gain, β (h_{FE}), is a function of I_C/I_B. When biased linearly, the EB junction is forward biased, and the CB junction is reverse biased. Under these conditions, the collector current is approximately 95% to 99% of the emitter current. The device is considered a conventional amplifier.

Nonlinear operation has both junctions forward biased or both reverse biased. Under these conditions the device is a switch, not an amplifier. Such devices are used in digital applications.

Silicon bipolar NPN devices have an upper cutoff frequency of about 25 GHz (varies with manufacturing improvements). At higher frequencies, most designers typically use field-effect transistors (described in section 7.6). The primary limitations to higher-frequency use are base and emitter resistance, capacitance, and transit time.

Modern silicon bipolar transistors tend to be NPN devices of either planar or epitaxial diffused design. These are shown in Figure 7.19. The planar, Figure 7.19a, has its elements

Figure 7.18 Conventional NPN bipolar transistor.

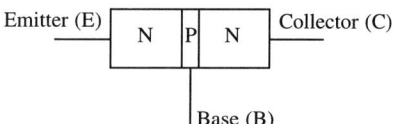

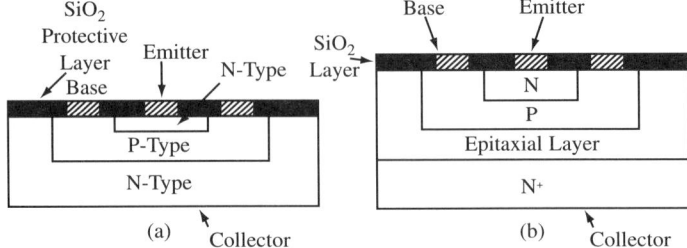

Figure 7.19 Types of silicon bipolar transistors. (a) Planar diffused NPN silicon transistor. (b) Epitaxial diffused NPN silicon transistor.

diffused under the surface of Si using photolithography techniques. The surface has a protective layer of silicon dioxide (SiO_2) that protects against contaminants.

The epitaxial transistor (Figure 7.19b) uses a thin, low-conductivity epitaxial layer for part of the collector region (the remainder is highly conductive, N^+ material).

Heterojunction bipolar transistors (HBTs) have been designed with much higher maximum frequencies. For high-frequency applications the trick is to scale down the size of the device. Narrower widths of the elements within the transistor are the key to superior high-frequency work.

One way to control widths and other problems associated with high-frequency work is to change the conventional construction geometry. One of the more popular geometries is the *interdigital* (Figure 7.20), which yields thin wide-area low-resistance base regions that increase the operating frequencies. The pitch, or emitter-to-emitter centerline spacing, controls the high-performance aspects of the transistor. Finer pitches result in more gain and a lower noise figure at higher frequencies. Devices with coarser pitches are typically easier to manufacture but are limited to lower-frequency applications. The number of emitter fingers controls the current-handling ability of the device and is a measure of output power capability. Devices with larger numbers of fingers are suitable for power applications such as transmitter stages; devices with small numbers of fingers operate at lower biases and are often the choice for battery-operated applications.

7.6 MICROWAVE FIELD-EFFECT TRANSISTORS

The field-effect transistor (FET) operates by varying the conductivity of a semiconductor channel through changes in the electric field across the channel. (It is the semiconductor equivalent of a triode vacuum tube.) Because the input voltage varies an output current ($\Delta I_D/\Delta V_{in}$), the FET is called a *transconductance amplifier*. The two basic forms of the FET are the *junction FET* (JFET) and the *metal-oxide semiconductor FET* (MOSFET). These are shown in Figure 7.21.

Figure 7.20 Interdigital construction geometry.

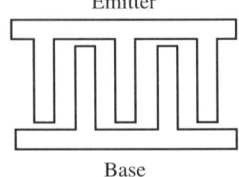

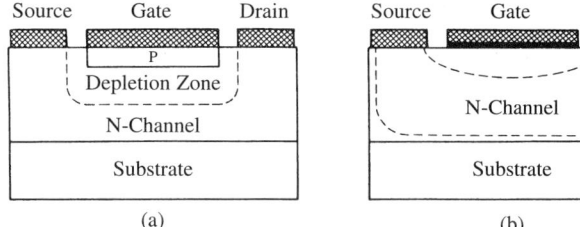

Figure 7.21 FET transistors. (a) Junction field-effect transistor (JFET). (b) Metal-oxide semiconductor field-effect transistor (MOSFET).

The JFET works on the depletion channel principle. The channel in this case is an N-type material, while the gate is of P-type material. The gate lead connects to the P-type material. Reverse bias is normally applied to the gate-source junction. The applied electric field extends into the channel and controls the level of current reaching the drain.

The MOSFET replaces the material in the gate with a layer of silicon dioxide insulating material. Gate metallization is applied over the insulator. This allows for the implementation of two types of channels: a depletion mode, in which channel carriers are controlled by the electric field, and an enhancement mode, in which the channel is formed by the electric field. Either forward or reverse bias may be used.

As stated earlier, microwave transistors need to be scaled down in size and have higher-mobility carriers. GaAs FETs offer this for improved high-frequency operation (above 4 GHz). They afford low-noise characteristics, improved temperature stability, and higher power levels. These devices can be used as low-noise amplifiers (LNAs), Class C amplifiers, oscillators, or in monolithic microwave integrated circuits (MMIC).

The principal player among the FET devices is the *metal semiconductor field-effect transistor* (MESFET), also called the *Schottky barrier transistor* (Figure 7.22).

The performance of a GaAs FET is determined primarily by the gate width and length. The gate length is the short dimension of the gate and sets the high-frequency aspects of performance. In general, shorter gate lengths result in superior performance. The gate length is the primary construction feature of the process. The gate width determines the active periphery and sets the transconductance (g_m) and saturated drain current (I_{DSS}) of the FET. In addition to establishing the power-producing capability, the gate width sets the S-parameters (chapter 9) of the device and, therefore, the optimum frequency of operation. The gate width is set by the geometry or mask design. It is fairly common for the same mask type to be used

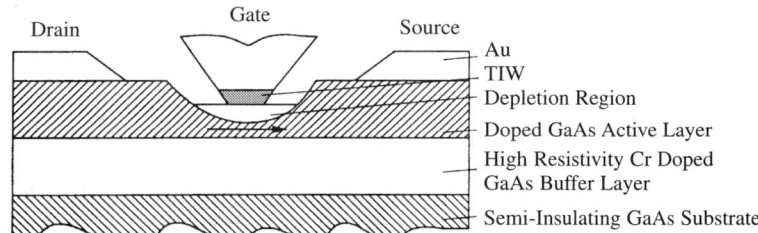

Figure 7.22 MESFET transistor.

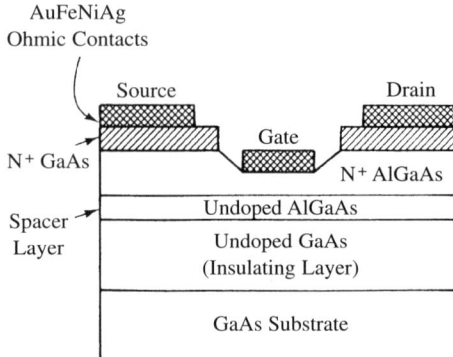

Figure 7.23   HEMT FET transistor.

with different processes to create FETs with the same gate width but different performance characteristics.

High performance FETs include the high electron mobility transistor (HEMT), shown in Figure 7.23, and the pseudomorphic high electron mobility transistor (PHEMT), shown in Figure 7.24. The HEMT has high power gains at frequencies approaching 100 GHz with low noise levels. This device is built using ion implantation, molecular beam epitaxy (MBE), or metal organic chemical vapor deposition (MOCVD).

The PHEMT uses the MBE material to create a GaAs-AlGaAs-InAlGaAs structure that results in superior mobility to standard MESFET or HEMT devices. This process is optimized for the lowest noise figure for critical receiver applications such as DBS (Direct Broadcast Satellite) block converters.

Higher-frequency applications will continue to be done via creative geometries, processes, and other manufacturing techniques. Compounds of bipolar and CMOS (BiCMOS), silicon and germanium (SiGe), and other materials will undoubtedly improve the performance of these devices. New techniques in photolithography and x-ray lithography allow gate lengths as small as 0.1 μm with silicon.

7.7   MICROWAVE INTEGRATED CIRCUIT AMPLIFIERS

The transistors described in the previous sections make up what are called *discrete amplifiers*. The word "discrete" means "made up of distinctive parts." There is a trend in low-power microwave applications to use the amplifiers in integrated form. (You are already

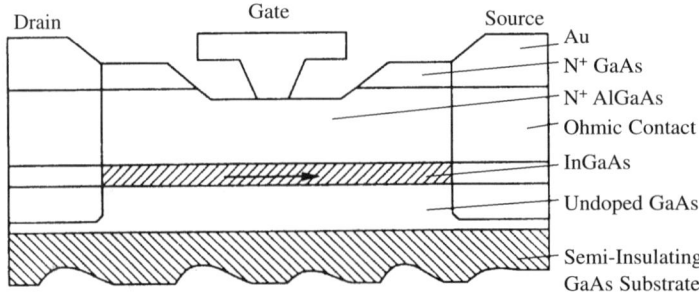

Figure 7.24   PHEMT transistor.

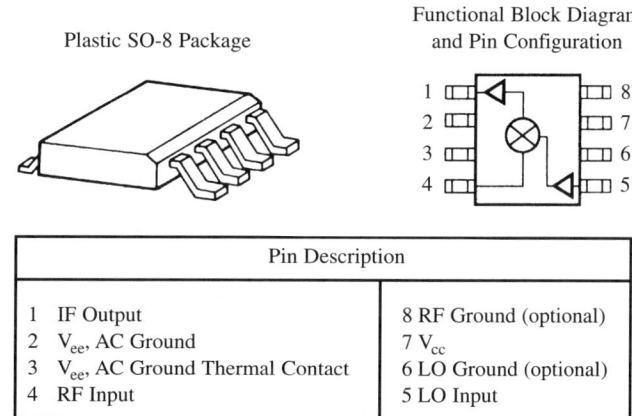

Plastic SO-8 Package

Functional Block Diagram and Pin Configuration

Pin Description	
1 IF Output	8 RF Ground (optional)
2 V_{ee}, AC Ground	7 V_{cc}
3 V_{ee}, AC Ground Thermal Contact	6 LO Ground (optional)
4 RF Input	5 LO Input

Figure 7.25   Hewlett-Packard IAM-82008 MMIC mixer/IF amplifier.

familiar with integrated forms of solid-state devices like the 741 op-amp and 555 timer.) This integrated circuit (IC) form includes the monolithic microwave integrated circuit (MMIC) and the hybrid microwave integrated circuit (HMIC). These devices can be made with either silicon or GaAs technology and in either bipolar or FET transistor form. For high-frequency applications, GaAs FET devices are the best choice.

An MMIC circuit has all the devices fabricated directly within the substrate. Discrete devices are not used. The word "monolithic" literally means "one stone," implying that the devices are within or under the surface of the IC material.

Typical MMICs are used as LNAs, as mixers, as modulators, in frequency conversion, in phase detection, and as gain block amplifiers. Silicon MMIC devices are designed to work optimally in the 100 MHz to 3 GHz frequency range.

One product (Hewlett-Packard IAM-82008) offers an active double-balanced mixer/IF amplifier in one eight-pin surface mount package. It is designed for narrow or wide bandwidth commercial and industrial applications having RF inputs up to 5 GHz. Typical applications include frequency conversion, modulation, demodulation, and phase detection. Markets include fiber optics, GPS satellite navigation, mobile radio, and communications transmitters and receivers. Its package, functional block diagram, and pin description are shown in Figure 7.25.

GaAs FET MMICs are available to work in the 1 to 18 GHz frequency range. These high-performance amplifiers are targeted for specialty applications in which excellent electrical performance is the primary consideration.

Figure 7.26 shows a Hewlett-Packard MGA-65100 two-stage GaAs FET MMIC. It offers 24 dBm output power at 14 GHz with high gain, 9.5 dB typical at 14 GHz. This figure shows the chip schematic and the substrate bonding diagram for this MMIC. Note that the input and output pads have matching networks for impedance matching purposes. Typical MMIC impedances are either 50 Ω or 75 Ω.

Another MMIC example is shown in Figure 7.27. Figure 7.27a shows a modular amplifier designed to meet the ultralinear transmitter output requirements of worldwide PCS (Personal Communications Services) systems. The amplifier is a class A, single-stage amplifier based on a GaAs MESFET transistor. The unit has a 15 dB gain from 1.7 GHz to

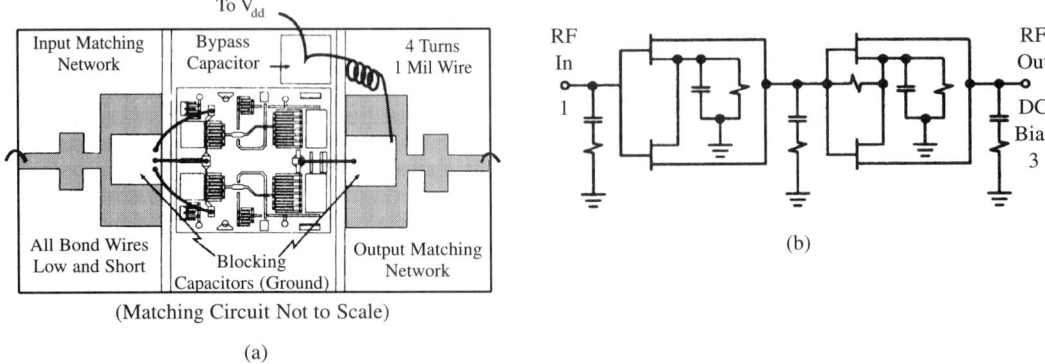

Figure 7.26   Hewlett-Packard MGA-65100 two-stage GaAs FET MMIC. (a) Substrate bonding diagram. (b) Chip schematic.

2.1 GHz. Figure 7.27b shows the application schematic for this stage. (Note the use of a 50 Ω microstrip transmission line.)

Hybrid ICs are a level closer to regular discrete circuit construction than MMICs. Passive components and planar transmission lines are laid down on various types of substrate

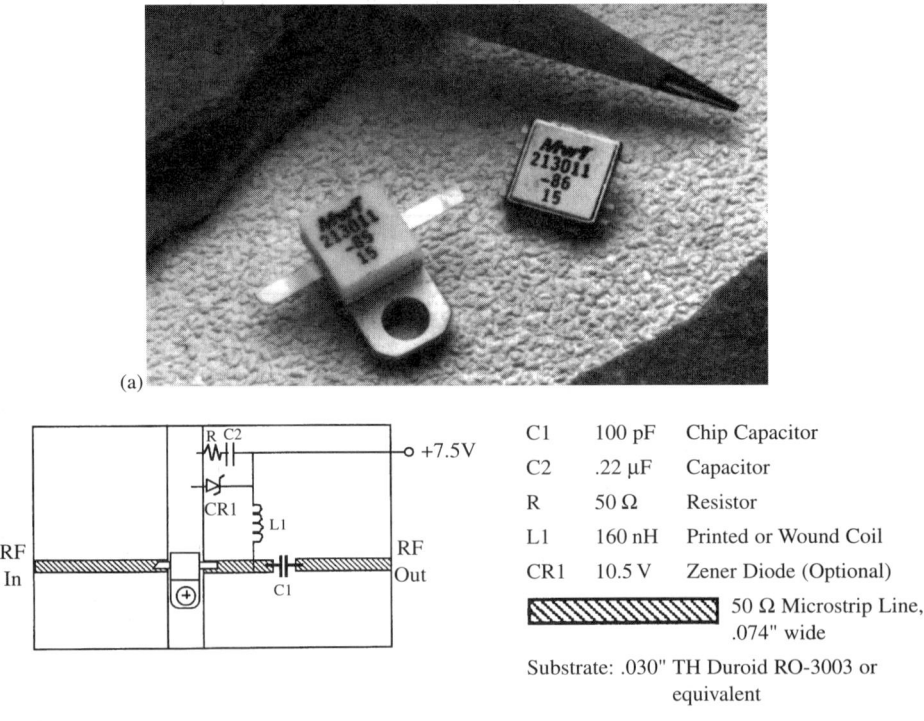

Figure 7.27   GaAs MESFET amplifier. (a) Modular GaAs MESFET MMIC transistor. (b) Schematic application. (Courtesy of Microwave Technology, Fremont, California.)

material. Transistors and unpackaged monolithic chip dies are bonded to the substrate and then connected to the substrate circuitry by mil-sized gold or aluminum bonding wires.

Packaging holds the key to further growth in MMIC technology. There has been a downward trend in the production costs of GaAs die and a corresponding increase in chip performance and reliability. Packaging designs reflect a high degree of customization and a wide variety of approaches. At the present time there is a lack of component standardization. This is seen as an impediment to MMIC market acceptance and product growth. However, the government is making a concerted effort to promote microwave packaging technology development and standardization.

In 1993 the San Diego, California, company Holzpak Corp. touted the use of polytetrafluoroethylene composite (PTFE) as a solution to the packaging dilemma. Glass microfiber-reinforced PTFE had never been used in microwave packaging before. The material displays a uniform dielectric constant, $\epsilon_r = 2.33$, over a wide frequency range. Its usefulness extends through the Ku band to 18 GHz. The packaging material is implemented in a stripline design, Ku-band single chip package.

This technology has been adapted to frequencies of 1 to 16 GHz currently. Current markets include personal communication networks (PCNs), digital European cordless telecommunications (DECT), wireless local area networks (WLANs), DBS satellites, terrestrial flight telephone systems (TFTFs), and digital short-range radio (DSRR).

The late 1990s will continue to see many improvements in packaging designs brought about by the various manufacturers of microwave integrated circuits.

7.8 SUMMARY

1. Semiconductors have four valence electrons. They are doped with either pentavalent or trivalent impurities to improve their conductivity and afford a means of control over their current carriers.
2. Gallium arsenide (GaAs) is a compound material that has higher electron mobility than either germanium or silicon. It is becoming the dominant high-frequency semiconductor material.
3. Passive microwave diodes are used in rectification, limiting, switching, and detection. PIN and Schottky barrier diodes are commonly used microwave passive diodes.
4. Active microwave diodes are used as signal sources to generate or amplify microwave frequencies.
5. The varactor diode is a reverse biased PN diode used as a variable capacitor.
6. Step-recovery diodes act as charge-controlled switches. They produce a series of current pulses that are rich in harmonics. They are used in comb generators and multipliers.
7. Negative resistance is a dynamic phenomenon occurring when V and I in a semiconductor device vary inversely.
8. The tunnel diode was the first semiconductor device shown to possess a negative resistance region. This phenomenon makes the device useful in oscillators and amplifiers.
9. The Gunn diode is a GaAs semiconductor that also exhibits negative resistance. It is operated in either a bulk (transit-time) or limited space-charge accumulation (LSA) mode. In either case, negative resistance occurs, and a cyclical operation occurs at gigahertz frequencies.

10. The IMPATT diode is another negative resistance diode, which works on avalanche breakdown forming current pulses. Variations of this diode include the TRAPATT and BARRITT diodes.

11. A parametric amplifier uses a varactor diode in a variable-capacitance amplifier. A pump oscillator is used to provide an input that causes an increase in output power. It is a low-noise amplifier.

12. Microwave bipolar transistors can be used to about 25 GHz. They use different construction geometries to scale down their dimensions and improve high-frequency response.

13. The most popular microwave transistors are based on the field-effect transistor model. The principal player is the metal semiconductor field-effect transistor (MESFET). It is made of GaAs and it affords higher-frequency performance.

14. Microwave IC amplifiers are either of the monolithic microwave integrated circuit (MMIC) or the hybrid microwave integrated circuit (HMIC) type. MMICs employ no discrete components. All devices are formed within the substrate material based on the same planar diffusion technology as bipolar transistors. HMICs employ passive and active components bonded to a substrate. Interconnections are made to the devices to form an amplifier stage.

15. Packaging holds the key to further growth in MMIC technology. The next decade will see many improvements in packaging designs and a move toward standardization.

Key Equation:

$$\mu = V_d/E \qquad\qquad \textbf{(Eq. 7.1)}$$

QUESTIONS

1. Describe the three energy level categories we divide all substances into.

2. What is the doping process? Why do we do it?

3. Describe the movement of free electrons in semiconductor materials.

4. What is meant by *mobility* of carriers? How does this affect high-frequency performance?

5. Describe the features that make semiconductor materials good for making various solid-state devices.

6. What is meant by the term *passive* in relation to semiconductor devices? What are passive devices used for?

7. Describe the features and operation of the PIN diode.

8. Describe the features and operation of the Schottky barrier diode.

9. What is meant by *active* devices? What are they used for?

10. Describe the features and operation of the varactor diode.

11. Describe the features and operation of the step-recovery diode.

12. Define *negative resistance*. How is it used by semiconductor devices?

13. Describe the features and operation of the tunnel diode.

14. Describe the features and operation of the Gunn diode.

15. Describe the features and operation of the IMPATT diode.

16. Describe the features and operation of a parametric amplifier.

17. How are bipolar transistors made?

18. How is a microwave bipolar transistor different from the conventional bipolar transistor?

19. Describe the features and operation of the various FET transistors.

20. Where are GaAs FETs used?

21. What else is being done to improve the frequency performance of transistors?

22. Describe how MMIC amplifiers differ from HMIC amplifiers.

23. What power levels and frequencies are current IC amplifiers able to achieve?

24. What does packaging have to do with further growth in MMIC technology?

8   ANTENNAS

OBJECTIVES

1. To describe how electromagnetic waves radiate into free space and the fields associated with this phenomenon.
2. To describe the properties of common antennas, such as the monopole, dipole, and arrays.
3. To describe the properties of microwave antennas, such as the horn, parabolic, slot, and microstrip.
4. To calculate antenna properties, such as gain, beamwidth, efficiency, and radiation resistance.
5. To define common antenna terms.

8.1   INTRODUCTION

To this point, we have considered the ways in which electromagnetic waves can propagate along transmission lines and through space. The antenna is the interface between these two media. It is a very important part of the communications path. An *antenna* can be defined as a conductor or system of conductors used either for radiating electromagnetic energy into space or for collecting electromagnetic energy from space. Wave travel between transmitter and receiver entails high losses. Therefore, to have adequate signal strength at the receiver, either the power transmitted must be extremely high, or the efficiency of the transmitting and receiving antennas must be high.

Antennas are passive devices. Therefore, the power radiated by a transmitting antenna cannot be greater than the power entering it from the transmitter. In fact, the power radiated from the antenna is less because of losses. When we speak of antenna gain, we must remember that gain in one direction results from a concentration of power, and is accompanied by a loss in other directions. Antennas achieve gain the same way a flashlight reflector increases the brightness of the bulb—by concentrating energy.

Recall from chapter 1 that the isotropic antenna represents a point source from which electromagnetic waves emanate. It disperses the energy equally in all directions. To have gain, an antenna must concentrate the energy in a preferred direction. Thus, the isotropic antenna is a theoretical antenna and is the reference antenna for comparison purposes. For example, when an antenna's gain is rated at 6 dB, it means that at a common reference point this antenna has 6 dB more power than an isotropic antenna could produce at the same point. It is common to label the value 6 dBi. The *i* means that this value is referenced to an isotropic antenna.

Another concept to keep in mind is that antennas are reciprocal devices; that is, the same design works equally well as a transmitting or a receiving antenna and, in fact, has the same gain. This does not mean that transmitting and receiving antennas are necessarily identical. For instance, the conductors in a transmitting antenna must be sized to handle larger currents. However, the designs are quite similar, and many of the calculations are identical.

The energy in the transmission line is contained in the electric field between the conductors and in the magnetic field surrounding them. All that is needed is to "launch" these fields, and the energy they contain, into space.

At the receiving antenna, the electric and magnetic fields cause current to flow in the conductors that make up the antenna. Some of the energy is thereby transferred from these fields to the transmission line connected to the receiving antenna.

8.2   ANTENNA RADIATION

Because the half-wave dipole antenna is the fundamental element in an antenna system, we will use it as a starting point for discussing the radiation of electromagnetic energy into space. A dipole antenna can be developed from an open-end transmission line one-quarter wavelength long. Figure 8.1a shows the quarter-wave section and the standing waves of current and voltage developed on the line. The current, as indicated, is flowing in opposite directions in the two conductors. The magnetic fields due to these currents cancel. The electrostatic fields, too, are in opposite directions.

Efficient radiation requires a very heavy field density; therefore, the fields must reinforce, or aid, each other. Suppose the two ends of the conductors are spread apart. The direction of current flow doesn't change, but the magnetic field is being "opened" into space and less cancellation occurs. The electrostatic field is opened out instead of being concentrated in the small space between the conductors. If the two conductors are moved until they are in line with each other, as shown in Figure 8.2b, the current flow in the two conductors is now in the same direction. The standing waves are exactly as they were before.

The magnetic (H) field about the wire is shown in Figure 8.1c. The direction of current flow is indicated by the arrows, from left to right. The magnetic field is in the direction indicated. Since maximum change of current takes place near the center of the dipole, the magnetic field is greatest at the center. Figure 8.1d shows the electrostatic (E) field about the conductor when the difference in potential between the ends of the conductor is greatest. This does not occur at the same time as the maximum magnetic field. Since the voltage and current on the dipole are 90° out of phase, the E and H fields surrounding the antenna are also 90° out of phase.

The fields associated with the energy stored in the resonant circuit (antenna) are called the *induction fields*. They are only local in effect and play no part in the transmission of electromagnetic energy. They represent only the stored energy in the antenna and are responsible only for the resonant effects that the antenna reflects to the source. The fields set up in the transfer of energy through space are known as the *radiation fields*.

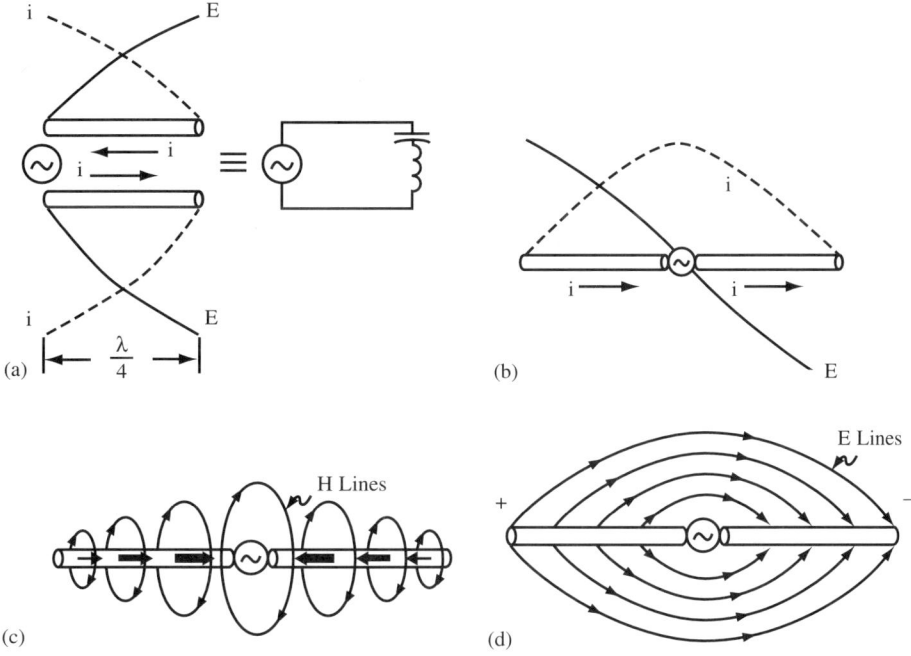

Figure 8.1 Development of half-wave antenna from open-end quarter-wave transmission line. (a) The open circuited two-wire RF line appears to the generator as a series resonant circuit. Currents flow in opposite directions, so fields cancel. The external field is small. (b) Moving one conductor around to opposite side causes currents to add to one another. (c) Distribution of H field around half-wave antenna. (d) Distribution of E field around half-wave antenna (one quarter cycle after c).

Another way to describe these fields is to call them the *near field* and the *far field*. The far field is assumed to be a plane wave when the distance between the fields is greater than 10λs. (The plane wave was described in section 1.7.)

Figure 8.2 gives a simple picture of an E field detaching itself from an antenna. The H field is not considered, although it is present. In Figure 8.2a the voltage is maximum and the E field has maximum intensity. The lines of force begin at the end of the antenna that is positively charged and extend to the end of the antenna that is negatively charged. Note that the outer E lines are stretched away from the inner ones. This is due to the repelling force between lines of force in the same direction. As the voltage drops, the separated charges come together, and the ends of the lines move toward the center of the

Figure 8.2 E field detaching itself from an antenna.

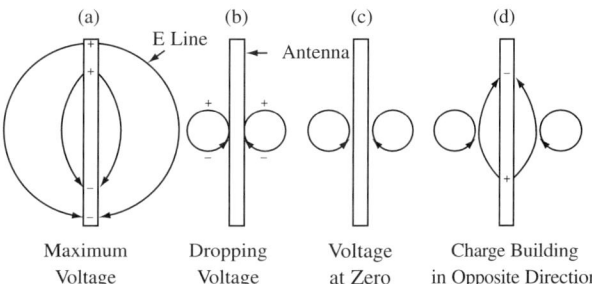

dipole (Figure 8.2b). But, since lines of force in the same direction repel each other, the centers of the lines are still held out.

As the voltage approaches zero, some of the lines collapse into the dipole, but the ends of other lines come together and form complete loops. Notice the direction of these lines of force next to the dipole in Figure 8.2c. At this point, the voltage on the antenna is zero. As the charge builds up in the opposite direction, electric lines of force again begin at the positive end and stretch toward the negative end. These lines of force, being in the same direction as the sides of the closed loops next to the antenna, repel the closed loops and force them out into space at the speed of light (Figure 8.2d). As they travel, they generate a magnetic field.

The ease with which radiation occurs varies with frequency. At lower frequencies, voltage on the antenna changes so slowly that the component of energy radiated is extremely small and is of no practical value. At higher frequencies, such as 50 kHz and up, the radiated energy is great enough to meet communication requirements.

8.3   DIPOLE ANTENNAS

The half-wave antenna described in the previous section was characterized as a dipole antenna. The word "dipole" simply means it has two parts, as was shown. A dipole antenna does not have to be one-half wavelength in length, but this length is handy for impedance matching because the feed can be applied at the center, at one end, or spaced apart near the center as required. In practice, the length of a dipole antenna is slightly less than one-half the free space wavelength, to compensate for the fact that energy on the antenna does not travel at the same speed as in space. A half-wave antenna is sometimes called a *Hertz* antenna.

In Figure 8.3 note that the voltage standing wave is high at the ends of the antenna and low at the center. As previously explained, this is also the case with the quarter-wave open-circuit two-wire line from which the half-wave antenna is developed.

An examination of the current-distribution curve shows that the current standing wave reaches maximum a quarter-cycle after the voltage reaches maximum. At this time the current is maximum at the center and zero at the ends.

Figure 8.4 shows the manner in which radiation fields are propagated from the antenna. The E field and the H field are shown as separate sets of flux lines about the antenna. The electric flux lines are closed loops on each side of the antenna. The magnetic flux lines are closed circular loops that have their axis around the antenna; in other words, the antenna is the center of each loop. The sine waves that are labeled the curves of radial variation of flux density indicate the relative field strength at various distances and angles from the antenna. These effectively define the radiation pattern of the antenna.

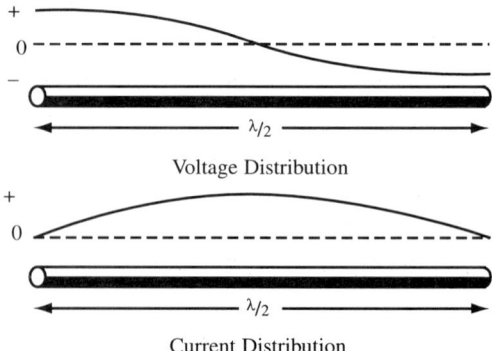

Figure 8.3 Current and voltage distribution in half-wave antenna.

Voltage Distribution

Current Distribution

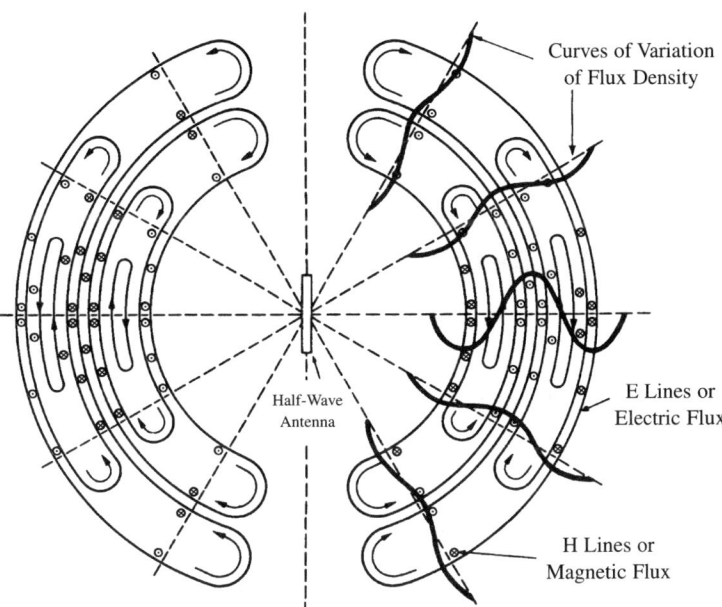

Figure 8.4 Fields in space around a half-wave antenna.

Due to the directional properties of the half-wave antenna, most of the radiated energy travels perpendicular to the antenna, while very little energy travels along its axis. Each field builds the other, and one cannot exist without the other.

The direction of motion is the travel of the fields away from the antenna. The current associated with the H field does not flow because in space there is no conductor to carry the current; the field exists, nonetheless. To visualize a field existing without current flow, think of a moving H field cutting a glass rod. A voltage (electrostatic stress) is induced in the rod, but there is little actual electron movement because the rod is a good insulator. Magnetic lines move in space and set up electric stresses in space in a similar manner.

The physical length of the dipole is about 5% shorter than a half-wave traveling in space. It can be expressed by the equation:

$$L = 0.95(\lambda/2) \qquad \textbf{(Eq. 8.1)}$$

where
L = length of dipole, meters
$\lambda = c/f$
$c = 3 \times 10^8$ m/s
f = operating frequency, Hz

EXAMPLE 8.1:

Calculate the length of a dipole antenna, when the operating frequency is 500 MHz.

$$\lambda = (3 \times 10^8)/(500 \times 10^6)$$
$$= 0.6 \text{ m}$$
$$L = 0.95(0.6/2)$$
$$= 0.285 \text{ m}$$

An antenna of the correct length acts as a resonant circuit and presents pure resistance to the excitation circuit. An antenna slighter longer than a half-wave acts inductive. To correct its length, it can be cut shorter, or a capacitor can be placed in series to the antenna. Conversely, an antenna slightly shorter than a half-wave acts capacitive. To correct its length, an inductor can be placed in series to the antenna.

8.4   ANTENNA PROPERTIES

Radiation Resistance

The radiation of energy from a dipole is quite apparent if we measure the impedance at the feed point in the center of the antenna. At a distance one-quarter wavelength from the open end, an actual open-circuit lossless line looks like a short circuit. At distances slightly greater than or less than one-quarter wavelength, the line appears reactive. There is never a nonzero resistive component to the feed point, since an open-circuit line has no way of dissipating power.

The half-wave dipole does not dissipate power either, assuming the material of which it is made is lossless, but it does radiate power into space. The effect on the feed-point impedance is the same as if a loss had taken place. Whether power is dissipated or radiated, it disappears from the antenna and, therefore, causes the input impedance to have a resistive component. The half-wave dipole, for instance, looks like a resistance of approximately 70 Ω (actually 73.1 Ω) at its feed point.

The portion of an antenna's input impedance that is due to power radiated into space is known, appropriately, as the *radiation resistance*. It is important to understand that this does not represent losses in the conductors that make up the antenna.

The idealized antenna just described radiates all the power supplied to it into space. Of course, a real antenna does have ohmic losses (including skin effect) in the conductor, so it has an efficiency less than 100%. Practical antenna efficiency is typically on the order of 90% or more. This efficiency can be defined as:

$$Eff = Pr/P_T \qquad\qquad \textbf{(Eq. 8.2)}$$

where P_r = radiated power

P_T = total power supplied to the antenna

Recalling that $P = I^2R$, we have:

$$Eff = I^2R_r/I^2R_T$$

$$Eff = R_r/R_T \qquad\qquad \textbf{(Eq. 8.3)}$$

where R_r = radiation resistance, as seen at the feed point

R_T = total resistance, as seen at the feed point

EXAMPLE 8.2:

Calculate the efficiency of a dipole antenna having a radiation resistance of 67 Ω and a loss resistance of 5 Ω, measured at the feed point.

$$Eff = 67/(67 + 5)$$
$$= 0.93 \text{ or } 93\%$$

Radiation Patterns

If we measured the radiation from a single point in space, we would find that this point radiates equally well in all directions. Since the strength of the radiated energy varies inversely as the distance, if we plotted all the points at which the energy was of the same strength, the points would form a sphere, with the radiating point at the center.

If radiation occurs from more than one point in space, the radiated signal from each point adds vectorially to produce the total radiation strength. When these facts are applied to a half-wave antenna, we can look at the conductor as a series of points arranged in a straight line. The radiation from points equally spaced from the center adds in directions perpendicular to a line through them but cancels along the line through them. A plot of all points of equal strength produces a doughnut-shaped figure. This three-dimensional figure is shown in Figure 8.5a. Note that a cross-sectional view resembles two ellipses that are adjacent to one another (Figure 8.5b).

The cross-sectional view is called a *polar diagram*. This sort of diagram is standard for delineating antenna performance on paper. The directional properties of the antenna radiation pattern are thus defined. Figure 8.6 shows a typical blank polar diagram that can be used for plotting the radiation pattern of an antenna.

Figure 8.7 shows the polar diagram for a half-wave antenna. The diagram holds good for any plane containing the antenna. The radiation patterns are only valid in the far-field region. This is at a distance of several wavelengths.

The distance out from the center on the graph paper represents the strength of the radiation in a given direction. The scale is usually in decibels with respect to some reference. Often, the reference is to an isotropic radiator. When the half-wave dipole is referenced to the isotropic, the dipole has a 2.14 dB value at its strongest point (while the isotropic has 0 dB). The gain of the lossless half-wave dipole is thus 2.14 dB compared to the isotropic. This gain is usually expressed as 2.14 dBi.

Beamwidth

Beamwidth is the angle between the points on the graph at which the electric field strength is 0.707 (–3 dB) of its maximum. For a horizontal half-wave dipole, this is about 78° in one plane and 360° in the other. Many antennas are much more directional than this, with a narrower beamwidth in both planes. Figure 8.8 shows where the beamwidth is measured for a typical directional antenna. Angle *a* defines the beamwidth in degrees. In addition, the various shapes

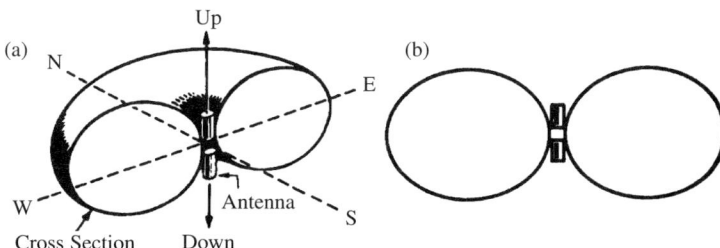

Figure 8.5   Radiation patterns of a half-wave antenna. (a) Three dimensional view of antenna pattern. (b) Cross-sectional view of pattern.

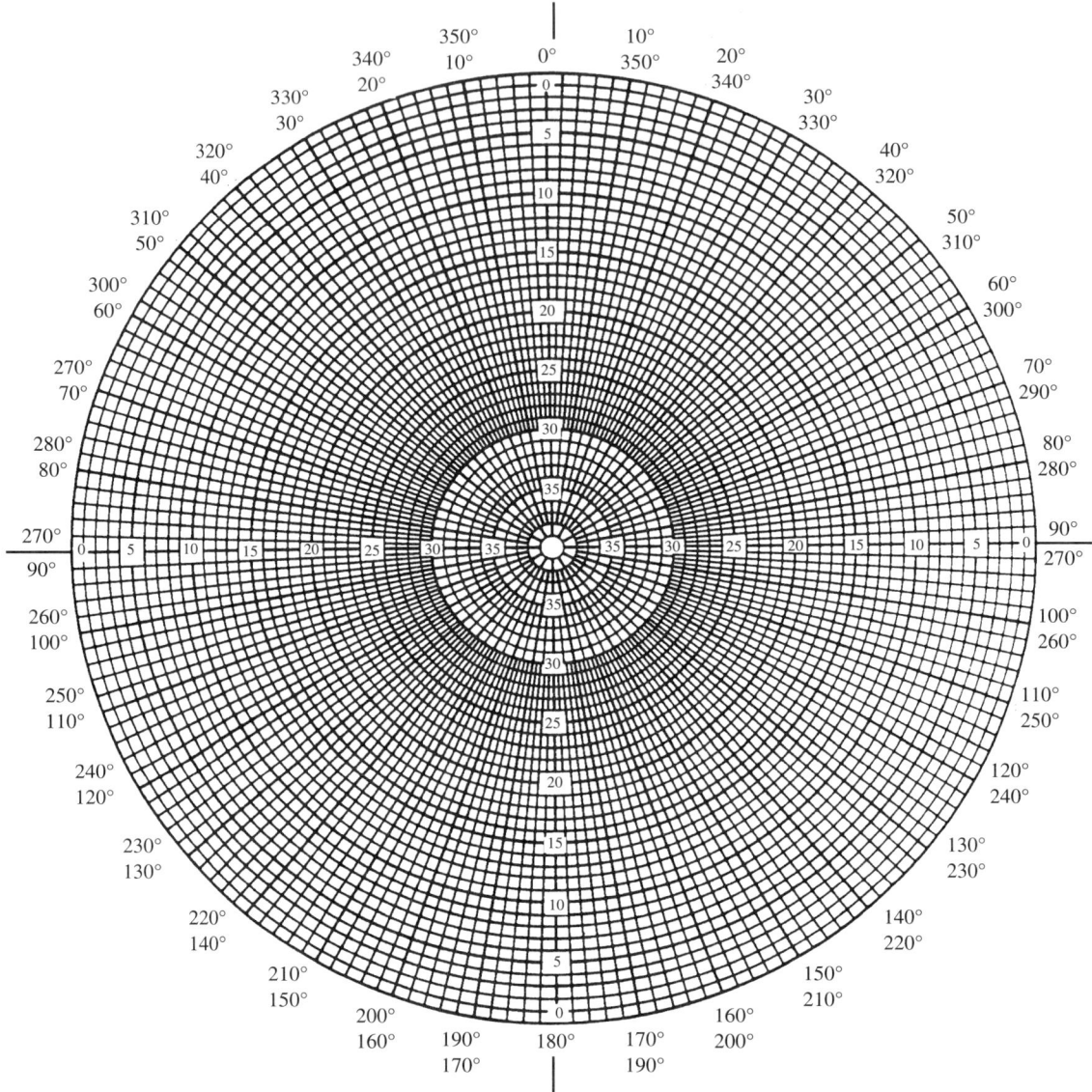

Figure 8.6 Polar diagram for plotting an antenna's radiation pattern.

of the radiation pattern are given the names *main lobe* (radiation in the main direction of travel), *side lobes* (radiation to the side of the antenna), and *back lobe* (radiation behind the antenna).

Gain and Directivity

Remember that for antennas power gain in one direction is at the expense of losses in others. Sometimes the term "directivity" is used. This is not quite the same as gain. *Directivity* is the gain calculated assuming a lossless antenna in a preferred direction at maximum

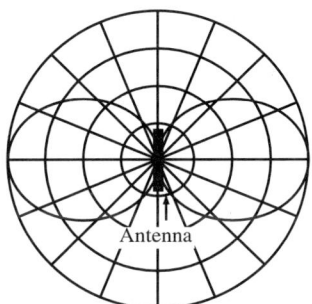

Figure 8.7 Polar diagram for half-wave antenna in any plane containing the antenna.

radiation. Real antennas have losses, and *gain* is simply the directivity multiplied by the efficiency of the antenna.

When an antenna is used for transmitting, the total power emitted by the antenna is somewhat less than that delivered to it by the feed line. It can be determined by rearranging equation 8.2:

$$P_r = P_T(Eff)$$

where
P_r = total radiated power

P_T = power supplied to the antenna

Eff = antenna efficiency

The figure of 2.14 dBi we used for the gain of a lossless half-wave dipole is also the directivity for *any* half-wave dipole. To find the gain of a real (lossy) dipole, it is necessary first to convert the directivity decibel to a power ratio, then to multiply this by the efficiency. This can be shown as:

$$D = \text{antilog dBi}/10 \qquad \textbf{(Eq. 8.4)}$$

where
D = directivity as a power ratio

dBi = the given directivity decibel

Figure 8.8 Antenna radiation patterns (lobes) and beamwidth measurements.

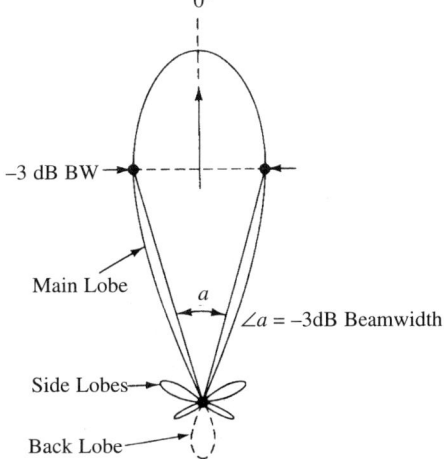

Gain can then be found by:

$$G = D \times Eff \qquad \text{(Eq. 8.5)}$$

where

G = gain as a power ratio

D = directivity as a power ratio

Eff = antenna efficiency

EXAMPLE 8.3:

A half-wave dipole antenna has an efficiency of 85%. Determine its gain as a power ratio.

$$D = \text{antilog } 2.14/10$$
$$= 1.638$$
$$G = 1.638 \times 0.85$$
$$= 1.39$$

(Note: The gain of 1.39 can be converted to decibels by taking 10 log 1.39 = 1.43 dBi.)

Effective Radiated Power

Effective radiated power (ERP) represents the power input multiplied by the antenna gain measured with respect to a half-wave dipole. *Effective isotropic radiated power* (EIRP) represents the actual power going into the antenna multiplied by its gain with respect to an isotropic radiator.

$$ERP = P_i \times G \qquad \text{(Eq. 8.6)}$$

where

ERP = effective radiated power, W

P_i = input power

G = gain, power ratio

EXAMPLE 8.4:

The input power to a half-wave dipole antenna is 50 W and the gain is 15. Find the ERP.

$$ERP = 50 \times 15$$
$$= 750 \text{ W}$$

8.5   MONOPOLE ANTENNAS

The monopole antenna is a quarter-wave antenna that is classified as an *image antenna*. Image antennas, unlike half-wave dipoles that are isolated from any conductor, utilize the reflecting properties of a conductor. The effect of a nearby conductor is illustrated in Figure 8.9.

Here, a real antenna is shown perpendicular to a horizontal perfectly conducting plane. When energy radiates from the antenna, one path moves into space, while the other moves

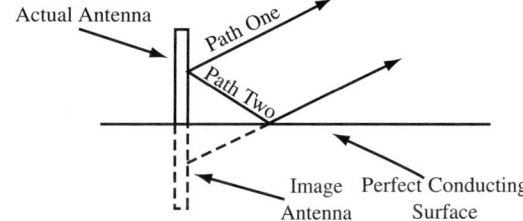

Figure 8.9 Image antenna.

toward the conducting plane. Upon striking the perfect conductor, the wave is reflected in reverse phase.

As with light waves, the angle of reflection is equal to the angle of incidence. The reflected wave also proceeds toward some point in space. The sum of the two waves makes up the total wave at a point in space. The action in the conducting plane can be replaced by another antenna that is a mirror image of the actual antenna.

A quarter-wave grounded antenna is called a *Marconi antenna*. In a Marconi antenna (Figure 8.10a), the vertical field strength pattern (polar diagram), shown in Figure 8.10b, is the same as that of a half-wave element, except the plane cuts it off at the center. The image is only effective above the plane, because no energy penetrates the conducting plane.

Figure 8.10 (a) The Marconi is a quarter-wave grounded antenna. (b) Vertical field strength pattern (polar diagram). (c) Horizontal polar diagram.

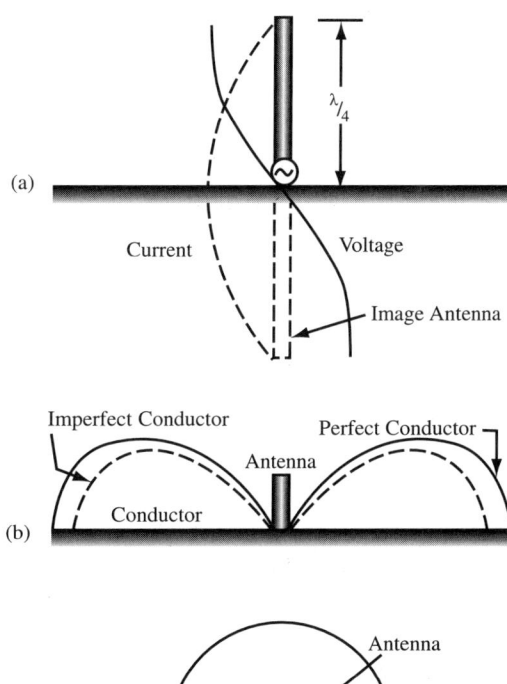

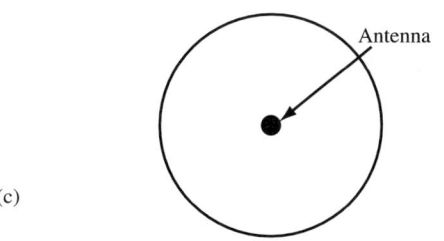

In the horizontal polar diagram (Figure 8.10c), the vertical field strength pattern can be rotated with the antenna as the axis to form the horizontal polar diagram of a Marconi. It is uniform in all directions in a plane perpendicular to the length of the antenna.

The input impedance to the Marconi is approximately 37 Ω when the antenna is fed at its base. In addition, a quarter-wave Marconi is resonant and displays zero reactance, just like a half-wave antenna.

If the conducting plane is not a perfect conductor, then some modification is needed to the size of the polar diagram, since the gain is lower by some percentage. The conducting plane can be the skin of an aircraft (for airborne antennas), the fender of a car, the earth's surface for grounded equipment, or an artificial ground called a *counterpoise*.

Conductors reradiate energy and affect the resultant field strengths via phase relationships. When they are deliberately placed near the antenna to distort or alter the field strength pattern they are called arrays.

8.6   ANTENNA ARRAYS

For some antenna uses it is desirable that the radiation pattern be uniform in all directions. For other uses the antenna must be very directional. In the latter case, the antenna system usually consists of two or more simple half-wave elements spaced so that the fields from the elements add in some directions and cancel in others. The set of antenna elements is called an *antenna array*. There are two types of antenna arrays—*driven arrays* and *parasitic arrays*.

Driven Arrays

The driven array consists of two or more elements with all elements connected to the source (feed). They are divided into two basic types—the *broadside array* and the *end-fire array*.

Broadside Arrays

When two or more half-wave elements placed one half-wave apart and parallel to each other are excited in phase, most of the radiation pattern is *broadside* to the plane of the elements, i.e., out of the page along the z-axis as shown in Figure 8.11a. The individual fields radiated from an array combine to form a directional field. The gain in the broadside direction, or the direction perpendicular to the plane of the antenna, is twice the gain of the individual elements. As more elements are added, the pattern becomes more directional and side lobes appear (Figure 8.11a). The antenna array is termed a broadside array (Figure 8.11b).

End-Fire Arrays

If half-wave dipole elements spaced a quarter-wave apart are fed 90° out of phase, the array resembles that of a broadside array, but a completely different radiation pattern results because the spacing between the elements is different. The radiation pattern of an end-fire antenna (Figure 8.12a) is different from that of a broadside in that its beamwidth is somewhat broader for the same number of elements, and the direction of the main lobe is in the plane of the array, not at right angles. Moreover, the antenna is essentially unidirectional, with only a slight minor lobe, as shown in Figure 8.12b. The end-fire array is, like the broadside, a resonant device with limited usefulness as a broadband receiving antenna.

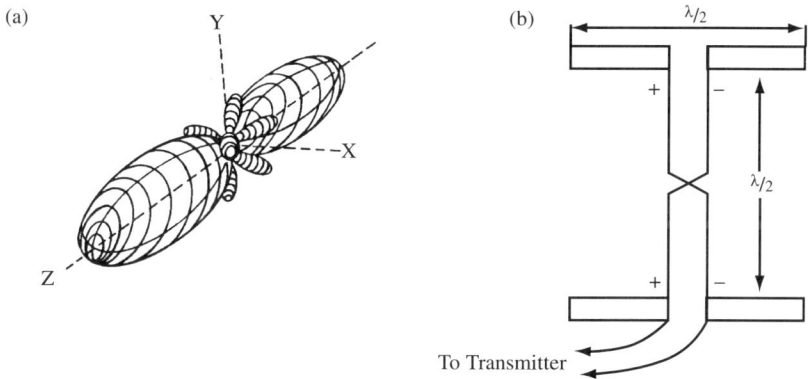

Figure 8.11   (a) Solid radiation pattern. (b) Broadside array.

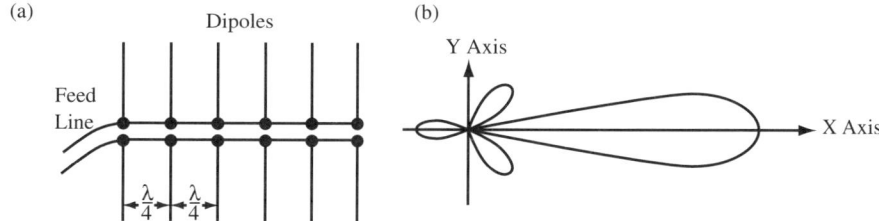

Figure 8.12   (a) End-fire array. (b) Radiation pattern.

Parasitic Arrays

A *parasitic array* is an antenna system that consists of two or more elements, one or more of which are excited by the induction and radiation fields produced by the driven element or elements. With such an array it is possible to obtain highly directional patterns without changing the length of the antenna.

The parasitic elements are placed parallel to the driven half-wave antenna. The spacing between these elements depends on the effect desired. Because of currents set up in the parasitic elements by induced voltage, they, too, radiate energy. At some points the energy reradiated by the parasitic element is in phase with that radiated directly by the driven element; at other points the fields are 180° out of phase, and there is cancellation. The unidirectional pattern can be found by plotting the signal strength measured in all directions about the array. When the radiation pattern is strongest in the direction of the driven element, the parasitic element is called a *reflector*. When the maximum radiation is in the direction of the parasitic element, this element is called a *director*.

The phase differences that occur between the driven element and the parasitic element are a function of the distance between them. For a reflector, the maximum gain over a half-wave antenna alone is obtained with a 0.15λ separation; for a director, a 0.1λ separation is required.

A common parasitic array antenna is the Yagi (Yagi-Uda) array (Figure 8.13). It combines a driven element with a reflector and several directors. In this way a more directional antenna can be built. Yagi-Uda arrays are commonly used as TV and FM reception antennas.

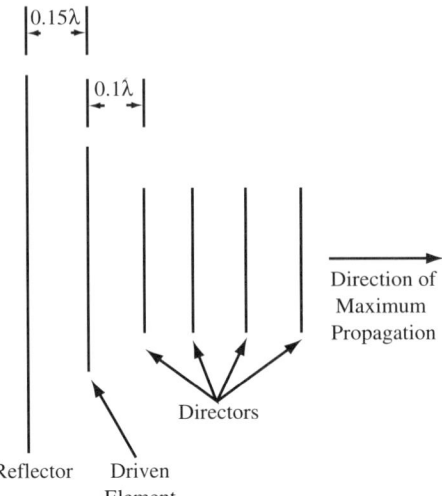

Figure 8.13   Yagi-Uda array.

Typical Yagi-Uda antennas have gains in the range of 7 to 17 dBi. Such antennas can be driven by a dipole, but for TV applications the *folded dipole* is most often used. The folded dipole is a full wavelength conductor that is folded to form a half-wave element. A better description is that it consists of a pair of half-wave elements connected at the ends. Its input resistance is four times the input resistance of the simple dipole. This means this antenna does not need any special matching elements to use it with 300 Ω twin-lead line. In addition, the folded dipole affords an increase in bandwidth. For 75 Ω coaxial lines, the simple dipole is used as the driven element, which also provides greater noise immunity via the braided shield.

8.7   REFLECTOR ANTENNAS

For radar and microwave links antennas must be highly directional. Beamwidths on the order of 2° or less are commonplace. At higher frequencies, the size of components is automatically reduced. This lowers the power-handling ability of the system. To offset this, the associated antenna must have a very high gain to achieve the required level of effective radiated power (ERP). There is more noise at higher frequencies, consequently the receiver signal must be as large as possible to achieve an adequate signal-to-noise ratio.

The most common antenna that meets these requirements is the *parabolic reflector antenna* (often called a *dish antenna*). The dish is not actually an antenna, but serves as a reflector. It must be driven by a radiating element at the focal point. A parabola is a mathematical curve such that its reflection property causes an incoming beam of parallel rays to all focus to one point (Figure 8.14a). Conversely, radiated waves from a point signal placed at the focal point are reflected by the surface to form parallel rays in the outgoing beam (Figure 8.14b). Thus, a parabolic antenna can be employed as a transmitter/receiving device.

The parabolic reflector may take many forms. The larger the reflector with respect to the wavelength, the narrower the beamwidth. This fact makes it possible to shape a beam by changing the dimensions of the reflector. Some of these shapes are shown in Figure 8.15.

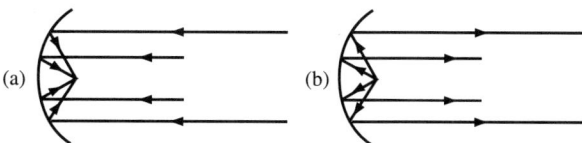

Figure 8.14 Lines of reflection (a) and radiation (b) in a parabolic dish.

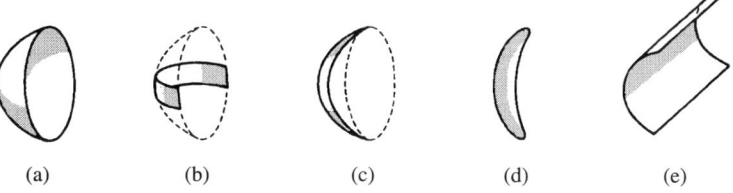

(a)   (b)   (c)   (d)   (e)

Figure 8.15 Construction dimensions of parabolic antennas. (a) Paraboloid. (b) Truncated paraboloid (surface search). (c) Truncated paraboloid (height finding). (d) Orange-peel paraboloid. (e) Cylindrical paraboloid.

The differences in these shapes are used primarily by radar antennas to determine elevation and height of targets. Each paraboloid shapes the beam for a specific function.

The gain of a parabolic antenna can be determined by equation 8.7. Note that k, the reflection efficiency, is provided to adjust for the fact that the feed does not perfectly illuminate the reflecting surface of the antenna. It is common for the illumination to fall away uniformly from the center to the edges of the reflector such that the power density at the edges is 10 dB lower than the center power density.

$$G = k(\pi D)^2/\lambda^2 \qquad \text{(Eq. 8.7)}$$

where G = gain over isotropic, ratio

D = diameter

λ = wavelength (same units as D)

k = reflection efficiency (typically 0.4–0.7)

The reflection efficiency is 0.6 for most antennas. Thus, gain can be approximated more easily by equation 8.8.

$$G = 6D^2/\lambda^2 \qquad \text{(Eq. 8.8)}$$

The beamwidth (−3 dB) of a uniformly illuminated parabolic reflector antenna is approximated by:

$$BW = 70\lambda/D \qquad \text{(Eq. 8.9)}$$

EXAMPLE 8.5:

Determine the gain and beamwidth of a parabolic antenna whose diameter is 3 m. The wavelength of the signal is 2 cm (15 GHz).

$$G = 6 \times (3/.02)^2$$
$$= 1.35 \times 10^5$$
$$BW = 70 \times (.02/3)$$
$$= 0.47°$$

The radiation pattern from a parabolic antenna looks very similar to that in Figure 8.8. Fifty percent of the energy radiated by the parabolic antenna is within the −3 dB beamwidth, and 90% is between the first null on either side of the main lobe.

Two methods for feeding energy into the parabolic antenna are common. In Figure 8.16a, we see the method in which the radiator element is placed at the focal point. This method is used in low-cost installations, such as home satellite TV receive-only (TVRO) antennas.

In Figure 8.16b we see the *Cassegrain feed* system. This system is modeled after the Cassegrain optical telescope. The radiator element is placed at an opening at the center of the dish. A hyperbolic subreflector is placed at the focal point and is used to reflect the wave fronts to the surface of the parabolic reflector. This antenna is characterized by low-noise operation.

If a dipole is used as the source of radiation, there is radiation from the antenna out into space as well as toward the reflector. The energy that is not directed toward the paraboloid has a wide beamwidth that would interfere with the narrow pattern from the reflector. To prevent this, a parasitic reflector may be used with the half-wave driven element. In this way all of the radiated energy is redirected into the parabolic reflector, and a narrow beam is produced.

Very often the *horn radiator* is used as the source of radiation. As can be seen in Figure 8.17, the flared opening is very directive. The *rectangular horn* (Figure 8.17a) has the guide flared in only one direction. If the flaring is in both directions, the horn is said to be *pyramidal* (Figure 8.17b). The *conical* horn is used with circular waveguide (Figure 8.17c). Additionally, the tapered termination provides the impedance transformation between the waveguide and free-space impedance.

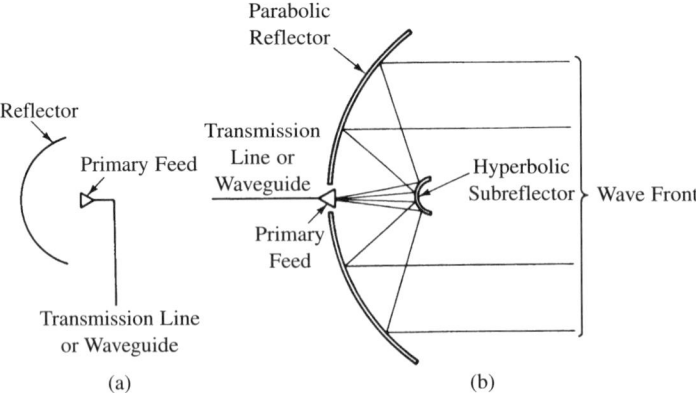

Figure 8.16 Parabolic antenna feed geometries. (a) Conventional feed. (b) Cassegrain feed.

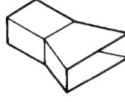 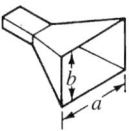 

Figure 8.17   Horn radiators.   (a) Rectangular   (b) Pyramidal   (c) Conical

The gain of the horn radiator is proportional to the area (*A*) of the flared open flange and inversely proportional to the square of the wavelength.

$$G = 10A/\lambda^2 \qquad\qquad\qquad\text{(Eq. 8.10)}$$

where   *A* = flange area

λ = wavelength (same units as *A*)

The −3 dB beamwidth for vertical and horizontal extents can be approximated from the following:

$$\text{Vertical } \phi_v = 51\lambda/b \qquad\qquad\qquad\text{(Eq. 8.11)}$$
$$\text{Horizontal } \phi_h = 70\lambda/a \qquad\qquad\qquad\text{(Eq. 8.12)}$$

where ϕ_v = vertical beamwidth in degrees

ϕ_h = horizontal beamwidth in degrees

b = narrow dimension of the flared flange

a = wide dimension of the flared flange

λ = wavelength

EXAMPLE 8.6:

A horn radiator uses a 4×9 cm open flange. What is its gain at 10.25 GHz? What are the vertical and horizontal field −3 dB beamwidths?

$$\lambda = c/f \qquad\qquad\qquad\text{(Eq. 1.4)}$$
$$= 3.0 \times 10^{10}/10.25 \times 10^9$$
$$= 2.93 \text{ cm}$$
$$G = (10 \times 36)/2.93^2$$
$$= 360/8.58$$
$$= 41.9$$
$$\phi_v = (51 \times 2.93)/4$$
$$= 37.4°$$
$$\phi_h = (70 \times 2.93)/9$$
$$= 22.8°$$

The wave front leaving the horn radiator is somewhat curved. To reshape this pattern to that of a plane (parallel) wave front, a dielectric lens may be employed. In microwave antenna applications, the lens functions as a *wave collimator*, as shown in Figure 8.18.

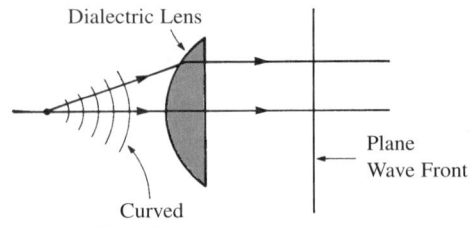

Figure 8.18 Wave collimator. A dielectric lens converts a curved wave front to a plane wave front.

Dielectric lenses are made of polystyrene and other dense materials. Such materials are chosen because they produce the greatest amount of diffraction for the smallest size and weight. However, such material also produces greater attenuation of the signal as it passes through the lens. To offset this problem, dielectric lenses are typically stepped or zoned as shown in Figure 8.19.

Slot array antennas (Figure 8.20) are made by cutting resonant slots into the walls of a waveguide. The slots are usually a half wavelength long and spaced half a guide wavelength ($\lambda_g/2$) along the waveguide. The slot cut into a waveguide is equivalent to a dipole antenna.

The phase of the signals to the individual elements can be changed. This causes the directional properties to vary electronically, thus producing a narrow, high-gain beam. Internally, there may be a wedge for impedance matching, flanges to enhance directivity, or phase-shifting stubs to fine-tune the antenna radiation pattern.

Slot array antennas are used for marine navigation radars, airborne radars, and telemetry systems. They are also used for the reception of microwave TV signals for the Multipoint Distribution Service (MDS).

8.8   MICROSTRIP ANTENNAS

When a microwave radiating or receiving element is formed on top of an MMIC dielectric substrate with a complete ground plate, it is a *microstrip antenna* or a *patch antenna* depending on the shape of the pattern of the active element on the substrate. If a large part of the ground plate corresponding to the active element is absent or the ground plate is completely missing, then it is a *planar antenna*.

Figure 8.19 Stepped dielectric lenses.

Figure 8.20 Slot array.

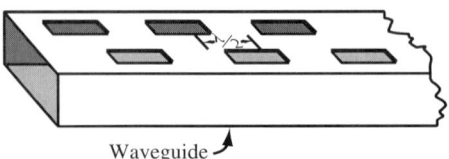

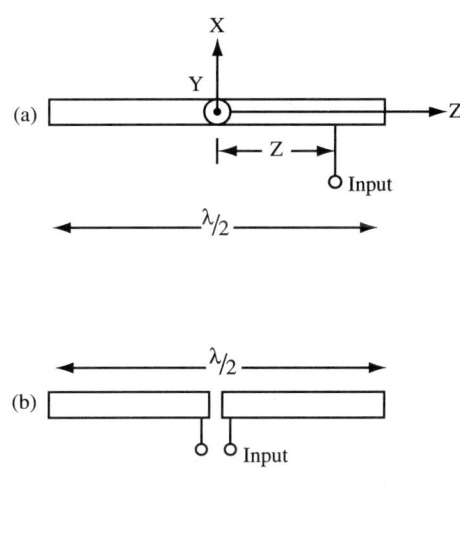

Figure 8.21 Microstrip dipole antennas. (a) Side-fed antenna. (b) Planar antenna. (c) Feed point at one end.

Figure 8.21a shows a side-fed microstrip half-wave dipole antenna. If the feed point is moved, the impedance varies as well. The feed is chosen (point Z) to be located such that the antenna impedance and the microstrip impedance are close to each other.

If the ground plate underneath the radiator shown in Figure 8.21a is removed, the antenna becomes a planar antenna (Figure 8.21b). The feed is different in this antenna. The microstrip planar antenna can be fed through a balanced parallel stripline or a conventional single microstrip line. When the conventional strip is attached to the feed point of one element, the feed point of the other element must be grounded to the ground plate. The radiation patterns from either of these antennas is the same as that from the familiar half-wave dipole antenna. When fed from one end (Figure 8.21c) the antenna impedance is usually higher that the characteristic impedance of the microstrip. Here, an impedance matching transformer is needed.

Typical microstrip loop antennas are shown in Figure 8.22. A single-ended microstrip loop is shown in Figure 8.22a, while a balanced microstrip loop is shown in Figure 8.22b.

Figure 8.22 Microstrip loop antennas.

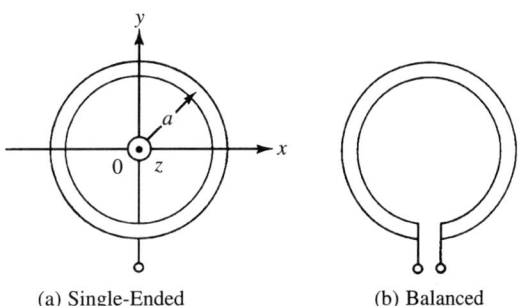

(a) Single-Ended   (b) Balanced

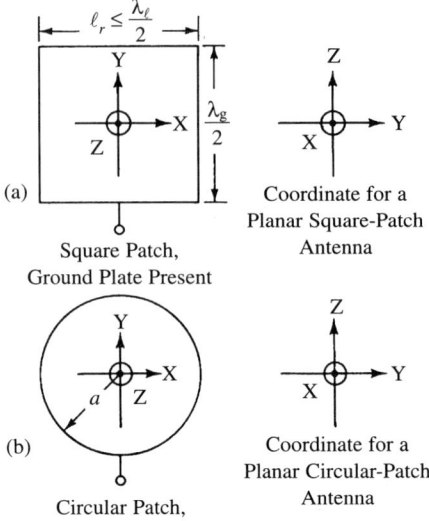

Figure 8.23   Microstrip patch antennas.

A conducting square or circular patch formed on a dielectric substrate, as shown in Figure 8.23, can be used as an antenna. If the ground plate is present, the radiation pattern is in both the Z and Y directions. When the ground plate is not present, the antennas are planar antennas. More specifically, they are considered end-fed $\lambda/2$ dipole antennas. Here, there is very little radiation in the Z direction.

Microstrip antenna arrays are shown in Figure 8.24. A large variety of patterns can be utilized. The figure shows four-element arrays; the actual number can be in the hundreds to

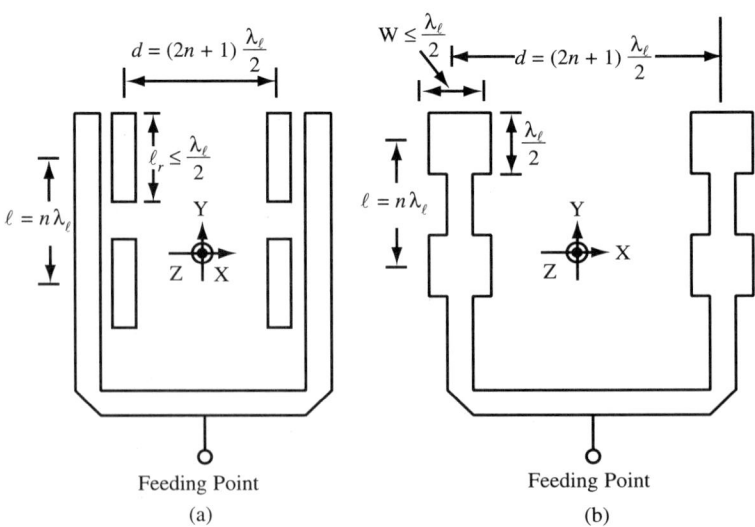

Figure 8.24   (a) Microstrip array antenna. (b) Micropatch array antenna.

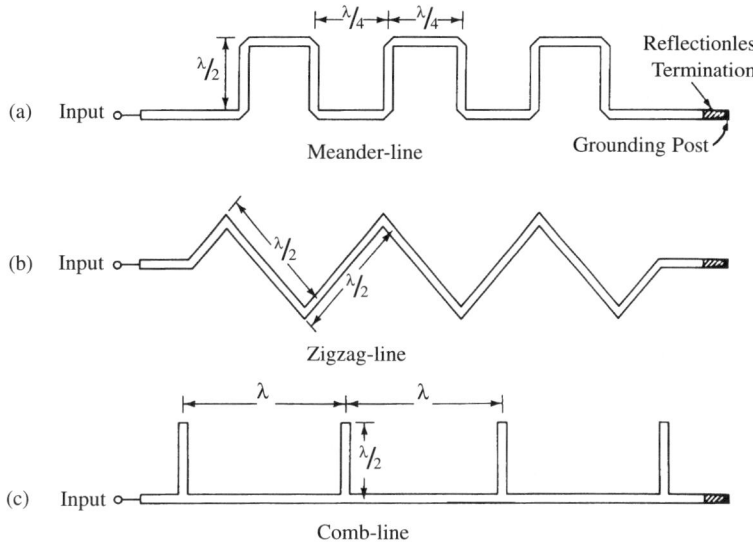

Figure 8.25 Microstrip traveling-wave antennas.

form a desired radiation pattern. Array spacing (dimension "d") is varied to suppress radiation in the X direction. The major direction of radiation is then in the Z direction, which produces a sharp directive beam. Radiation patterns can be varied further by changing the radiator spacing "l" as well as the array spacing "d".

Microstrip traveling-wave antennas are shown in Figure 8.25. The load ends of the microstrip lines are terminated in reflectionless terminations by using a matched resistive strip grounded to the ground plate at the end. Traveling waves are produced on the line, and the radiating elements are sequentially excited.

Radiation patterns vary by the pattern of the microstrip line having shapes noted as meander-line, zigzag-line, and comb-line. Radiation exits from the corners in the meander-line and zigzag-line patterns (Figures 8.25a and 8.25b), and produces a broadside radiation pattern.

The traveling-wave comb-line microstrip antenna of Figure 8.25c is basically an array of end-fed, half-wave antennas. The radiation pattern is broadside. The open-stub line produces a high impedance at the feed point.

Most microstrip antennas are formed on a substrate of high permittivity. This makes the transmission line wavelength short. The physical size of the antenna is, therefore, made small. In an array structure, this compactness requires a large number of elements to be put together in a limited area. The result is a sharply directed high-gain antenna array in a compact structure.

As further developments occur in MMIC technology, with applications in personal communication systems (PCS), these antennas will see even more applications.

8.9 SUMMARY

1. The antenna is the interface between the transmission line and free space.
2. Antenna gain refers to the directional property of the antenna in transmitting energy in a preferred direction. The reference is usually made with respect to an isotropic antenna.

3. Antenna radiation consists of E and H fields at right angles to each other. One cannot exist without the other, and each builds the other.
4. Radiation fields are known as induction and radiation fields. The terms near field and far field are also used to describe radiation.
5. The dipole (Hertz) antenna is a half-wave antenna. It can be fed from the center, from one end, or from various points in the middle, each of which has a different impedance.
6. Radiation resistance is the portion of an antenna's impedance that is due to power radiated into space.
7. Common antenna properties that can be calculated include efficiency, gain, directivity, beamwidth, and effective radiated power (ERP).
8. The polar diagram is often used as a plotting reference to describe the radiation pattern of an antenna.
9. The Marconi antenna is a quarter-wave monopole antenna that uses the image (reflective) effects of a ground plane.
10. Antenna arrays (driven or parasitic) are made with the addition of extra elements. Parasitic elements are either directive or reflective. In any case, the extra elements aid in making the antenna more directional in nature. This improves the gain of the antenna when referenced to the isotropic antenna.
11. The parabolic reflector antenna is the most common microwave antenna. It uses the reflection properties of the parabola to confine the field into a very directive beam. It can have a very high gain.
12. The horn radiator is often used to feed the parabolic antenna. It is made by flanging the end of a waveguide.
13. Microstrip antennas are formed on the surface of the dielectric substrate of MMICs. They can be made in a variety of configurations. These include the dipole, monopole, array, and traveling-wave types.

Key Equations:

$L = 0.95(\lambda/2)$	(Eq. 8.1)
$Eff = P_r/P_T$	(Eq. 8.2)
$Eff = R_r/R_T$	(Eq. 8.3)
$D = $ antilog dBi/10	(Eq. 8.4)
$G = D \times Eff$	(Eq. 8.5)
$ERP = P_i \times G$	(Eq. 8.6)
$G = k(\pi D)^2/\lambda^2$	(Eq. 8.7)
$G = 6D^2/\lambda^2$	(Eq. 8.8)
$BW = 70\lambda/D$	(Eq. 8.9)
$G = 10A/\lambda^2$	(Eq. 8.10)
Vertical $\phi_v = 51\lambda/b$	(Eq. 8.11)
Horizontal $\phi_h = 70\lambda/a$	(Eq. 8.12)

PROBLEMS

1. Calculate the length of a dipole antenna when the operating frequency is 900 MHz.

2. Calculate the efficiency of a dipole antenna having a radiation resistance of 77 Ω and a loss resistance of 8 Ω.

3. Using the efficiency of problem 2, determine the total radiated power when the power supplied to the antenna is 1 kW.

4. Determine the directivity of an antenna when its gain is 3.25 dBi.

5. Calculate the gain of the antenna in problem 4 when the efficiency is 95%.

6. Determine the ERP for the antenna in problem 5 when the input power is 1000 W.

7. Calculate the length of a monopole antenna at 100 MHz.

8. Determine the gain of a parabolic antenna whose diameter is 3 m. ($f = 6$ GHz; $k = 0.55$)

9. Convert the gain of problem 8 into dB.

10. Determine the gain ($k = 0.6$) for a parabolic antenna whose diameter is 1 m. ($f = 12.25$ GHz)

11. Determine the beamwidth for the antenna in problem 10.

12. Determine the gain of a horn radiator whose dimensions are 2 cm $\times$ 4 cm.

13. Determine the gain of the horn radiator in problem 12 if the flange is increased to 4 cm $\times$ 7 cm.

14. What is the vertical field beamwidth for the horn in problem 13?

15. What is the horizontal field beamwidth for the horn in problem 13?

QUESTIONS

1. Describe the radiation properties of an antenna.

2. Describe how energy is radiated from a dipole antenna.

3. Define *antenna radiation resistance.*

4. How can you "lengthen" the length of an antenna? How can you shorten the length?

5. What does the shape of a radiation pattern describe?

6. Describe the radiation pattern of a dipole antenna.

7. Define *antenna beamwidth.*

8. Why do microwave antennas need a narrow beamwidth?

9. Describe how an antenna has gain.

10. What is a typical antenna efficiency?

11. Describe how image antennas work.

12. What are the characteristics of antenna arrays?

13. Describe how a parabolic reflector antenna works.

14. Describe the feed types and mechanisms for a parabolic antenna.

15. Describe how microstrip antennas are formed. Compare their characteristics.

9 MICROWAVE MEASUREMENTS

OBJECTIVES

1. To describe how attributes of a microwave signal may be measured.
2. To define noise and describe its common measuring parameters.
3. To describe attributes of the common detection devices used in microwave measurements.
4. To describe the operation of basic microwave test instruments.
5. To compare and contrast applications of basic microwave test instruments.
6. To classify microwave power levels.
7. To describe the properties of attenuation and insertion loss.
8. To define the basic S-parameters and compare them to the coefficients of transmission and reflection.

9.1 INTRODUCTION

This chapter looks at tests and measurements performed primarily in a lab setting. It involves much more precise measurement techniques than you did in low-frequency course work. Basic measurements of voltage and current are extremely difficult, or impossible, at microwave frequencies; therefore, it is much more convenient to measure power. The measurement of microwave power requires that you know how to operate power detectors and indicating instruments and how to apply techniques that minimize errors and increase the accuracy of the measurement. For power measurements (or any other parameter) to have any significance, the instruments used must be calibrated to specifications. Training may include the concept of *uncertainty analysis*: "How does one account for errors?" There are really three different power levels in a system: the power generated by the source, the amount on the transmission line, and that absorbed by the load. Evaluating these power differences involves a concept that is quite mathematical in nature and beyond the scope of this text. It is more appropriately taught in a precision measuring equipment lab (PMEL), like the military service schools have, or as part of an engineering program. Nonetheless, frequency, detection, power, attenuation, and VSWR measurements are explained here as they relate to the skills of a bench technician or a technologist.

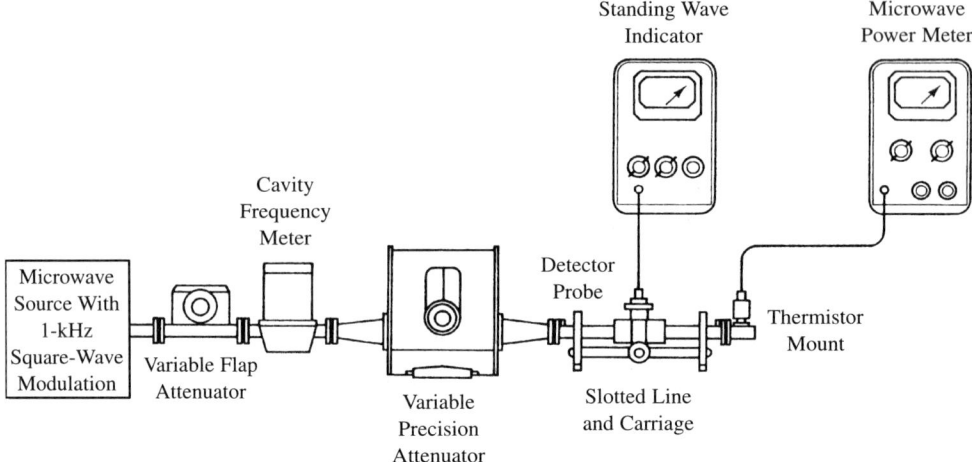

Figure 9.1   Typical setup for microwave measurements.

The RF sources found in the lab include almost any that can be used as oscillators. These include thermionic devices, such as the reflex klystron or the backward wave oscillator. More often the reflex klystron is used as the source. Microwave lab equipment manufactured by Hewlett-Packard has been around since the late 1960s. This equipment utilizes the X-13 reflex kystron, an X-band source from Varian. Figure 9.1 shows the typical setup using this equipment.

Today, most training programs use solid-state devices such as the Gunn diode oscillator, the YIG oscillator/filter, or the IMPATT diode as their RF source. Many new equipment vendors now offer complete microwave lab setups that meet the needs of an educational institution as well as the needs of industry.

We have seen how phase and impedance measurements can be done with the use of the Smith chart and the slotted line. Now we will look at some microwave test equipment analyzers—the spectrum analyzer and the network analyzer.

An introduction to scattering parameters (S-parameters) is included. S-parameters are normally taught in engineering schools. Many microwave components, particularly active devices, have associated S-parameter values in their documentation or specifications. Technicians and technologists who operate network analyzers find that the output data is in S-parameter terminology. Many texts fail to cover this topic, so some techs have not seen the correlation between the coefficients they were taught and the S-parameter values used in industry.

Before we begin our study of measurement techniques, we need to define and evaluate noise parameters. Noise is a part of any electronic circuit and includes both the transmission and the reception of a microwave signal. With the vast distances microwave signals are propagated, and the low power many of these signals contain, it is extremely important to compare noise and the microwave signal. This includes analyzing the signal-to-noise ratio *(S/N)*, noise values in terms of dB or as a noise figure, and equivalent noise temperature (T_e).

9.2   NOISE

It is helpful to divide noise into two types: *internal noise*, which originates within the microwave component or equipment, and *external noise*, which is a property of the channel. The channel is the link through which the signal travels.

At any temperature above absolute zero (0°K or –273°C), electrons in any material are in constant random motion. Although this random motion does not produce a current flow in any direction, it does produce current pulses that are the source of noise. This signal noise is called several names: *thermal agitation noise, thermal noise,* or *Johnson noise.*

Shot noise is caused by the interception of the electron beam by electrodes within active devices. There are other sources of noise also, but we are mainly concerned with those defined.

Most electronic systems are evaluated on the basis of a signal-to-noise ratio (*S/N* or *SNR*). Noise resulting from thermal agitation of electrons is measured in terms of noise power (P_n) and carries the units of power. Noise power is found by:

$$P_n = kT(BW) \qquad \text{(Eq. 9.1)}$$

where
P_n = noise power in watts

k = Boltzman's constant (1.38×10^{-23} J/K)

T = temperature in degrees Kelvin (°K)

BW = bandwidth in Hz

EXAMPLE 9.1:

Calculate the noise power from a resistor when a receiver bandwidth is 10 kHz and the resistor temperature is 300°K.

$$P_n = (1.38 \times 10^{-23})(300)(10 \times 10^3)$$
$$= 4.14 \times 10^{-17} \text{ W}$$

What is important about this equation is that noise is considered equally across an entire bandwidth, so that a 20 MHz bandwidth centered on 2 GHz produces the same thermal noise level as a 20 MHz bandwidth centered on 6 GHz, and so on.

The noise of a system or network can be defined in three different but related ways: signal-to-noise ratio (*S/N*), noise figure (*NF*), and equivalent noise temperature (T_e).

Signal-to-Noise Ratio

We need to know the effect that noise has on system performance. It is not really the amount of noise that concerns us, but rather the amount of noise compared to the level of the desired signal; that is, it is the *ratio of signal to noise power.* This signal-to-noise ratio (*S/N*) can be expressed in decibels. Note that the ratio is given in power, not voltage.

The formula for *S/N* in dB is given by:

$$S/N_{(dB)} = 10 \log P_s/P_n \qquad \text{(Eq. 9.2)}$$

where
$S/N_{(dB)}$ = the ratio expressed in dB

P_s = signal power

P_n = noise power

This ratio is difficult to measure, because although it may be possible to measure the noise power by turning off the signal, it is not possible to turn off the noise in order to measure the signal power alone. Consequently a variant of *S/N*, called (*S + N*)/*N*, is often found in receiver specifications. This stands for the *ratio of signal-plus-noise power to noise power alone.* It is found by equation 9.3.

$$(S + N)/N = (\text{Signal} + \text{Noise power})/\text{Noise power alone} \qquad \textbf{(Eq. 9.3)}$$

EXAMPLE 9.2:

A receiver produces a noise power of 200 mW with no signal. The output level increases to 5 W when a signal is applied. Calculate $(S + N)/N$ as a power ratio.

$$(S + N)/N = 5/0.2$$
$$= 25$$

To convert this ratio into dB, take $10 \times \log$ of the ratio.

$$(S + N)/N_{(dB)} = 10 \log 25$$
$$= 14 \text{ dB}$$

Noise Figure

Noise figure (*NF*), also called *noise factor*, is a figure of merit, indicating how much a component, stage, or series of stages degrades the signal-to-noise ratio of a system. The noise figure is, by definition:

$$NF = (S/N)_i/(S/N)_o \qquad \textbf{(Eq. 9.4)}$$

where
$$(S/N)_i = \text{input signal-to-noise ratio}$$
$$(S/N)_o = \text{output signal-to-noise ratio}$$

When both *S/N* and *NF* are expressed in decibels, we have:

$$NF_{(dB)} = 10 \log NF_{(ratio)} \qquad \textbf{(Eq. 9.5)}$$

or,
$$NF_{(dB)} = (S/N)_{i(dB)} - (S/N)_{o(dB)} \qquad \textbf{(Eq. 9.6)}$$

EXAMPLE 9.3:

The signal power at the input to an amplifier is $100 \ \mu\text{W}$ and the noise power is $1 \ \mu\text{W}$. At the output, the signal power is 1 W and the noise power is 30 mW. What is the noise figure as a ratio of the amplifier?

$$(S/N)_i = 100 \times 10^{-6}/1 \times 10^{-6}$$
$$= 100$$
$$(S/N)_o = 1/30 \times 10^{-3}$$
$$= 33.3$$
$$NF = 100/33.3$$
$$= 3.0$$

EXAMPLE 9.4:

The signal at the input of an amplifier has an *S/N* of 40 dB. If the amplifier has a noise figure of 6 dB, what is the *S/N* at the output (in dB)?

$$NF_{(dB)} = (S/N)_{i(dB)} - (S/N)_{o(dB)}$$
$$(S/N)_{o(dB)} = (S/N)_{i(dB)} - NF_{(dB)}$$
$$= 40 \text{ dB} - 6 \text{ dB}$$
$$= 34 \text{ dB}$$

Equivalent Noise Temperature

The equivalent noise temperature (T_e) is a means of specifying noise in terms of an equivalent temperature. The *equivalent noise temperature* is not the physical temperature of the circuit, but is the *absolute temperature* that would produce the same noise power. The equivalent noise temperature is related to the noise figure as:

$$T_e = (NF-1)T_o \qquad \textbf{(Eq. 9.7)}$$

where $\qquad T_o$ = a reference temperature of 290°K
(approximately room temperature)

Equivalent noise temperature of low noise amplifiers (LNAs) are quite low, often less than 100°K. LNAs are used quite often in many types of transmitter/receiver applications, particularly in satellite communications.

EXAMPLE 9.5:

An amplifier has a noise figure of 2 dB. What is its equivalent noise temperature? (Note: You must first convert $NF_{(dB)}$ into $NF_{(ratio)}$.)
    Using equation 9.6, we have:

$$NF_{(dB)} = 10 \log NF_{(ratio)}$$
$$NF_{(ratio)} = \text{antilog } NF_{(dB)}/10$$
$$= \text{antilog } 2/10$$
$$= \text{antilog } 0.2$$
$$= 1.585$$

Now from equation 9.7, we find T_e as:

$$T_e = (NF - 1)T_o$$
$$= (1.585 - 1)290°\text{K}$$
$$= 169.6°\text{K}$$
$$\approx 170°\text{K}$$

Since the noise figure is so often expressed in dB, the spiral chart (Figure 9.2) was developed to quickly read the equivalent noise temperature.

When cascading stages, the first stage becomes the most important in analyzing the overall noise effects. Any noise in the first amplifier is re-amplified by each succeeding stage. Noise produced in later stages is amplified less, and noise generated in the last stage is amplified least of all. It is possible to derive an equation that relates the total noise figure to the gain and noise figure of each stage. Note that the noise figure and gains are expressed in ratios, not in decibels.

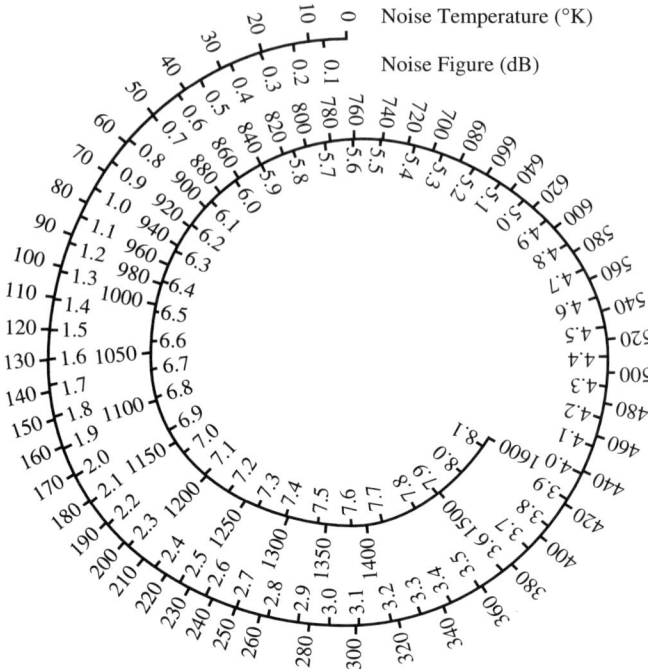

Figure 9.2   Conversion chart between noise temperature and noise figure.

This equation is known as Friis' formula and can be generalized to any number of stages as:

$$NF_T = NF_1 + (NF_2 - 1)/G_1 + (NF_3 - 1)/G_1G_2 + (NF_4 - 1)/G_1G_2G_3 + ... \textbf{(Eq. 9.8)}$$

EXAMPLE 9.6:

Determine the total noise figure for a three-stage amplifier with the following specifications.

Stage	Power Gain	Noise Figure
1	10	2
2	20	4
3	30	6

$$NF_T = 2 + [(4 - 1)/10] + [(6 - 1)/(10)(20)]$$
$$= 2.325$$

9.3   FREQUENCY MEASUREMENTS

Many microwave procedures require a measurement of frequency. There are basically two approaches to the measurement of microwave frequencies. The first approach, and the most accurate, is to measure the frequency directly using a frequency counter. Direct frequency measurements are made by comparing an unknown signal to a reference frequency, the crystal oscillator.

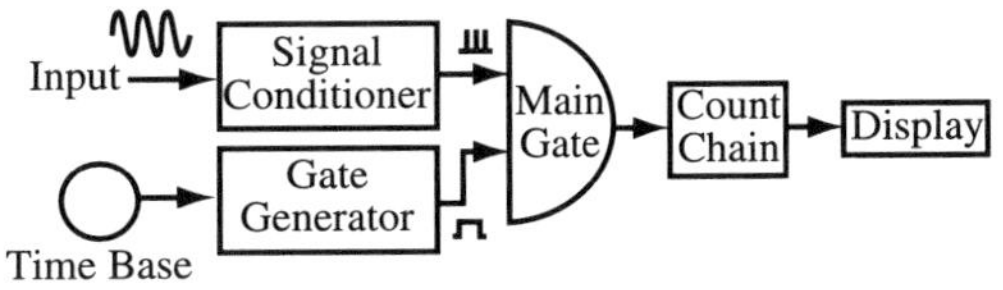

Figure 9.3 Direct frequency counter block diagram.

The input signal is first conditioned into a series of pulses, then passed to the main gate (Figure 9.3). The frequency is measured by generating a gate time, consisting of a number of cycles of the reference clock, during which the input signal is counted. The frequency is calculated by dividing the number of cycles by the gate time.

To make frequency measurements at microwave frequencies various down-conversion techniques are used to convert the microwave input to an IF so that the resultant signal can be directly counted. The three basic techniques for down-conversion are prescaling, transfer oscillator, and harmonic heterodyne.

Prescaling

Prescaling uses a divider circuit to reduce the frequency of the input signal down to a lower frequency that can then be counted by the direct counter circuit. However, this technique has frequency limitations.

Transfer Oscillator

The transfer oscillator method may be thought of as a microwave prescaler since it locks a harmonic of a low-frequency oscillator to the incoming signal. This technique offers high sensitivity but suffers from a performance trade-off in terms of immunity, FM signals, and noise.

Harmonic Heterodyne

This technique uses a mixer and a local oscillator (LO) signal to down-convert the incoming signal. The resulting IF is then counted directly by the basic counter techniques.

The second approach is to measure the wavelength associated with the frequency of interest and then convert this wavelength to frequency through the following equation:

$$f = (3 \times 10^8)\sqrt{[(1/\lambda_g)^2 + (1/2a)^2]} \qquad \textbf{(Eq. 9.9)}$$

where
$$3 \times 10^8 = \text{speed of light, m/s}$$
$$a = \text{the width of the waveguide, m}$$
$$\lambda_g = \text{wavelength in the guide, m}$$

Most microwave lab exercises include this second type of frequency measurement. It involves the use of the slotted line with its built-in crystal detector and probe. This is done by measuring the distance between successive minima of a standing wave. It is generally preferable to measure the distance between standing-wave minima rather than between maxima, since the amplitude variation is more pronounced near the minima. The presence of large reflections in the waveguide is also advantageous. For this reason, a short circuit is often used as the load. A short circuit does not absorb any power, so the amplitude of the reflected wave is equal to the amplitude of the incident wave. It is possible to obtain a measurement precision of about 1% when determining the guide wavelength in this manner. This technique cannot be used to measure the microwave signal

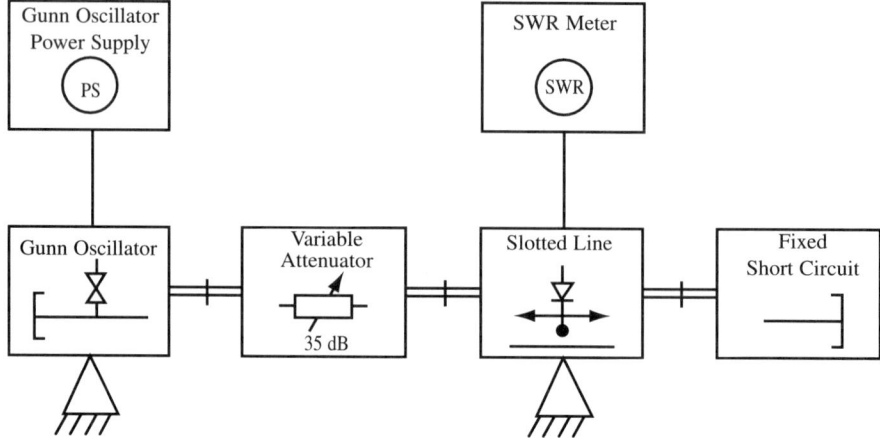

Figure 9.4 Typical setup for measuring frequency via wavelength in the waveguide.

frequency in a waveguide terminated in a matched load, because in that case no standing wave is generated.

As the carriage is moved on the slotted line, the detector's output signal is fed to an SWR meter. The reading is directly related to the microwave voltage at the probe's input and, therefore, to the intensity of the electric field in the waveguide. A typical setup for this type of measurement is shown in Figure 9.4.

EXAMPLE 9.7:

Using the slotted line technique to determine the frequency, the distance between two adjacent minima is found to be 1.9 cm. Since two adjacent minima are $\lambda_g/4$ apart, you must double this to know λ_g. If the guide's a dimension is 2.29 cm, calculate the value of the frequency.

$$\lambda_g = 2 \times 1.9 \text{ cm} = 3.8 \text{ cm} = 0.038 \text{ m}$$
$$2a = 2 \times 2.29 \text{ cm} = 4.58 \text{ cm} = 0.0458 \text{ m}$$
$$f = (3 \times 10^8)\sqrt{(1/.038)^2 + (1/.0458)^2}$$
$$= (3 \times 10^8)\sqrt{1169.24}$$
$$= 3 \times 10^8 \times 34.19$$
$$= 10.26 \text{ GHz}$$

For measurements of frequency that are not as critical, passive frequency measurement techniques may be employed. These techniques use passive frequency measuring devices such as *transmission wavemeters* and *absorption wavemeters*. Both types of wavemeters employ a cylindrical cavity as the frequency determining device.

The absorption cavity method of frequency measurement is the simplest of all methods to perform. The absorption cavity frequency meter is connected in the waveguide transmission line (see Figure 9.1). An equivalent series resonant circuit is presented to the line at resonance, allowing energy to pass into the cavity. Energy is in turn lost to the wave-

Figure 9.5   Resonant cavity frequency meter for coaxial lines HP Q532A. (Photo courtesy of Hewlett-Packard Co.)

guide cavity as it is being dissipated within the cavity walls. This represents a loss of power to the transmission line. When tuned to resonance, a fraction of the power is absorbed; consequently, there is a dip of about 1 dB in the transmitted power. The frequency at which this dip occurs may be read directly from the scale on the frequency meter. Figure 9.5 shows a typical resonant cavity frequency meter, the HP-Q532A from Hewlett-Packard. It employs coaxial adapters. It is tunable from 33 to 50 GHz with an accuracy near 0.12%.

The transmission cavity wavemeter is connected in parallel with the transmission line. The indicating circuit does not show power until the cavity is tuned to resonance. At that point the meter reads maximum power.

9.4   DETECTION DEVICES

Only a few detection devices convert power to a signal that the meter itself can read. Among those detection devices are bolometers (thermistors and barretters), thermocouples, and diode sensors.

The most common detection methods of the older type (but still in use) power meters are classified as bolometer mounts. This category includes the thermistor and barretter. The barretter is generally not used for absolute power measurements due to its high susceptability to burnout. The thermistor, on the other hand, is much more rugged and therefore very popular.

Barretters have a more predictable square law response than thermistors and respond quicker to a power change. Thermistors have a negative temperature coefficient and barretters have a positive temperature coefficient.

The bolometer mount consists of resistive elements that make up one leg of a bridge (generally a Wheatstone bridge). In less-expensive power meters, a single thermistor is often used.

RF power applied to the detection thermistor is dissipated in the elements, which give a corresponding impedance change. This change is read out on the metering circuit and is related back to power.

Crystal detectors are widely used in microwave measurements because these devices have high sensitivity and wide frequency response. The crystal is used to detect the standing waves of voltage in VSWR measurements. The microwave signal is converted to a DC current or voltage whose amplitude is a function of the amplitude of the microwave signal. By converting the microwave signal to a DC signal, conventional low-frequency instruments can be used to measure many of the parameters of the microwave signal, or just to detect its presence.

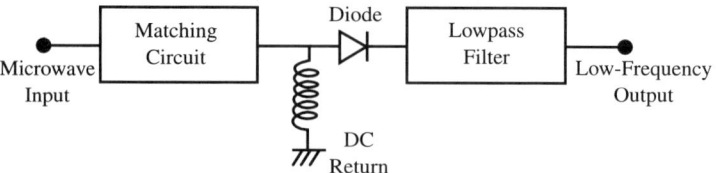

Figure 9.6 Simplified crystal detector circuit.

Figure 9.6 represents a simplified crystal detector circuit. Crystal detectors are basically a type of diode rectifier. Most older devices are of the familiar "cat whisker" diode type, in which a piece of tungsten in the form of a whisker touches a piece of germanium or silicon. Because a diode is a nonlinear element, harmonic components of the input signal are generated in the detection process. These harmonic components are attenuated by the lowpass filter, leaving only the DC output signal.

If the microwave signal's amplitude is sufficiently low, the output of the detector is proportional to the square of the microwave signal voltage and, therefore, proportional to the microwave signal power (since $V^2 = P$). Figure 9.7 shows the detector sensitivity curve. When the voltage is low, the detector is said to be operating in its quadratic, or square-law region.

When the microwave signal power is greater than about –15 dBm, the voltage of the detector's output signal tends to be directly proportional to the microwave signal voltage. The detector is said to be operating in its linear region; that is, it rectifies the applied signal.

When followed by a low-noise high-gain amplifier, as illustrated in Figure 9.8, a crystal detector can be used to detect weak signals. The output level, amplified to a level high enough to be easily measured, is then applied to an indicator.

This technique is quite general and can be applied to nonmodulated as well as amplitude-modulated (AM) signals. Usually, the microwave signal is amplitude modulated by a 1 kHz square wave. This permits the use of instruments and circuits with lower noise levels and allows the detection of even weaker signals. The weakest signal that can be detected with this technique is limited by the diode's noise level. The *tangential sensitivity*

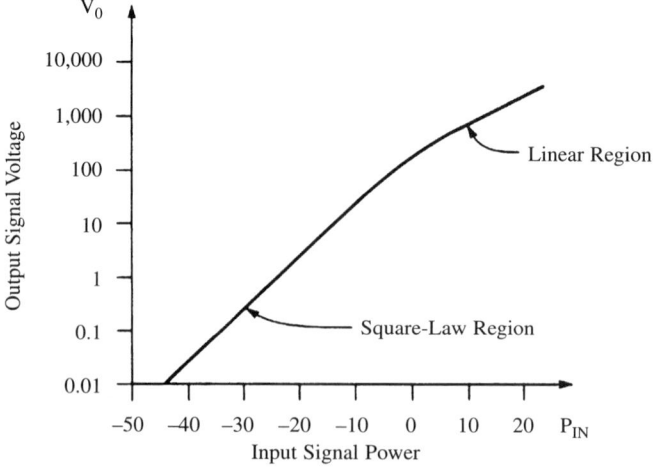

Figure 9.7 Typical crystal detector sensitivity curve.

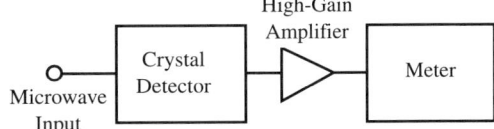

Figure 9.8 System for detecting low-power signals.

of a detector is a measure of a detector's ability to detect a low-level signal. It is the power of the weakest microwave signal that can be detected above the noise level.

Most crystal detectors have a tangential sensitivity on the order of –50 dBm for a bandwidth of 1 MHz. Since the noise power at the amplifier output is proportional to the amplifier's bandwidth, the tangential sensitivity is better when the bandwidth is reduced.

Newer power meters use GaAs barrier or Schottky barrier diodes. The GaAs diodes offer very flat frequency response versus both frequency and temperature. They have excellent square-law characteristics, and have an upper-bound frequency range of more than 100 GHz.

Table 9.1 is a comparison chart of some detectors used in microwave measurements.

The thermocouple is another device for detecting power that is widely used and implemented for use with new equipment. A *thermocouple* is a pair of dissimilar metal wires joined together at one end (sensing end) and terminated at the other end (reference end). If a temperature difference exists between the two junctions, voltage is produced, causing current to flow in the circuit. Compared with bolometers, thermocouples have a low sensitivity and are seldom used as detectors; however, thermocouples are more reliable in the measurement of absolute power levels. Thermocouples are often used in conjunction with a diode detector.

9.5   POWER MEASUREMENTS

At microwave frequencies, the amplitudes of the voltages and currents vary along a waveguide. However, the power transmitted by a lossless waveguide is independent of the location along the waveguide. Therefore, it is more convenient to measure power instead of voltage and current. Moreover, at RF and microwave frequencies, power measurement techniques are simpler and more precise than voltage and current measuring techniques.

For microwave measurements, power levels are often broken down into three classes. Levels under 1 mW are considered to be low power. Medium power levels are from 1 mW to 10 W. Powers over 10 W are considered to be high power levels.

Table 9.1 Comparison of Detectors

Characteristics	Crystals	Barretters	Thermistors
Response Time	Extremely fast	$\approx 350\ \mu s$	≈ 1 sec
Square-law Response	$\approx 10\ \mu W$	$\approx 200\ \mu W$	$\approx 200\ \mu W$
Resistance to Burnout	Determined by design	≈ 12 mW	≈ 25 mW
Resistance to Shock	Poor	Fair	Good
Temperature Coefficient	None	Positive	Negative
Minimum Discernable Signal	$1.8 \times 10^{-6}\ \mu W$	$1.0 \times 10^{-4}\ \mu W$	$1.0 \times 10^{-4}\ \mu W$
Method of Operation	Rectifies Voltage	Absorbs EM energy	Absorbs EM energy

When doing general microwave work, it is more convenient to express power levels in dBm rather than in watts (W) or milliwatts (mW). To convert a power expressed in milliwatts to dBm, equation 9.10 is used. (Remember that 1 W = 1,000 mW. See Appendix A for a review of logarithms and dBs.)

$$P_{(dBm)} = 10 \log (P_{mW}/1 \text{ mW}) \qquad \textbf{(Eq. 9.10)}$$

Another power unit, used for higher power levels, is the dBW, referenced to 1 watt and calculated as follows:

$$P_{(dBW)} = 10 \log (P_W/1 \text{ W}) \qquad \textbf{(Eq. 9.11)}$$

Using the logarithmic properties of the preceding equations, dBW are easily converted into dBm using the relation:

$$P_{(dBm)} = P_{(dBW)} + 30 \qquad \textbf{(Eq. 9.12)}$$

Thus, 10 dBW is equivalent to 40 dBm.

EXAMPLE 9.8:

Determine the appropriate dB level for both a 30 mW and a 5 W microwave signal.

$$P_{(dBm)} = 10 \log 30 \text{ mW}/1 \text{ mW}$$
$$= 10 \log 30$$
$$= 14.8 \text{ dBm}$$
$$P_{(dBW)} = 10 \log 5 \text{ W}/1 \text{ W}$$
$$= 10 \log 5$$
$$= 6.99 \text{ dBW}$$

Most commonly used power measurement techniques use thermocouples, diode detectors, or thermistors. Methods using thermocouples or diode detectors require fairly elaborate instrumentation and a calibrated oscillator. However, diode detectors can be used to measure power levels as low as –70 dBm. The thermistor method is simpler, but its sensitivity is limited to about –20 dBm. Figure 9.9 shows a simplified block diagram of a thermistor power meter. The thermistor, represented by resistance R_T, is connected to one leg of a Wheatstone bridge and inserted into the waveguide via a mount. The three other resistors of the bridge are mounted in the power meter.

The power meter operates in the following manner. Before the microwave signal is introduced, the bridge is balanced using a 10 kHz audio signal applied between points A and B. The audio signal warms up the thermistor, causing its resistance to decrease. This signal is adjusted until the thermistor resistance R_T equals R. When this condition is met, there is no current through the ammeter connected between points C and D, and the Wheatstone bridge is said to be at equilibrium. Then the audio signal power (P_0) that is dissipated in the thermistor can be calculated using the voltage across the bridge (between points A and B) and the known value of R of the resistors.

When microwave power is injected into the waveguide, the bridge goes out of balance. This is caused by the additional heating of the thermistor by the microwave signal. Equilibrium is restored by reducing the audio signal power at point A, and the audio signal

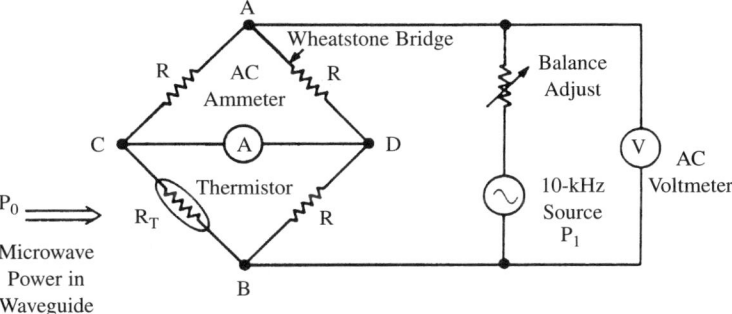

Figure 9.9 Simplified block diagram of a thermistor power meter.

power P_1 fed to the thermistor is measured again. The microwave power is simply the difference between audio powers P_0 and P_1.

Some power meters are designed to automatically balance the bridge by means of a feedback loop. Prior to measuring the microwave power, the display must be adjusted to zero. When the microwave signal is introduced, the display indicates the power of the microwave signal.

The automatic balanced bolometer bridge power meter can be summarized as follows:

1. Measurement is rapid, and no auxiliary equipment is required other than the bolometer mount.
2. The bolometer operates with one resistance, making a good impedance match at all power levels within the dynamic range of the system.
3. Accuracy of substituted power is better than 5% of full scale.
4. Dynamic range is about 20 dB, which is limited by bolometer burnout at the high end and zero drift on the low power end. At the low end, slight ambient temperature variations at the bolometer cause large variations in meter readings. The system has the inability to distinguish applied power changes from environmental temperature variations. This is true of all non-temperature-compensated bolometer bridges.

To circumvent this ambient temperature problem, there is a type of power meter that meets the requirements for low-level power measurements. This is a dual bolometer bridge that is an automatically balanced temperature-compensated power meter. The HP 432A analog power meter (Figure 9.10) is an example of this type of meter.

Newer digital readout power meters are available that can interface with automatic test equipment (ATE) measurement stations. Some have a built-in GPIB (General Purpose

Figure 9.10 Analog power meter HP 432A. (Photo courtesy of Hewlett-Packard Co.)

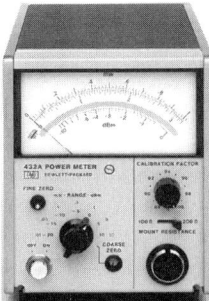

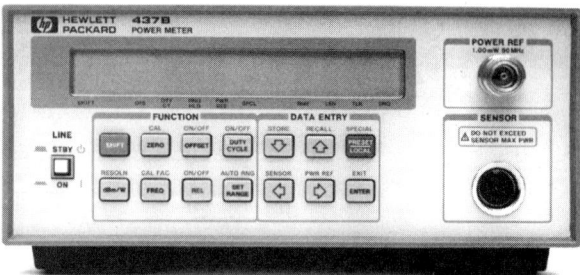

Figure 9.11   High performance digital power meter, HP 437B. (Photo courtesy of Hewlett-Packard Co.)

Interface Bus) to properly interface with the computer. They are compatible with thermocouple and diode power sensors. Figure 9.11 shows an example of this type of power meter, the HP 437B. The HP 437B has a very fast measurement speed at frequencies up to 110 GHz.

9.6   ATTENUATION MEASUREMENTS

Transmission measurements is a very broad term that takes in such areas as insertion loss, gain, and attenuation. *Attenuation* is a general term describing the decrease of signal amplitude in transmission from the input to the output of a device.

An attenuation measurement is determined by the decrease in power level at the load caused by inserting a device between a matched source and load. Under this condition, the measured value is a property of the device alone, so this is the ideal system in which to make measurements.

Insertion loss is the term used to describe the attenuation caused by inserting a component into a transmission system. More specifically, the insertion loss is the ratio of the power (P_1) provided to the load before the component insertion to the power (P_2) provided to the load after the component insertion (see Figure 9.12).

Figure 9.12   Insertion loss measurements. (a) Before insertion. (b) After insertion.

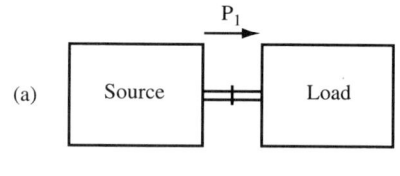

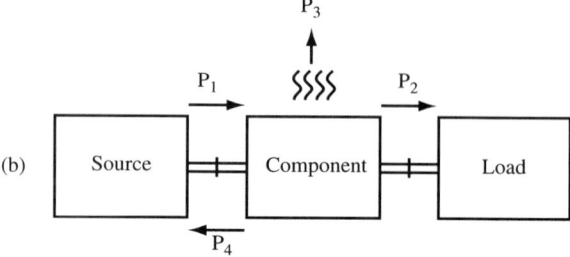

The insertion loss expressed in dB is:

$$IL_{(dB)} = 10 \log (P_1/P_2) \qquad \text{(Eq. 9.13)}$$

or, alternately:

$$IL_{(dB)} = P_{1(dBm)} - P_{2(dBm)} \qquad \text{(Eq. 9.14)}$$

EXAMPLE 9.9:

Determine the insertion loss when the power provided to the load before insertion of a component is 15 mW and the power provided after insertion is 11 mW.

$$IL_{(dB)} = 10 \log 15 \text{ mW}/11 \text{ mW}$$
$$= 10 \log 1.36$$
$$= 1.35 \text{ dB}$$

EXAMPLE 9.10:

Determine the insertion loss when the power provided to the load before insertion of a component is 12 dBm and the power provided after insertion is 9.5 dBm.

$$IL_{(dB)} = 12 \text{ dBm} - 9.5 \text{ dBm}$$
$$= 2.5 \text{ dB}$$

Note in Figure 9.12 that a number of phenomena combine to limit power P_2. Some power, P_3, is dissipated through conductor or dielectric losses, while reflected power, P_4, also reduces the power transferred to the load. Insertion loss is the total effect of both power reflected and power attenuated (dissipated).

In general, the insertion loss of a component is a function of the source and load impedances, and their degree of matching. If there is no mismatch, insertion loss and attenuation are the same.

There are four basic types of attenuation: residual, incremental, intrinsic, and reflective. The first two are properties of fixed attenuators, and the last two are properties of variable attenuators. *Residual attenuation* is the attenuation that is measured when the variable attenuator is set to minimum (0 dB) and inserted in an ideal matched system. The *incremental attenuation* is the amount of attenuation between the positions (minimum and any other position) on the variable attenuator. The residual and incremental attenuation make up the total attenuation of a variable attenuator. *Intrinsic attenuation* is the actual dissipative (internal) attenuation of the device itself, whereas the *reflective attenuation* is the loss of power due to the mismatch of impedances.

In most insertion loss measurements, variable attenuators are the standard, with fixed attenuators used to extend the range of the measurement. There are four recognized methods of insertion loss (attenuation) measurements. These are the DC substitution, RF substitution, IF substitution, and power ratio methods. The most common of these are the RF substitution and the power ratio methods.

The power ratio method is a simple method that provides relatively precise measurements of attenuation. First, a load is connected directly to a microwave source, and the

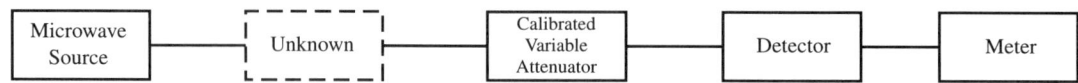

Figure 9.13   Attenuation measurement by the RF substitution method.

power provided to the load is measured. Then the component under test is inserted between the source and load, and the power in the load is measured again. The ratio of these two powers gives a measure of the attenuation or insertion loss.

The advantage of this method is that it requires relatively little equipment and no modulating signal. This contributes to the simplicity of the method. The precision of the measurement in highly dependent on the accuracy of the power meter. When the power detector is a thermistor, the main disadvantage is that detection is limited to an attenuation range of about 30 dB. Moreover, this technique is not suitable for low power levels.

The RF substitution method of measuring attenuation, illustrated in Figure 9.13, provides a high degree of precision. In this method, the component whose attenuation is to be measured (the unknown) is inserted after a source and before a calibrated variable attenuator and a detector. Then the variable attenuator is adjusted to produce a convenient signal level at the detector's output. Finally, the unknown is removed from the setup, and the variable attenuator is adjusted to produce the same signal level at the output of the detector as was measured with the unknown in the system. The difference in the two attenuation settings of the calibrated attenuator determines the attenuation caused by the unknown.

The dynamic range offered by this technique is a function of the detector used. The use of a power meter and a thermistor results in a rather low dynamic range (as seen in the power ratio method). A diode detector and an SWR meter provide a dynamic range of about 60 dB.

The SWR meter consists of a variable gain amplifier tuned to 1 kHz and a meter. The source is amplitude modulated by a 1 kHz square wave. The SWR meter is calibrated for use with a square-law detector at its input.

An advantage of this method is that it eliminates errors that could be caused by the detector. Another advantage is that the measuring process is not restricted to any particular type of detector.

The main sources of error are the calibration of the attenuator, the reflections at component junctions, the presence of harmonic signals, and the fact that any measuring instrument has a limited stability.

9.7   VSWR MEASUREMENTS

It has already be shown that when the load is not equal to the characteristic impedance of the transmission line, standing waves result. A standing wave is periodic and is characterized by its minima and maxima. The ratio of the voltage amplitude at a maximum to the voltage amplitude at a minimum is called the voltage standing-wave ratio, VSWR.

Maxima correspond to the points at which the incident and reflected waves add in phase. Minima correspond to the points at which the reflective and incident waves add 180° out of phase. The standing-wave ratio is then the ratio of E max to E min.

VSWR measurements are made using a slotted line in a setup like the one shown in Figure 9.14. The slotted line is inserted right next to the load and the VSWR is measured as close to the load as possible. This minimizes the effect of the insertion loss of the slotted line.

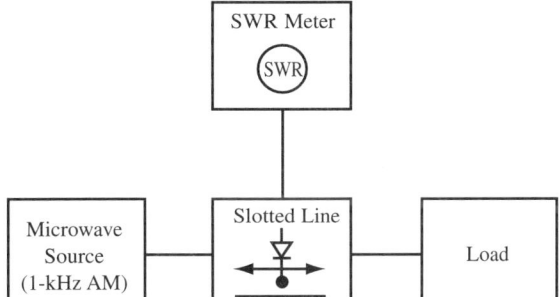

Figure 9.14 VSWR measurement.

In general, VSWR and $SWR_{(dB)}$ measurements are made in the same way. First, the slotted line's probe is located over a maximum, preferably as close to the load as possible. Then, the needle of the SWR meter is adjusted to a reference position. When the VSWR is to be read off the scales, the needle must be aligned with the VSWR = 1.0 position. For dB measurements, the setting is only a reference, with the 0 dB position a convenient position. Next, the probe is located over a minimum, the appropriate meter scale is selected, and the reading is taken. If the reference position for VSWR was VSWR = 1.0, the scale reading *is* the value of the VSWR. For dB measurements, the $SWR_{(dB)}$ is the difference in dB reading at the reference and the dB reading at the minimum.

Particular attention must be given to the adjustment of probe depth inside the slotted line. If the probe penetrates too far into the slotted line, the field distribution may be distorted, and moreover, the signal may be too strong and drive the crystal detector out of its square-law region.

Figure 9.15 shows a typical SWR meter, the Lab-Volt Model 9502, designed for use with square-law detectors.

9.8 INTRODUCTION TO S-PARAMETERS

S-parameters are a two-port analysis tool, much like the engineering parameters *h, y,* and *z,* but S-parameters are used to describe microwave components. The name "S-parameters" was derived from the word "scattered" (from reflected waves).

Recall from transistor theory that engineers use *h* (hybrid) parameters to describe the characteristics of the bipolar transistor. H-parameters are measured under open and short con-

Figure 9.15 A typical SWR meter, model 9502. (Photo courtesy of Lab-Volt systems.)

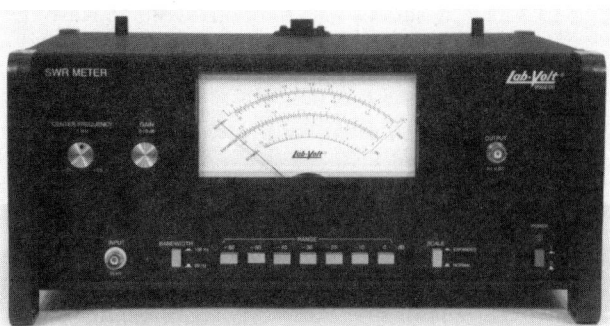

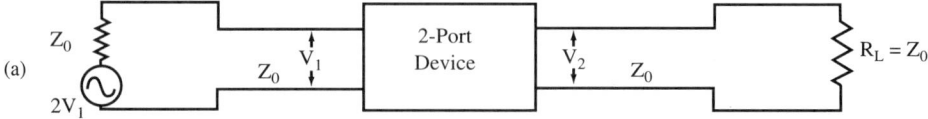

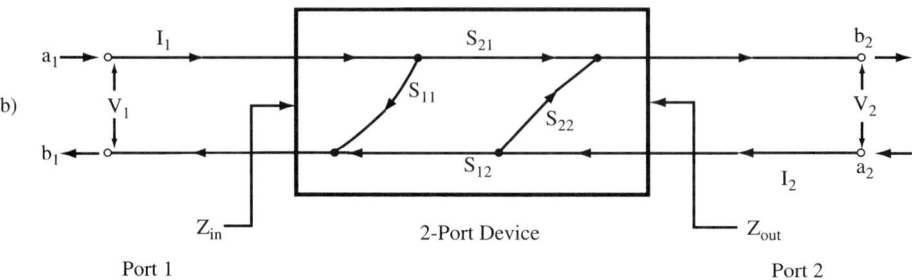

Figure 9.16   S-parameters. (a) Typical setup. (b) Flowgraph.

ditions. At microwave frequencies opens and shorts cannot be used for testing, because they cause reflections and act reactively. Therefore, S-parameters are better interpreted with a pure resistance matched termination.

Most network analyzers give their output data in terms of S-parameters for various coefficients. Technical students are taught forward or reverse voltage coefficients, such as Γ and ρ. In S-parameter terms these are S_{11}, S_{22}, S_{12}, or S_{21}. Obviously, we need to know the terminology to work in an environment that uses network analyzers.

Figure 9.16a shows the typical setup for S-parameter measurements, while Figure 9.16b shows the development of the various S-parameters via a flowgraph.

Along with the S-parameter set used to identify the flowgraph branches, subscripted lowercase letters are used to identify the flowgraph nodes. The a denotes a voltage incident to a port, and b denotes the emergent node. For example, a_1 represents the node at which voltage is incident to port 1, and b_1 represents the node at which voltage is emergent from port 1. When using S-parameters to define the flowgraph branches, keep in mind that these parameters are node coefficients. There are two type of node coefficients, transmission and reflection.

Consider a signal applied to port 1 (node a_1), that is transmitted through the network along branch S_{21} and emerges from the network at port 2 (node b_2). The direction of signal flow within the network is described by the directed branch, S_{21}. The S-parameter S_{21} describes the effect on a voltage wave traveling to port 2 coming from port 1. The value of b_2 depends on the magnitudes of a_1 and S_{21}, as shown in the following equation:

$$b_2 = (a_1)(S_{21})$$

Solving for S_{21}, we have:

$$S_{21} = b_2/a_1$$

S_{21} is referred to as the *forward transmission coefficient*. Remember, to measure and define S-parameters so that their results are meaningful, we must terminate any unused port by a pure resistance equal to Z_o. Therefore:

$$S_{21} = b_2/a_1$$

when port 2 is match terminated ($a_2 = 0$). An alternate solution for S_{21} is $S_{21} = V_2/V_1$.

If a signal is applied to port 2, with port 1 match terminated, we can see the value of b_1 depends on a_2 and S_{12}.

$$b_1 = (a_2)(S_{12})$$

S_{12} describes the reverse signal flow within the network and is often referred to as the *reverse voltage transmission coefficient*. Solving the above equation for S_{12}, we have:

$$S_{12} = b_1/a_2$$

when port 1 is match terminated ($a_1 = 0$). An alternate solution to S_{12} is $S_{12} = V_1/V_2$.

When port 2 is Z_o terminated, any voltage emerging from b_1, upon application of a voltage incident to a_1, is a reflected wave. The branch depicting this signal path is S_{11}. The value of the node b_1 is dependent on the value of a_1 and S_{11}, as shown by the following:

$$b_1 = (a_1)(S_{11})$$

Solving for S_{11}, we have:

$$S_{11} = b_1/a_1$$

when port 2 is match terminated ($a_2 = 0$).

S_{11} refers to a voltage traveling to port 1 coming from port 1. Therefore, S_{11} is the familiar voltage reflection coefficient, Γ, of port 1, or the *input voltage reflection coefficient*, Γ_{in}, of the two-port network. An alternate solution to Γ_{in} is $\Gamma_{in} = (Z_{in} - Z_o)/(Z_{in} + Z_o)$.

If we terminate port 1 of the two-port network with a Z_o impedance and examine the reflected voltage from port 2 in relation to an incident voltage to port 2, we see that the value of b_2 depends upon a_2 and S_{22}, or:

$$b_2 = (a_2)(S_{22})$$

Solving for S_{22}, we have:

$$S_{22} = b_2/a_2$$

when port 1 is match terminated ($a_1 = 0$).

S_{22} is also a ratio of reflected voltage to incident voltage, since it relates values going to and coming from port 2. It is called the *output voltage reflection coefficient*, or the Γ_{out} of port 2. An alternate solution to Γ_{out} is $\Gamma_{out} = (Z_{out} - Z_o)/(Z_{out} + Z_o)$.

S_{21} squared, $|S_{21}|^2$, is interpreted as the power gain of the network, since V_2 equals the output voltage, while V_1 is the input voltage to the network. [$(V_2/V_1)^2$ = power gain]

To summarize the S-parameters:

S_{11} and S_{22} are reflection coefficients.

S_{12} and S_{21} are transmission coefficients.

$$S_{11} = \frac{b_1}{a_1} \ (a_2 = 0) \text{ or } \Gamma_{in} = \frac{Z_{in} - Z_o}{Z_{in} + Z_o}$$

$$S_{22} = \frac{b_2}{a_2} \ (a_1 = 0) \text{ or } \Gamma_{out} = \frac{Z_{out} - Z_o}{Z_{out} + Z_o}$$

$$S_{21} = \frac{b_2}{a_1} \ (a_2 = 0) \ \text{or} \ \frac{V_2}{V_1}$$

$$S_{12} = \frac{b_1}{a_2} \ (a_1 = 0) \ \text{or} \ \frac{V_1}{V_2}$$

9.9   MICROWAVE TEST EQUIPMENT ANALYZERS

Spectrum Analyzer

Single-frequency measurements performed at many points are not satisfactory when making measurements on wide frequency band devices, because it takes too long to make them, and even then not all the frequencies in question are covered. However, voltage controlled oscillators are capable of providing a swept-frequency signal source. One such device, the backward wave oscillator (BWO), makes a reasonable swept-frequency oscillator. Newer sweep generators employ the YIG oscillator as the source. The YIG oscillator has a DC loop that sets up a magnetic field around the sphere; the strength of the field determines the output frequency. When we apply a sawtooth signal to the DC loop of the YIG, the output is a varying frequency. We call this varying frequency range "sweeping," because it sweeps a range of frequencies.

The heart of the spectrum analyzer is the sweep generator. A *spectrum analyzer* is an instrument that graphically presents a plot of signal amplitude as a function of frequency for a selected portion of the spectrum. It is essentially a frequency-selective, peak-responding voltmeter calibrated to display the root-mean-square (rms) value of a sine wave. This is known as *frequency domain* presentation. Signals are displayed as a spectrum on a CRT screen with signal energy plotted on the vertical axis against frequency on the horizontal axis. (The oscilloscope presents its display in a *time domain* format, amplitude vs. time.)

A spectrum analyzer display provides the following information: the presence or absence of signals, their frequencies, frequency drift, noise, relative amplitude of the signals, and the nature of modulation, if any, plus many other characteristics. The spectrum analyzer was originally designed to look at the output of radar transmitters. A pulse radar signal is a train of RF pulses with a constant repetition rate, constant pulse and shape, and constant amplitude. By looking at the characteristic spectra, all important properties of the signal can be evaluated.

Figure 9.17 is a basic block diagram of a spectrum analyzer that has circuitry similar to both an oscilloscope and a superheterodyne receiver.

The sawtooth produced by the sweep generator provides horizontal sweep voltage for the CRT and at the same time provides a frequency sweep voltage for the voltage-tunable sweep oscillator. The sawtooth action also provides for synchronization of the amplitude versus frequency display.

Input frequencies beat with the sweeping signal from the voltage-tunable oscillator in the input mixer stage. When the different frequencies produced are within the bandwidth of the narrowband IF amplifier, they are sent to the video section. The detector changes the IF video pulses into DC pulses and sends them to the vertical amplifier, which drives the vertical deflection plates. Thus, the input frequencies are displayed as vertical deflections that are synchronized with the horizontal sweep so that the lower frequencies appear at the start of the sweep and the higher frequencies at the upper end of the sweep. By varying the amplitude of the sawtooth, the pulses can be brought closer together or farther apart for proper analysis.

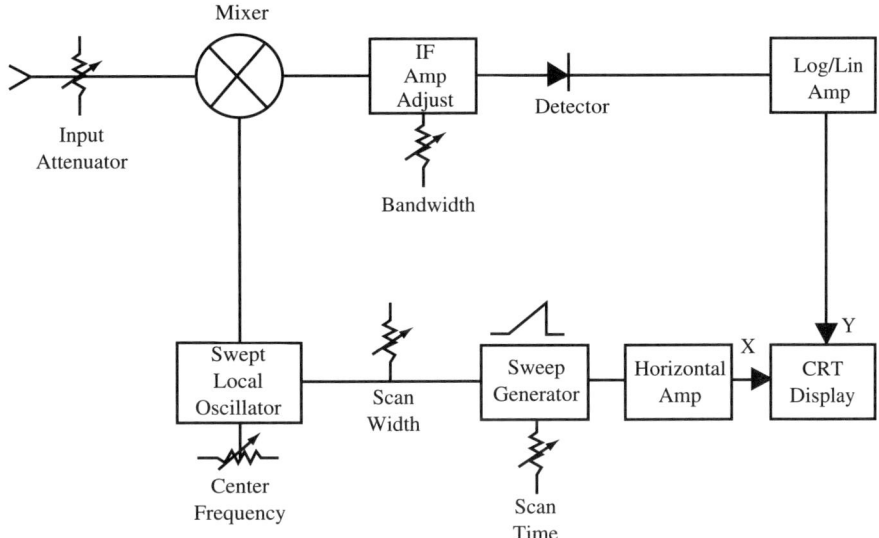

Figure 9.17 Simplified block diagram of spectrum analyzer.

Figure 9.18 shows a spectrum analyzer display for a typical amplitude modulated (AM) carrier signal modulated by a single signal frequency. The center deflection (f_c) represents the carrier frequency and its associated power level. The deflections to either side represent the frequency and power of the upper and lower sidebands (f_1, f_2). The sidebands are displaced from the carrier by an amount equal to the modulation frequency. When more than one audio frequency modulates the carrier, additional sidebands occur on the CRT. The presentation by the spectrum analyzer is periodic, while an oscilloscope presentation is continuous.

Distortion in the spectrum analyzer is primarily produced by the analyzer itself. Distortion is caused by nonlinearities in the mixing process, the level of the local oscillator, and the level of the input signal.

The types of distortion include harmonic distortion (caused by the mixer producing harmonics), second-order intermodulation (two high-frequency signals mix to produce harmonics), and third-order intermodulation (a combination of harmonic distortion and second-order intermodulation). Third-order intermodulation is the worst type of distortion; it is caused by the distortion products being closely spaced to the original signals.

Figure 9.18 Spectrum analyzer display for AM carrier modulated by a single signal frequency.

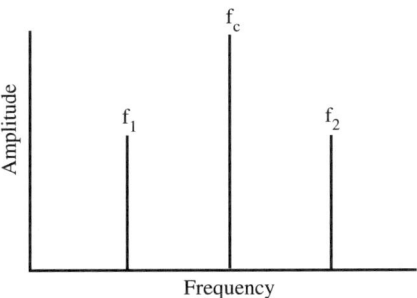

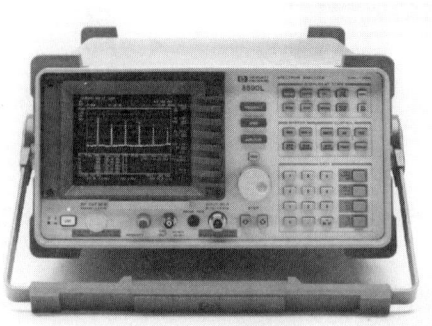

Figure 9.19 Portable spectrum analyzer, HP 8590L. (Photo courtesy of Hewlett-Packard Co.)

Figure 9.19 shows an example of a portable microwave spectrum analyzer, the HP 8590L. It offers a frequency range of 9 kHz to 22 GHz.

Network Analyzers

Network analysis is the process of creating a data model of the transfer and/or impedance characteristics of a linear network (active or passive). This is done through stimulus-response testing over the frequency range of interest. Some analyzers do this with point-to-point frequency testing, while others do this by sweeping the frequency band at one time.

Network analysis is generally limited to the definition of linear networks. Sine wave testing is an ideal method to characterize magnitude and phase response as a function of frequency. *Network analyzers* are instruments that can measure the transfer and/or impedance functions through sine wave testing. Since transfer and impedance function are ratios of various voltages and currents, a means of separating the appropriate signals from measurement ports of the device under test is required. The analyzer must detect the separated signals, form the desired signal ratios, and display the results. At microwave frequencies, where standing waves might occur on the transmission line, the analyzer must be capable of separating the signal from the traveling waves.

Automatic network analyzers (ANA) are commonplace for doing these precise forms of measurements. Scalar network analyzers (magnitude only) and vector network analyzers (both magnitude and phase) are available.

Two types of detection methods are usually employed by network analyzers. Broadband detection accepts the full frequency spectrum of the input signal, while narrowband detection involves tuned receivers that convert continuous wave (CW) or swept RF signals to a constant intermediate signal (IF). Each detection scheme has its advantages.

Scalar analyzers usually employ broadband detection techniques. Broadband detection reduces instrument cost by eliminating the IF section required by narrowband analyzers. This sacrifices noise and harmonic rejection. Broadband systems can make measurements when the input and output frequencies are not the same, as in measurements of the insertion loss of mixers and frequency doublers. Narrowband systems cannot make these measurements.

Vector network analyzers normally employ narrowband detection techniques. This makes for a more sensitive low-noise detection of the constant IF. This also increases the accuracy and dynamic range for frequency selective measurements (as compared to broadband systems).

There is a good range of adaptability and flexibility in analyzer systems. Impedances can be shown on a Smith-chart overlay for a polar display. An S-parameter test set can be attached to perform S-parameter measurements. Computer controlled network analyzers

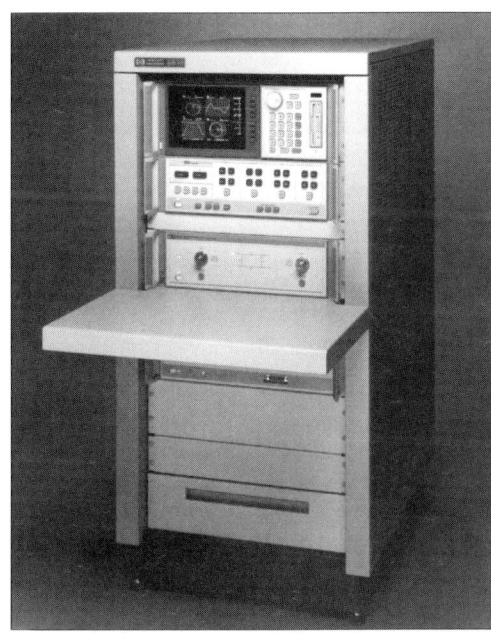

Figure 9.20 Vector network
analyzer, HP 85107B. (Photo
courtesy of Hewlett-Packard Co.)

can be programmed to set up and make many measurements automatically. The measurement process is further accelerated by the computer's ability to store, transform, summarize, and output data in a variety of formats to a number of peripherals. Functions that are normally displayed in the frequency domain can be converted to the time domain for additional analysis.

Figure 9.20 shows an HP 8510 series vector network analyzer (the HP 85107B). This analyzer was first produced in 1984 and has become an industry benchmark. It can characterize linear behavior of either active or passive networks over the 45 MHz to 40 GHz range. For millimeter wave measurement needs, systems can be configured to 100 GHz.

Table 9.2 shows the typical scattering parameters data from a 2 to 18 GHz low-noise PHEMT transistor. This data was output from a vector network analyzer.

9.10   SUMMARY

1. Common sources for microwave measurements include the BWO and the reflex klystron (thermionic), with newer lab equipment using the Gunn diode, the YIG, or the IMPATT diode.
2. Noise can be divided into internal and external forms. Thermal noise is the most common form of component-generated noise.
3. Noise is evaluated in terms of signal-to-noise ratio, noise figure, and equivalent noise temperature.
4. Frequency measurements are done by a direct frequency measurement or by measuring the wavelength and then converting that to frequency. The latter involves using a slotted line and an SWR meter.
5. Passive frequency determining devices include the absorption and transmission wavemeters.

Table 9.2   Typical Scattering Parameters for a PHEMT Transistor

Typical Scattering Parameters: Common Source, $Z_o = 50\ \Omega$											$T_A = 25°C, V_{DS} = 1.5\ V, I_{DS} = 20\ mA$	
Frequency	S_{11}		S_{21}			S_{12}			S_{22}			
GHz	Mag	Ang	dB	Mag	Ang	dB	Mag	Ang	Mag	Ang		
2.0	.97	−33	13.89	4.95	147	−29.62	.033	68	.41	−24		
3.0	.95	−45	13.76	4.88	135	−26.56	.047	60	.40	−33		
4.0	.91	−62	13.57	4.77	119	−24.44	.060	50	.37	−45		
5.0	.86	−79	13.20	4.57	103	−22.73	.073	38	.33	−57		
6.0	.80	−95	12.75	4.34	88	−21.72	.082	28	.30	−69		
7.0	.76	−110	12.34	4.14	74	−20.82	.091	19	.27	−82		
8.0	.74	−121	12.07	4.01	65	−20.09	.099	14	.24	−90		
9.0	.71	−135	11.75	3.87	51	−19.66	.104	5	.22	−102		
10.0	.67	−150	11.35	3.70	37	−19.17	.110	−5	.20	−113		
11.0	.62	−164	10.95	3.53	24	−18.86	.114	−14	.17	−123		
12.0	.59	−178	10.63	3.40	11	−18.42	.120	−23	.15	−138		
13.0	.57	167	10.47	3.34	−2	−18.20	.123	−33	.14	−152		
14.0	.56	156	10.32	3.28	−9	−17.99	.126	−37	.12	−164		
15.0	.54	140	10.10	3.20	−23	−17.92	.127	−47	.10	−173		
16.0	.52	125	9.82	3.10	−37	−17.65	.131	−58	.06	−179		
17.0	.49	107	9.70	3.05	−51	−17.39	.135	−70	.03	165		
18.0	.50	88	9.72	3.06	−66	−17.08	.140	−82	.03	53		

6. The most common power detection devices include the bolometer, thermocouple, and diode sensor.
7. Bolometers are in the form of barretters or thermistors.
8. Crystal detectors operated in their square-law region make good power detectors.
9. The three classes of power levels are: low power (under 1 mW), medium power (from 1 mW to 10 W), and high power (greater than 10 W).
10. Power meters come in many forms. Some are manually balanced, while others are automatically balanced. Some have temperature compensation for the bolometer mounts.
11. Attenuation is a decrease in signal amplitude. Insertion loss is a function of impedance matching and is the total effect of both reflected power and dissipated power. If there is no impedance mismatch, insertion loss and attenuation are the same.
12. The four basic types of attenuation are residual, incremental, intrinsic, and reflective.
13. The most common insertion loss measurements are done via the power ratio method or by RF substitution.
14. SWR meters show both the VSWR and the $SWR_{(dB)}$. They are used in conjunction with a slotted line and a crystal detector.
15. S-parameters are a two-port analysis tool to describe the reflection and transmission coefficients. They are node coefficients.
16. S_{11} and S_{22} are reflection coefficients. S_{12} and S_{21} are transmission coefficients.
17. A sweep generator sweeps a range of frequencies. It is the heart of a spectrum analyzer.
18. A spectrum analyzer plots signal amplitude as a function of frequency for a selected portion of the spectrum.

19. A spectrum analyzer has circuitry similar to an oscilloscope and a superheterodyne receiver.
20. Automatic network analyzers (ANA) measure the transfer and/or impedance functions of a linear network through sine wave testing. Scalar (magnitude only) and vector (both magnitude and phase) are common network analyzers.

Key Equations:

$$P_n = kT(BW) \qquad \text{(Eq. 9.1)}$$

$$S/N_{(dB)} = 10 \log P_s/P_n \qquad \text{(Eq. 9.2)}$$

$$(S + N)/N = (\text{Signal + Noise power})/\text{Noise power alone} \qquad \text{(Eq. 9.3)}$$

$$NF = (S/N)_i/(S/N)_o \qquad \text{(Eq. 9.4)}$$

$$NF_{(dB)} = 10 \log NF_{(ratio)} \qquad \text{(Eq. 9.5)}$$

$$NF_{(dB)} = (S/N)_{i(dB)} - (S/N)_{o(dB)} \qquad \text{(Eq. 9.6)}$$

$$T_e = (NF - 1)T_o \qquad \text{(Eq. 9.7)}$$

$$NF_T = NF_1 + (NF_2 - 1)/G_1 + (NF_3 - 1)/G_1G_2 + (NF_4 - 1)/G_1G_2G_3 + \cdots \text{ (Eq. 9.8)}$$

$$f = (3 \times 10^8)\sqrt{(1/\lambda_g)^2 + (1/2a)^2} \qquad \text{(Eq. 9.9)}$$

$$P_{(dBm)} = 10 \log (P_{mW}/1 \text{ mW}) \qquad \text{(Eq. 9.10)}$$

$$P_{(dBW)} = 10 \log (P_W/1 \text{ W}) \qquad \text{(Eq. 9.11)}$$

$$P_{(dBm)} = P_{(dBW)} + 30 \qquad \text{(Eq. 9.12)}$$

$$IL_{(dB)} = 10 \log (P_1/P_2) \qquad \text{(Eq. 9.13)}$$

$$IL_{(dB)} = P_{1(dBm)} - P_{2(dBm)} \qquad \text{(Eq. 9.14)}$$

PROBLEMS

1. Determine the noise power generated by a component at 250°K when the receiver bandwidth is 100 kHz.

2. If the bandwidth doubles in problem 1, determine the new noise power.

3. If just the carrier frequency changes from 2 GHz to 6 GHz in problem 1, determine the noise power.

4. If the signal power is 100 mW, and the noise power is 2 mW, determine the SWR.

5. The carrier power with noise is 10 W, the noise power alone is 50 mW. Determine the $(S + N)/N$ ratio.

6. Convert the $(S + N)/N$ ratio of problem 5 into dB.

7. Determine the noise figure (NF) when the input signal power is 200 µW and the noise power is 2 µW. The output signal power is 200 mW, and the noise power is 4 mW.

8. Convert the noise figure of problem 7 into dB.

9. The noise figure of a system is 5, the output signal to noise ratio is 90. What is the input signal to noise ratio?

10. Determine the equivalent noise temperature for the NF of problem 7.

11. Using the spiral chart, convert the NF of problem 7 into T_e.

12. Determine the total noise figure for the following amplifier stages.

Stage	Power Gain	Noise Figure
1	10	1.5
2	15	3
3	30	4

13. Convert the NF_T of problem 12 into T_e.

14. Determine the frequency when the guide wavelength is 2.2 cm, and the width of the waveguide is 3 cm.

15. Convert 100 mW into dBm.

16. Convert 40 dBm into power in mW.

17. Convert 50 W to dBW.

18. Convert the wattage of problem 17 into dBm.

19. Determine the *IL* when the power before component insertion is 20 mW, and after insertion is 16 mW.

20. Determine the *IL* when the power before component insertion is 16 dBm, and after insertion is 14 dBm.

QUESTIONS

1. What parameter is usually measured at microwave frequencies? Why?

2. What are common signal sources for the lab?

3. Describe thermal noise.

4. What is the most important parameter of noise power?

5. Define *noise figure.*

6. Define *equivalent noise temperature.*

7. Which stage contributes the most to noise in a system? Why?

8. Describe how frequency is determined using the slotted line.

9. How do passive wave meters work?

10. Describe the attributes of bolometers.

11. Which detector has the fastest response time? The greatest sensitivity?

12. Describe how the analog power meter works.

13. Define *attenuation*; *insertion loss.* When are they the same?

14. Describe the four basic types of attenuation.

15. Describe how insertion loss measurements are done.

16. What is meant by VSWR?

17. What are S-parameters? How do they relate to the coefficients of reflection and transmission?

18. Describe how a spectrum analyzer works.

19. What is a sweep generator? Where are they used?

20. Describe how a network analyzer works. What is the difference between a scalar and vector network analyzer?

10 MICROWAVE COMMUNICATIONS APPLICATIONS

OBJECTIVES

1. To describe how microwaves can be used in communications applications.
2. To describe basic features of radar systems.
3. To describe features and characteristics of terrestrial communications.
4. To describe characteristics of geosynchronous satellites and how they are used.
5. To compare and contrast characteristics of C-band and Ku-band satellites.
6. To describe characteristics of the high-power DBS satellites.
7. To define acronyms of various wireless technologies.
8. To describe characteristics and features of the new wireless technologies.

10.1 INTRODUCTION

As was shown in chapter 1, microwaves are finding ever expanding areas of applications. This is especially true in the area of communications. As the United States changed from a post-industrial society to the information society, microwave technology has been one of the most important vehicles of this transformation. Microwave, with its vast bandwidth capability, has made telecommunications, and communications in general, that much more feasible. As the world's need to share information and communicate continues to grow, microwave technology will continue to demonstrate it is an effective means of providing for this growth.

We will look at radar in section 10.2. Radar made possible the birth of microwave technology. World War II saw the growth of this technology accelerate rapidly.

At the end of World War II, terrestrial communications evolved with microwave radio links gradually replacing copper wire in the telephone industry. During the 1960s we saw satellite communications begin. Satellite technology (sections 10.4 and 10.5) has been growing and expanding ever since. This is especially true with electronic news-gathering and distribution. We will see how this is accomplished with a look at the C and Ku bands in section 10.5.

In addition, we saw many entertainment and sports channels evolve as direct broadcast satellite (DBS) service to the home became available. Further growth in the Ku band has occurred via VSAT (Very Small Aperture Terminals). High power, small antenna, direct broadcasts (section 10.6) became commercially successful with DirecTV's DSS™ (Digital Satellite System). DBS had already been used around the world, but in the United States, DBS was never commercially successful until the DirecTV system appeared.

Probably the biggest growth is occurring in what is termed PCS (Personal Communication Services) or wireless services. The wireless standards or lack thereof are covered in section 10.7. In section 10.8 we will look at another new growth area, such as the GPS (Global Positioning System).

10.2   RADAR

The word *radar* is an acronym for the description *r*adio *d*etection *a*nd *r*anging coined by the U.S. Navy in 1942. Radar is basically a means of gathering information about distant objects, or targets, by sending electromagnetic waves at them and then analyzing their echoes (reflections). It evolved during the years just before World War II. It was radar that gave birth to microwave technology, as the early workers quickly found that the highest frequencies gave the most accurate results. Higher frequencies produce the best echoes, make it possible to detect smaller targets, and permit the use of smaller antennas.

Figure 10.1 shows typical simplified radar systems. Note that if two antennas are used (one for transmitting and another for receiving), the radar system is called bistatic. On the other hand, if just one antenna is used for both transmitting and receiving, the system is

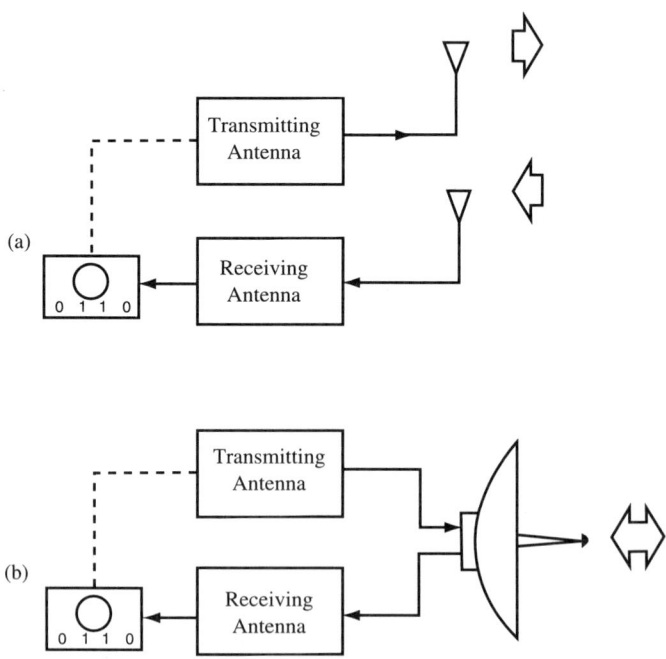

Figure 10.1   Simplified radar block diagrams. (a) Bistatic radar. (b) Monostatic radar.

Table 10.1 Standard Radar-Frequency Letter Band Designations

Band Designation	Nominal Frequency Range	Specific Radar Bands Based on ITU Assignments
L	1,000–2,000 MHz	1,215–1,400 MHz
S	2,000–4,000 MHz	2,300–2,500 MHz
C	4,000–8,000 MHz	5,250–5,925 MHz
X	8,000–12,500 MHz	8,500–10,680 MHz
Ku	12.50–18.00 GHz	13.40–14.00 GHz
		15.70–17.70 GHz
K	18.00–26.50 GHz	24.05–24.25 GHz
Ka	26.50–40.00 GHz	33.40–36.00 GHz

called monostatic. This is the usual form of radar. In general, radars consist of a transmitter, a receiver, a display, and antennas.

The standard radar-frequency bands are shown in Table 10.1. Note that within the common letter-band nomenclature fall the radar bands as designated by the ITU (International Telecommunications Union).

Radar equipment can be divided into two main categories. First is pulsed radar, which works by transmitting a short burst of microwave energy called a *pulse.* The time between transmitted and received pulses gives distance information. The second is continuous wave (CW) radar, which transmits a continuous wave and compares the frequency of the returned echo with that of the transmitted signal. Relative motion between the radar and the target causes a change in frequency, from which velocity information can be obtained. This is called the *Doppler effect.* The Doppler effect can be experienced while standing near a train track. A change in frequency (pitch) of the train whistle occurs as the train approaches and then moves away. There are also radars that combine both of these effects.

Figure 10.2 shows a typical radar system block diagram that utilizes the more sophisticated form of radar, a pulsed radar transmission.

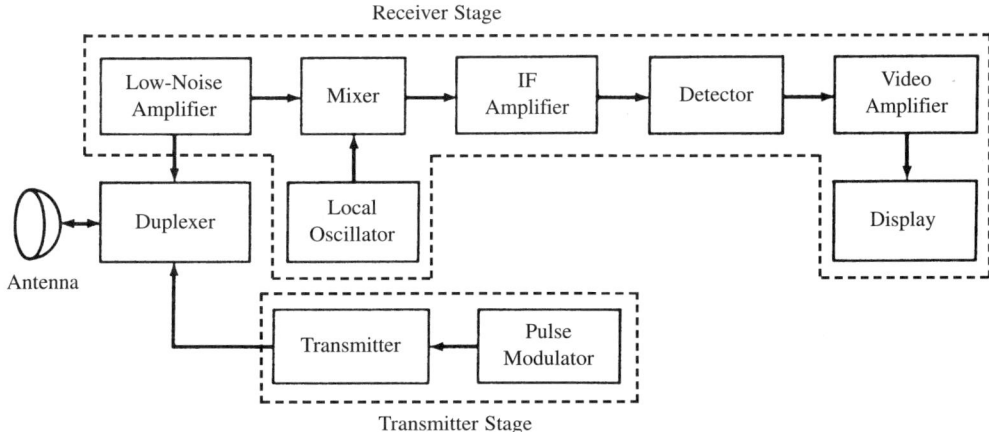

Figure 10.2 Pulsed radar system block diagram.

The blocks perform the following functions:

Duplexer. A circulator (ferrite duplexer) that allows microwave energy to pass from the transmitter to the antenna and from the antenna to the receiver, but not from the transmitter to the receiver.

Transmitter. Generates and amplifies the microwave signal.

Pulse Modulator. Switches the transmitter on and off, causing pulses of microwave power to be transmitted.

Low-Noise Amplifier. Amplifies the weak received echo signal.

Mixer, Local Oscillator. Convert the microwave signal to a lower, more convenient IF.

IF Amplifier. Amplifies the intermediate frequency signal.

Detector. Eliminates the IF, leaving only baseband information.

Video Amplifier. Amplifies the baseband signal.

Display. Presents the received radar echo in some form (usually visual) that is useable to the operator.

Recall that the propagation signal strength diminishes as the square of the distance. Since a target echo takes a total path length that is twice the transmitted signal, its signal strength is a function of the fourth power of the distance and is even weaker.

The power of a return signal is also affected by the size, shape, and composition of the target. *Radar cross sections* (RCS) are defined as the area of a perfectly conducting flat plane, facing the source that would reflect the same amount of power. They can be determined for various targets, as Table 10.2 shows.

The RCS of a target involves multiple scattering processes that are dependent on the operating frequency and the dimensions, geometry, and orientation of the target. Because the RCS of a complex target varies rapidly with small changes in orientation, the RCS is generally characterized by statistical functions.

The United States has two aircraft (the B-2 bomber and F-117A fighter) that employ stealth technology to make the aircraft "practically invisible" to radar. Stealth technology has been under development for more than twenty years and includes a combination of construction materials (radar absorbing) and shapes to make the aircraft difficult to detect.

Table 10.2 Radar Cross Section (RCS) of Familiar Targets

Object	RCS, m^2
Pickup truck	200
Automobile	100
Jumbo jet airliner	100
Large bomber or commercial jet	40
Cabin cruiser boat	10
Large fighter aircraft	6
Small fighter aircraft	2
Adult male	1
Conventional winged missile	0.5
Bird	0.01
Insect	0.00001

The information discussed so far can be quantified and expressed in a propagation equation called the *radar equation*.

$$P_R = \frac{\lambda^2 P_T G^2 \sigma}{(4\pi)^3 r^4} \qquad \text{(Eq. 10.1)}$$

where
P_R = received power in watts

λ = free-space wavelength

P_T = transmitted power in watts

G = antenna gain as a power ratio

σ = radar cross section of the target in m^2

r = range (distance to the target) in m

EXAMPLE 10.1:

A radar transmitter has a power of 20 kW and operates at a frequency of 10 GHz. Its signal reflects from a target 18 km away with a radar cross section of 12.5 m^2. The gain of the antenna is 20 dBi. Calculate the received power.

First:
$$\lambda = c/f$$
$$= 3 \times 10^8 / 10 \times 10^9$$
$$= 0.03 \text{ m}$$
$$G = \text{antilog (dB/10)}$$
$$= \text{antilog (20/10)}$$
$$= \text{antilog 2}$$
$$= 100$$

Then:
$$P_R = \frac{(0.03)^2 (20 \times 10^3)(100^2)(12.5)}{(4\pi)^3 (18 \times 10^3)^4}$$
$$= 1.08 \times 10^{-14} \text{ W}$$

The direction and range to a target are simple to determine. One method is to simply rotate the antenna in such a direction that the main lobe is strongest. Distance can be found by measuring the time between a transmitted pulse and its return. The range can be calculated using equation 10.2.

$$R = ct/2 \qquad \text{(Eq. 10.2)}$$

where
R = distance to target, m

$c = 3 \times 10^8$ m/s

t = time taken for echo to return, s

EXAMPLE 10.2:

A pulse sent to a target returns 12 µs later. How far away is the target?

$$R = \frac{(3 \times 10^8)(12 \times 10^{-6})}{2}$$

$$= 1800 \text{ m}$$
$$= 1.8 \text{ km}$$

Other considerations to be taken into account are time delays in the electronics, the pulse width of the transmitted signal, and whether the period between pulses is less than the time taken for a pulse to return.

As was stated earlier, the Doppler effect can be used to determine the velocity of the target. Equation 10.3 shows how to solve for the velocity of the target when the radar source is stationary.

$$V_r = f_d / 3.9 f_o \qquad \text{(Eq. 10.3)}$$

where V_r = velocity of target (toward or away from radar source), mph

f_d = Doppler shifted frequency, Hz

f_o = transmitted radar frequency, GHz

EXAMPLE 10.3:

A 5 GHz radar receives an echo that is shifted up in frequency by 10 kHz. Calculate the target's velocity in mph.

$$V_r = 10 \times 10^3 / (3.9)(5)$$
$$= 512.8 \text{ mph}$$

or

$$= 825.6 \text{ km/h}$$

EXAMPLE 10.4:

Determine the Doppler shift in Hz when the velocity of the target is 60 mph and the transmitted frequency is 10 GHz.

Rearranging equation 10.3:

$$f_d = 3.9 f_o V_r$$
$$f_d = 3.9(10)(60)$$
$$= 2,340 \text{ Hz}$$

The drawback to Doppler radar is that it cannot account for range. Range can be determined by frequency modulating the continuous wave of Doppler radar (FM-CW). The major application of FM-CW radar is the altimeter (height determination). The FM modulation allows one cycle to be distinguished from another; that is not be possible with CW alone. If the rate of change of frequency with time due to the FM process is known, the time difference between sent and received signals can be readily calculated, as can the height of the aircraft.

The Doppler effect can be utilized in a pulsed radar system. By combining special delay line techniques, the system can be made to determine target velocity and to distinguish moving targets from stationary targets. This improved system is called a moving target indicator (MTI) radar system.

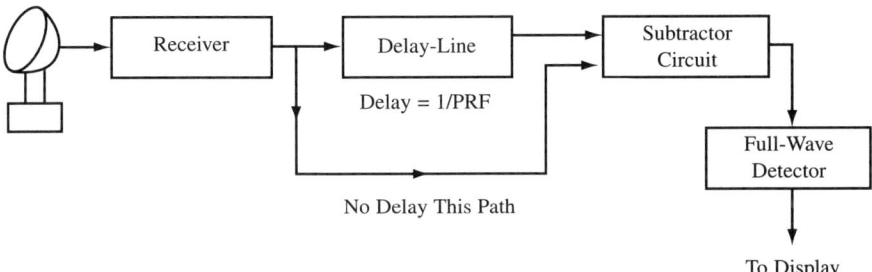

Figure 10.3 MTI radar block diagram.

The phase and amplitude of the echo of stationary targets do not change from pulse to pulse. However, the phase and amplitude of moving target echoes do change. Now, by applying a method in which the echo signal from one pulse is subtracted from the echo return of the previous pulse, everything cancels out *except* the moving targets. Performing this subtraction-cancellation process and displaying only the remainder, the moving target part, constitutes the basis of MTI radar. An MTI radar block diagram is given in Figure 10.3.

10.3   TERRESTRIAL COMMUNICATIONS

Terrestrial communications have evolved since World War II, when the spectrum of higher frequencies for radar was used by communication engineers as well. As stated in chapter 1, by the mid-1960s microwave communications had replaced 40% of the telephone circuits between major cities.

Microwave links perform the same functions as copper or fiber-optic cable, but in a different manner, by using point-to-point microwave transmission between repeaters. Most links operate in the 4 GHz and 6 GHz regions. Propagation is by means of space waves and therefore limited to line-of-sight distances. Typical repeater spacings are close to 50 km, unless located atop a special tower or on a hill or mountain.

A microwave link terminal has a number of similarities to a coaxial cable terminal. The multiplex equipment is similar, if not identical, as is the channel capacity. The similarities are in what is done, and the differences lie in the specific detail of how it is done. A simplified block diagram of a microwave repeater is shown in Figure 10.4. The repeater receives a modulated signal from one repeater and transmits it to the next one, and an identical chain is provided for working in the other direction. The frequencies are kept somewhat different in order to avoid interference. The frequency difference is typically a few hundred MHz at the 4 GHz or 6 GHz operating frequencies.

Note that there is no amplification of the received RF signal. Rather, there is conversion down to an IF (70 MHz) where the bulk amplification takes place. The reason for this down-conversion is that it is easier to low-noise amplify at the IF, rather than at a higher RF. This circuit exemplifies older equipment. Newer transistors have been developed that reduce noise figures, and so RF preamplifiers are now being used. Digital links in microwave are also being used.

With a low-noise-sensitive receiver, the power amplification can be reduced, which ultimately reduces system costs. The familiar parabolic reflector is the most common

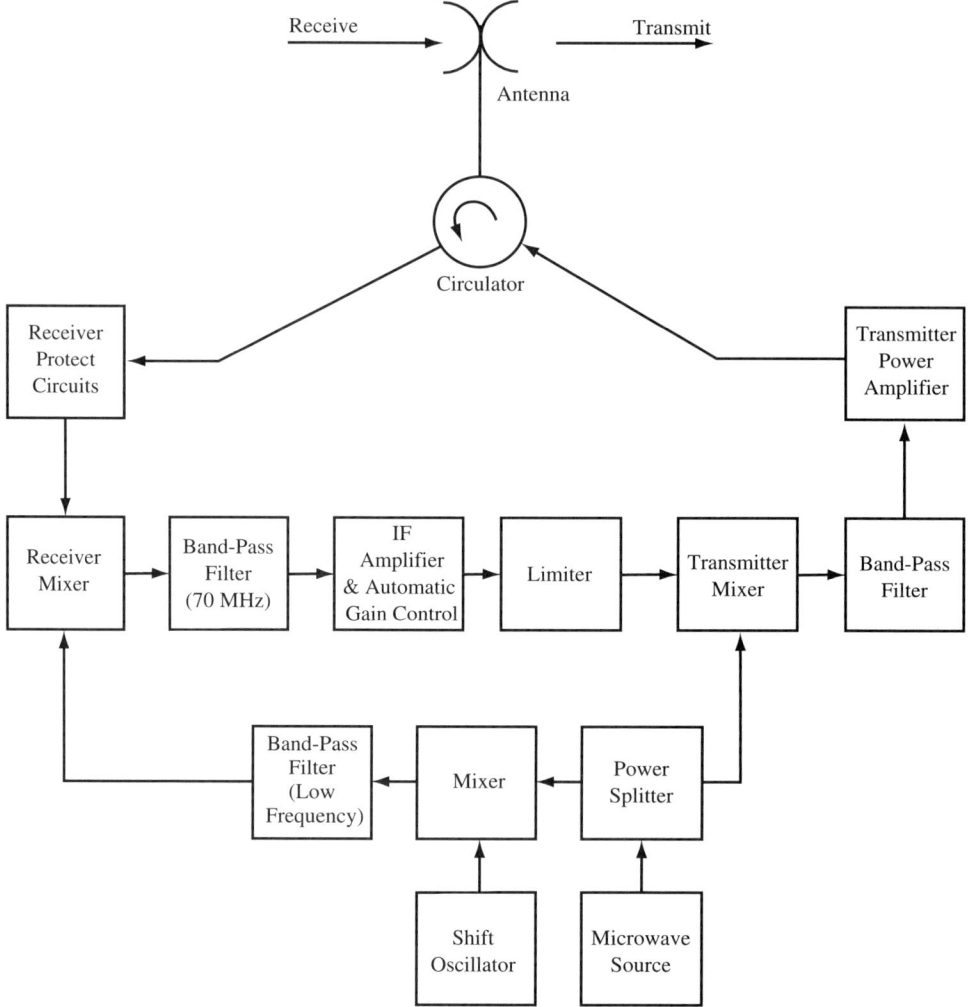

Figure 10.4   Simplified block diagram of microwave repeater.

antenna used with microwave links. They are often covered with a fiberglass cover (radome) to protect the antenna from the elements (Figure 10.5).

The most common antenna for high-density links is the hoghorn antenna, since it is a broadband and low-noise antenna (see Figure 10.6). It uses a tapered horn antenna to feed and impedance match the parabolic reflector. Frequency reuse is accomplished by separating signals through vertical and horizontal polarization. Hoghorn antennas are widely used in the United States for many types of microwave links.

The receiver mixer is usually a Schottky barrier diode, a very low-noise device. The IF amplifier is a low-noise, ultralinear, very broadband transistor amplifier. Varactor diodes are often used in the transmitter mixer, to convert the 70 MHz IF back up to the transmitting RF microwave frequency. The active element of the output power amplifier of the link varies from a reflex klystron in older equipment to a Gunn diode or an IMPATT diode in

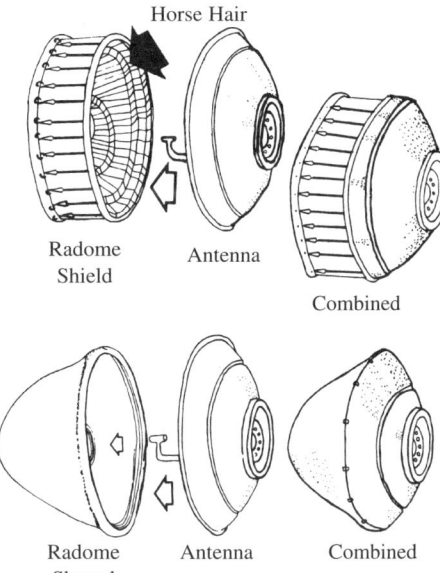

Figure 10.5 Antenna equipped
with radome shield or shroud.

Figure 10.6 Hoghorn parabolic antenna. (Photo courtesy of Andrew Corporation.)

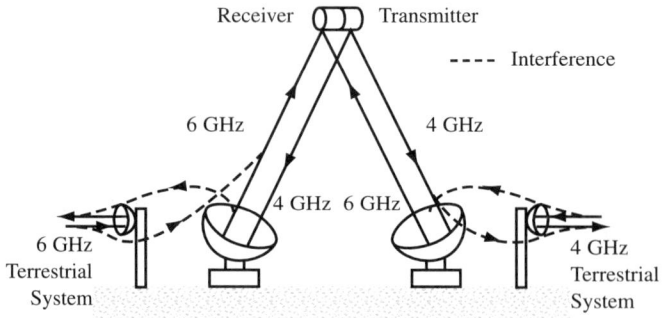

Figure 10.7   Possible interference between satellite and terrestrial links.

more modern equipment. MESFET power amplifiers are also common. Above 6 GHz, and for greater power, the TWT is used.

The typical number of carriers (in each direction) in a microwave link is at least four and sometimes as many as twelve. There are normally 600 to 2,700 channels per carrier.

The towers for the links are typically 25 meters high, depending on the terrain. Rooftops are often used. The links contain their own power supplies to power the onboard electronics. Antenna alignment is one of the many items checked at periodic maintenance visits. Often one repeater location can be used to check out other repeater locations via onboard diagnostic software. This remote testing reduces maintenance costs.

One possible interference is from the ground-to-satellite signal path in the C-band range of 4 to 6 GHz. These C-band systems obviously interfere with terrestrial line-of-sight repeaters and vice versa (see Figure 10.7). As satellite frequencies are pushed into the Ku band, this terrestrial interference problem becomes less of a concern.

10.4   SATELLITE COMMUNICATIONS

Arthur C. Clarke, in an article that appeared in *Wireless World* in February 1945, first postulated the potential of satellites for communications. This class of satellites became known as *geostationary* (also called *geosynchronous*). The orbit for such a satellite is shown in Figure 10.8.

The period of the satellite's rotation depends on its distance from the earth. The closer it is, the more rapidly it must rotate the earth, to offset the effects of gravity. Satellites that orbit in lower orbital paths are being used increasingly in the new PCS (Personal Communication Services). However, they disappear from the horizon rather quickly and must be tracked. (More on this in section 10.7.)

A satellite located 36,000 km above the earth's surface (at the equator) rotates at the same rate (angular velocity) as the earth. Hence, it appears to be stationary or fixed above one spot on earth. Thus the terms *geosynchronous* (its rotation equals the earth's) and *geostationary* (it appears fixed in location). Satellites parked in this orbit are available to all Earth stations within their shadow 100% of the time (excluding sun spots, weather, and the like). The shadow of a satellite includes all Earth stations that have a line-of-sight path to the satellite and within the radiation pattern of the satellite's antennas. Even though the distance to these satellites is relatively great in terms of signal loss and time delay, the advantages of this application outweigh any drawbacks.

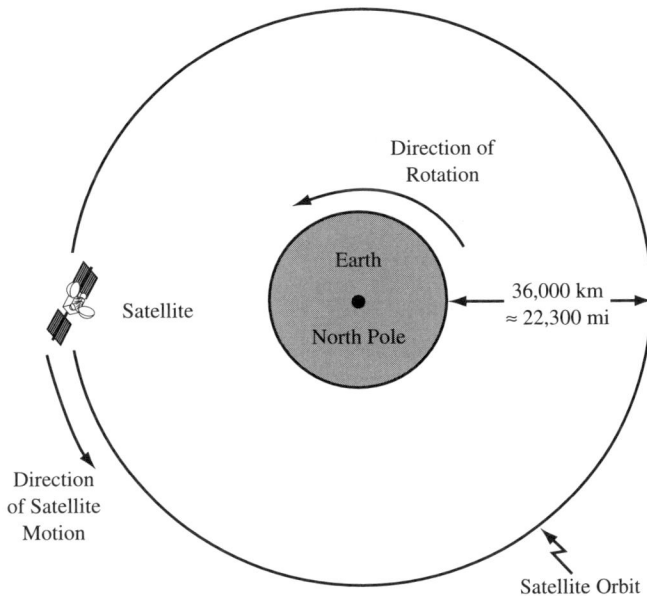

Figure 10.8   Geostationary satellite orbit.

Communications satellites act as repeaters. That is, they receive signals in one frequency band and retransmit in another. Typically these satellites (those in the C band) have relatively low-power amplifiers (5–10 W) with the newer Galaxy V satellite having 16 W amplifiers. The lower power is required to conserve the solar energy on board the satellite. To compensate for the lower power, they have high-gain, highly directional antennas, and low-noise receivers. An Earth-based satellite receiver sees a very "cold" noise source when looking at the satellite, therefore, it is technically possible for the satellite to output lower power while the Earth receiver still maintains a good signal-to-noise ratio. Ku-band satellites have even higher-power amplifiers on board. In this case the higher power allows for smaller Earth receiving antennas.

The use of satellites as a means of relaying information has matured very rapidly. Applications include communications (telephone, radio, television, data, and facsimile); navigation; weather forecasting; Earth resource management; and defense (reconnaissance, aircraft and missile detection, guidance, and control).

One of the largest users of satellite technology is INTELSAT (International Telecommunications Satellite Organization). INTELSAT is a nonprofit cooperative of 114 nations. It owns and operates a global system of communications satellites that provides international telecommunications services to more than 172 countries and territories. It also provides domestic telecommunications services to more than 30 nations.

The INTELSAT satellites are in geosynchronous orbit (equatorial orbit) over the Atlantic, Pacific, and Indian Oceans. These satellites make INTELSAT a major provider of transoceanic telephone, data, and television service.

Since its creation in 1964, INTELSAT's satellites have brought history-making events to televisions and radios all around the world. More than half of all international telephone calls and nearly all transoceanic television transmissions are carried by this system. Their first satellite was INTELSAT I launched in 1965. It had a capacity of 240 circuits or one TV circuit. INTELSAT VII was launched in 1992. It has frequency bands in both the C and Ku bands. It uses solid-state power amplifiers in the C band and TWTs for power in the Ku band. It has a capacity of 18,000 circuits and three TV circuits.

The INMARSAT (International Maritime Satellite Organization) provides a similar role for ships at sea. More than 20 INMARSAT Earth stations are now in service.

In addition to the aforementioned, there are numerous regional systems. They operate as a kind of mini-INTELSAT system in Indonesia, the Philippines, Europe, and the Middle East.

In North America, Telesat Canada was established in 1969. The United States followed soon afterward with the Westar system (1974) and then the competing Comstar, Satcom, SBS, STC, and Telstar networks. These are known as domestic satellite services.

Many other countries now have domestic satellite systems using their own satellites. Among these are the countries of the former U.S.S.R., China, Indonesia, India, the Scandinavian countries, Colombia, Japan, and Australia.

10.5   DOMESTIC SATELLITE BANDS (C AND KU)

C Band

The ordinary terrestrial broadcasting system is, of course, not the only way to transmit television signals. Communications satellites have become the major means by which networks send programs to their affiliate stations and by which specialized services such as news, sports, and movie channels send material to cable-television operators. They are also used for direct-to-home broadcasting.

The C band (4–6 GHz) was selected because at the time it was the most technically feasible way to go. C-band terrestrial links had already been in use, so the necessary technical base for electronics and microwave components was already in place.

The satellite's receiving and transmitting antennas are directional and have a propagation pattern. The center of this pattern, where maximum gain occurs, is called the *boresight*. The pattern of the transmitting antenna becomes particularly important when attempting to determine the strength of a satellite signal reaching the earth. As the transmission leaves the satellite, it forms a beam that covers a specific area of the earth. This pattern, referred to as a *footprint,* is generally shown on a map with contour lines that connect equal levels of EIRP (effective isotropic radiated power). The levels of EIRP (from the spacecraft) are expressed in "decibels above one watt" (dBW). The ground antenna offers an additional gain factor.

In a satellite like Hughes Communications Galaxy V, a newer C-band satellite, the gain provided by the antenna along with the power amplifier yields a 37 dBW signal over most of the 48 contiguous states (CONUS). This provides a signal level that is more than 5,000 times greater than a theoretical isotropic antenna could produce. A footprint map of a Galaxy V vertical polarized signal is shown in Figure 10.9.

The Galaxy V replaced Westar V at 125° W longitude. It has 24 C-band transponders operating at 16 W each. The power amplifiers are traveling wave tubes. There are actually 30 power amplifiers on board the Galaxy V but only 24 are used at any given time. This provides available spares in case of amplifier failure. The satellite is a Hughes HS 376 model spacecraft. The HS 376 satellite model has established an industry benchmark and is the most-purchased satellite model in the world.

Peak EIRP = 40.3 dBW

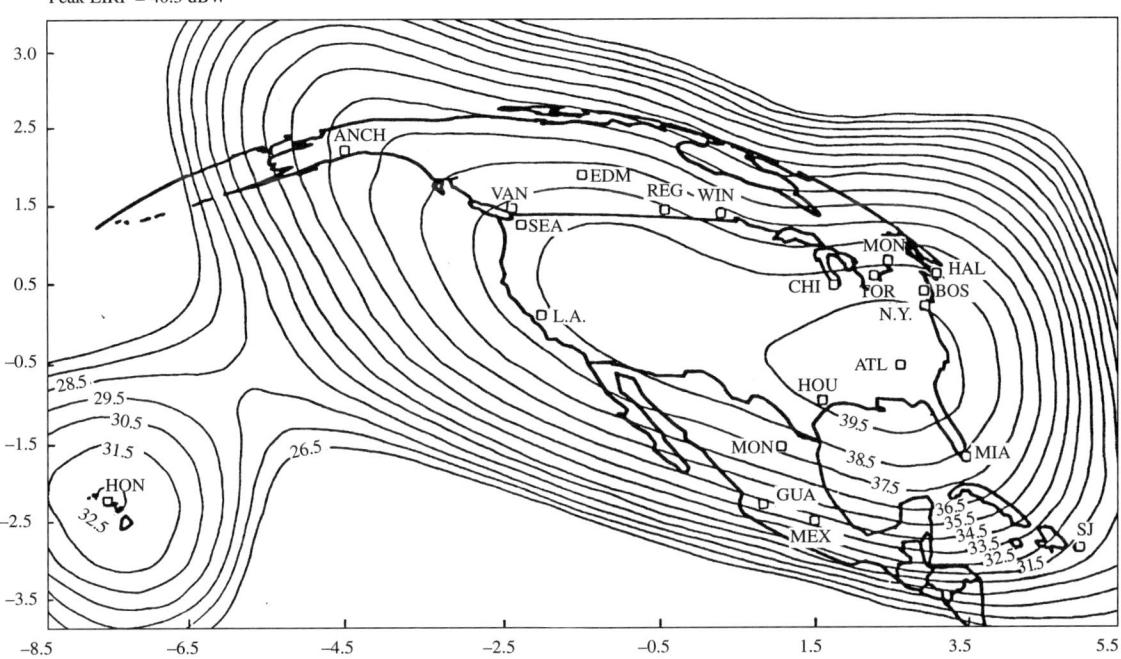

Figure 10.9   Vertical polarization footprint of Hughes Communications Galaxy V at 125° W longitude.

The typical receiving system, a TVRO (television receive only), includes the antenna, a low-noise amplifier (LNA), and a video receiver/demodulator or integrated receiver demodulator (IRD). The IRD is commonly called the "indoor" electronics. The uplink frequency is near 6 GHz, to help circumvent interference with terrestrial links and provide for duplex operation, and the downlink is near 4 GHz.

Table 10.3 shows the typical downlink transponder frequencies. Each has a bandwidth of 36 MHz with a 4 MHz guard band between transponders. Because orthogonal polarization (vertical and horizontal) is used, with odd and even channel numbers having their polarization 90° apart, the frequency ranges can overlap (duplex operation).

Table 10.3   C-Band Downlink Transponder Frequencies

Channel	Center Frequency (MHz)	Channel	Center Frequency (MHz)	Channel	Center Frequency (MHz)
1	3720	9	3880	17	4040
2	3740	10	3900	18	4060
3	3760	11	3920	19	4080
4	3780	12	3940	20	4100
5	3800	13	3960	21	4120
6	3820	14	3980	22	4140
7	3840	15	4000	23	4160
8	3860	16	4020	24	4180

Note, as an example, that channels 1 and 3 (odd channels) are separated by 40 MHz, the required 36 MHz plus the 4 MHz guard band. At the same time, channel 2 is utilizing a center frequency of 3740 MHz, in the middle of channel 1's bandwidth. No interference occurs, however, since channel 2 is of opposite polarization. This is true for all odd-even pairs.

The first active element in a receiver, the LNA (low-noise amplifier), also generates unwanted signals that contribute to system noise. The angle of elevation of the antenna contributes to noise as well. The lower the angle, the more Earth noise is picked up.

All of these items add up to become the *G/T* (antenna gain minus antenna system temperature), which provides a measure of merit of an antenna system. Systems are rated with such a ratio. By utilizing lower-temperature (°K) (but more costly) LNAs, the size of the antenna can be made smaller and still have optimum results. This all relates to the signal-to-noise ratio (*S/N*) discussed in chapter 9.

Figure 10.10 shows the diagram of a typical down-converter for a TVRO system. The LNA and mixer converts the C-band signals down to an IF in the L-band 950 to 1450 MHz range. (In all newer systems the LNA and mixer are found in one unit known as the LNB, or low-noise block, down-converter). The down-converter electronics are located at the dish after the feedhorn pickup. The IF is connected to the indoor electronics via a coaxial cable for further processing by a set-top unit (IRD) for viewing on a TV. Low-noise pre-amplification is best accomplished by the use of low-noise GaAs PHEMT or MESFET devices. The mixers can use either matched Schottky diode pairs in a passive balanced mixer or a silicon MMIC self-oscillating mixer. IF amplification can be provided by discrete bipolar transistors in a feedback amplifier or by silicon MMIC amplifiers. PIN diodes or GaAs MMICs are used for RF and IF switching in the dual polarization configurations.

Table 10.4 shows a representative sample of the programming offered on the C band. The first alpha character and number represent the satellite that offers this programming (G = Galaxy, T = Telstar, F = Satcom, S = SpaceNet, and so forth). The last number shows the specific channel (of the 24) on which the programming is found. (Note that this table is provided only as an example for interpreting where programming is located. The satellite

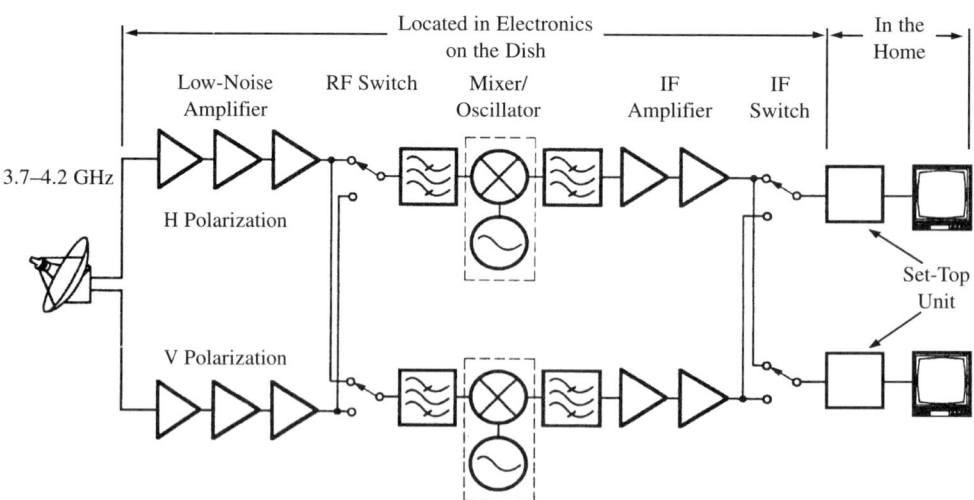

Figure 10.10   Typical C-band TVRO receiver.

Table 10.4 Representative Video Services on the C Band (All subject to change)

Video Service	Location	Video Service	Location
Cinemax 1	G1-19	ESPN	G5-14
Cinemax 2	T2-21	Prime Ticket	F1-07
Disney East	G5-01	SportSouth	S3-19
HBO 1 East	G5-15	Nashville Network	G5-18
HBO 2 West	T2-14	ABC N.Y.	F2-04
Action PPV	F4-02	CBS Denver	F1-02
CNN	G5-05	FOX West	T3-23
CNBC	G5-13	NBC Denver	F1-14

spacecraft have a finite lifespan, so new satellites are placed in orbit as replacements. Also, channel assignments may vary over the years.)

Figure 10.11 shows the azimuth and elevation angles for various Earth stations located in the northern hemisphere. To know where to point a dish during its installation, you need to know the longitude and latitude of the city where the installation is taking place. Next, you need to know the longitude of the satellite from which you wish to receive programming.

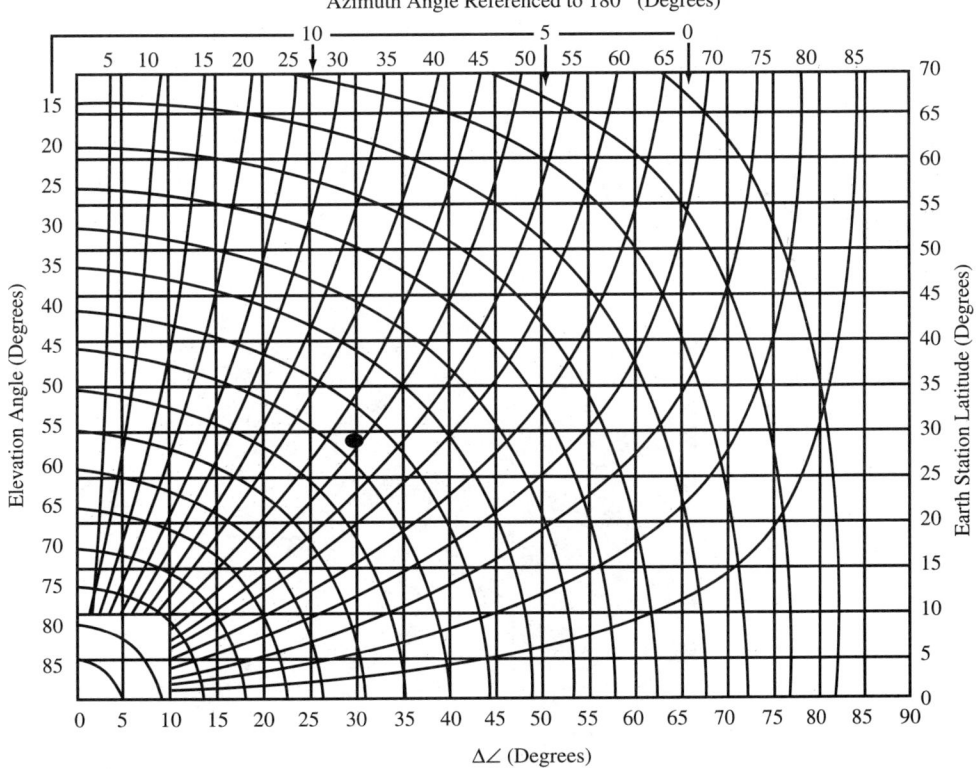

Figure 10.11 Azimuth and elevation angles for Earth stations located in northern hemisphere (referenced to 180°).

Suppose you are located in Houston, Texas (longitude 95.5° W and latitude 29.5° N), and wish to align your dish to Galaxy V located at longitude 125° W. The steps to find the azimuth and elevation angle are as follows:

1. Find ΔL, the difference in longitude, as $\Delta L = 125 - 95.5 = 29.5°$.
2. Locate the intersection of ΔL (29.5°) and the latitude of the Earth station (29.5° N). This is marked on Figure 10.11.
3. Read off the angle of elevation as 43°, and the azimuth as 49° W of south.

Ku Band

As was shown in the previous section, most home satellite TV and cable company systems receive satellite channels broadcast in the C band. Now satellite channels in the Ku band (using 12 to 14 GHz) are being used almost to the same extent. The Ku band was primarily used for DBS (direct broadcast system) to the home and nonvideo transmissions during the 1980s. Now all of that has changed.

In the United States, NBC (the National Broadcasting Company) was one of the first of the major players to opt to use the Ku band for distribution of their programming to affiliates. Most of their programming is on the Satcom K2 Ku-band satellite. Satellite news-gathering (SNG) and electronic news-gathering (ENG) agencies have used this band for some time. These systems can be mounted on mobile trucks for use by news-gathering agencies (Figure 10.12). The system also comes in a transportable mode that can be taken on an airplane for use in news-gathering in the field.

In addition, Europe, Japan, and Australia have used the Ku band for DBS to the home. Educational television is being done over the Ku band. Retail outlets, banks, and the like are using the band for two-way transfer of data and video.

Besides the obvious difference in frequencies used, there are other discernable differences between the Ku band and the C band. With the higher frequency, antennas can be

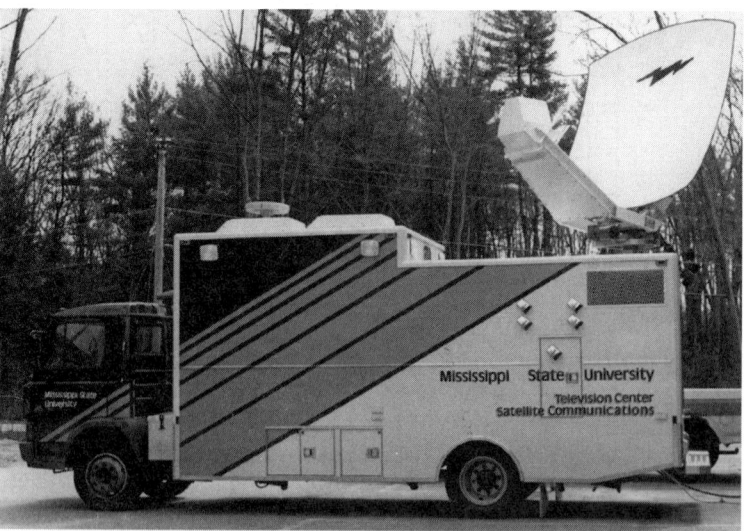

Figure 10.12 Truck-mounted Ku-band satellite news-gathering system. (Photo courtesy of Andrew Corporation.)

made smaller. (No more 8- to 10-foot dishes in the backyard.) Dishes can now be 3 feet or even smaller with the higher power DBS satellites. Another reason for the smaller antennas is that Ku transponders have higher-power amplifiers than their C-band counterparts. The typical power for Ku-band systems is 40 to 60 watts. The higher-power DBS system uses power amplifiers from 100 to 120 watts. This higher power is required, because the drawback to the Ku frequency band is that its wavelength is extremely small. This causes the signal to be severely attenuated by water molecules. Thus, it suffers from what is known as *rain fade.* This problem can be compensated for by increasing amplifier power, using larger receiving antennas, using spot beams to put more power into areas where there is an obvious concentration of rain (for example, the Southeastern United States), or by using circular polarization (rather than orthogonal). The higher-power DBS systems use circular polarization, since it is less affected by rain.

Because of the higher uplink frequency (around 14 GHz) and the higher downlink frequency, there is no interference with C-band terrestrial links. The downlink is around 12 GHz with 32 channels (32 transponders) using orthogonal polarization for frequency overlapping. The LNB and other receiver elements for this band are very similar to those shown in Figure 10.10.

VSAT

VSAT (Very Small Aperture Terminal) is a service also being done in the Ku band. VSAT systems employ small antennas (about 3-foot diameters) that sit atop businesses and retail stores (see Figure 10.13). They are used by convenience stores, auto parts distributors, banks, and the like. They offer two-way satellite communications, usually back to a hub

Figure 10.13   VSAT antenna. (Photo courtesy of Andrew Corporation.)

or headquarters. Retail information or data transfer can then be processed without the need for conventional phone lines.

Video conferencing can be done this way. Employer training can be done as a hub can distribute the training to its retail outlets in an asymmetrical form (one source feeding many sites).

The Public Broadcasting System (PBS) uses VSATs in public schools. In this way, schools across the country can exchange information and ideas.

10.6   DIRECT BROADCAST SATELLITES (DBS)

DBS has been available since the 1980s, but the political atmosphere in the United States was not conducive to the creation of a TV subscription service designed to compete with U.S. cable system operators. Because of this and other problems, early DBS ventures such as Sky Cable and Sky Pix failed. The Cable Act of 1992 may have helped DBS operators negotiate rates in parity with those offered to the cable industry.

In 1991 Hughes Communications announced its intention to enter into the direct-to-home (DTH) television business. In turn, DirecTV was formed as a unit of GM Hughes Electronics. With the launch of a Hughes HS 601 satellite in December 1993, high-power DBS arrived. A second DBS satellite (DBS-2) was launched in August 1994. A request for a third DBS satellite has been asked of the FCC by DirecTV.

Recall that other Ku-band satellites employ orthogonal polarization for frequency overlapping. This effectively doubles the channel capacity. By going to a digital architecture, which DirecTV does, the digital satellite system (DSS™) can utilize video and audio compression techniques with laser-disc quality. DirecTV's digital architecture allows for video compression that complies with the MPEG-2 standards (Motion Picture Experts Group). By using compression ratios of 5 to 7, 150 channels or more of programming are available from the two satellites collocated at 101° W longitude.

The first satellite, DBS-1, shares part of its 16 transponders with USSB (United States Satellite Broadcasting). DirecTV utilizes 11 transponders for 50 to 60 channels of programming, while USSB uses five transponders for 25 to 30 channels. DBS-2 is used exclusively by DirecTV to provide the service with its full capacity.

The HS 601 satellite (Figure 10.14) is a 4 kW body-stabilized satellite specially designed for high-power communications applications. For U.S. DBS, the HS 601 uses technology similar to the 50 to 60 watt traveling wave tube amplifiers for the Fixed Satellite Service (FSS) satellites. The TWTs for DirecTV have been scaled up in power level, but otherwise employ m-cathode tubes with cooling via heat pipes. The tubes are not radiantly cooled to space as in previous high-power designs.

These satellites contain sixteen 120-watt TWTs that can be combined to form eight pairs at 240 watts of power. This higher power can also be utilized for high-definition television (HDTV) transmission. They also utilize "shaped aperture" antenna technology to precisely control the transmit signal distribution over CONUS (the 48 continental states). More power is delivered to the Eastern United States than to the drier areas west of the Mississippi River. More power is also directed to the Southeast than the Northeast. The downlink EIRP varies over CONUS from about 48 dBW to 54 dBW, with the beam peak near Tallahassee, Florida.

Uplink frequency is 17.3 to 17.8 GHz, which is then translated to 12.2 to 12.7 GHz and retransmitted via a CONUS beam for the downlink in the Broadcast Satellite Service (BSS) band. The uplink power is accomplished with high-power klystron tubes feeding 13-meter antennas over a short waveguide run.

Figure 10.14   Hughes HS 601 DBS satellite. (Photo courtesy of DirecTv, Inc.)

Each satellite transponder carries a single quadrature phase shift keying (QPSK) carrier with a Reed-Solomon (RS) outer code and a convolutional inner code. This "concatenated" coding combination gives some of the most powerful error-control performance ever achieved in consumer products and permits the use of small dishes while maintaining high availability levels.

Table 10.5 shows a comparison of the DBS-1 to the Galaxy IV satellite. Note that each satellite is rated at 4,000 watts of power.

Table 10.6 shows the characteristics of the two DBS satellites. Note that DBS-1 utilizes left-hand circular polarization (LHCP), while DBS-2 uses right-hand circular polarization (RHCP). Circular polarization is less affected by rain than orthogonal polarization.

Table 10.5   DBS-1 Satellite Compared to Galaxy IV

	Galaxy IV (4 kW)	DBS-1 (4 kW)
C-band Transponders:		
Active (Spare)	24 (6)	0 (0)
Ku* Transponders:		
Active (Spare)	24 (6)	16 (16)
Ku Transponder Power	50 w	120 w

*DBS band 500 MHz above the Galaxy band

Table 10.6   DBS Satellite Characteristics

	DBS-1	DBS-2
Launch	December 1993	August 1994
Satellite Ready	January 1994	August 1994
Polarization	LHCP	RHCP
Channels	2,4,6,....,32	1,3,5,....,31
Orbit Location	101.2°	100.8°

Figure 10.15 DirecTV's 18-inch dish with "smart card" IRD. (Photo courtesy of DirecTv, Inc.)

On the receiving end is an 18-inch dish, an IRD using "smart card" technology and modern cryptographic techniques, with full remote capability. This is shown in Figure 10.15. These units are being made by Sony and GE as well.

In addition to DirecTV, other competing DTH services are available or planned. Primestar beams entertainment programming to viewers via GE Americom's Satcom K1, and utilizes 45 watt transponders. Space Systems/Loral was awarded a contract from Tempo, a subsidiary of Tele-Communications, Inc., to build two high-powered DTH broadcast TV satellites.

Tempo will have its satellites delivered in orbit in June and October of 1996. The satellites will be designed to use digitized and compressed video signals to provide a ten-fold increase in the number of channels. Each Tempo satellite will carry 32 high-powered 107 W transponders, switchable to 16 transponders at 200 W.

10.7   WIRELESS SERVICES AND STANDARDS

The new wireless voice and data communications technologies are the most exciting advances in mobile communications since the analog cellular telephone technology was introduced in the late 1960s. It is predicted that 60% of U.S. households will be using wireless services by 2005.

Wireless networks consist of microcells that connect people with truly global, pocket-size communication devices—telephones, pagers, personal digital assistants (PDA), and modems. A typical cellular cell has a 0.5 to 10 mile radius with a 100 watt transmitter. The handheld transmitter has a power of less than 3 W. The interconnect is done via microwave.

A PCN/PCS (Personal Communications Network/Personal Communications Service) cell has a radius less than 0.25 miles with a 0.01 to 1 W transmitter. The handheld transmitter power is less than 10 mW. The interconnect could be done via twisted-pair lines by local exchange carriers (LEC). The LECs would like to be the carriers of choice; however, cable TV's parallel broadband network could be used as well. It is still too early to tell how this link will be provided.

PCS will allow for each structure to employ several of their microcells. Many buildings or structures would thus become "cell sites"; that is why much lower power is required by PCS.

When these systems are fully implemented, individuals will be able to place or receive a call to and from anywhere in the world through a personal radio telephone about the size of a deck of cards. They will also be able to access a wide range of advanced intelligent network features, including numerous data and message features.

In 1994, the FCC allocated more than 160 MHz of the spectrum in the 900 MHz band and the 1.8 to 2.2 GHz band for a range of PCS that will supplement the existing cellular, paging, and data services. More than 220 firms requested experimental licenses to begin PCS. With the explosion and growth of wireless services come new markets, new technologies, new competitors, and multiple competing standards. Understanding this multiplicity of standards is one of the greatest challenges facing technical personnel in wireless communications.

Wireless personal communications standards for North America were once simple: Cordless was CT-0 (an analog 46/49 MHz standard) and cellular was AMPS (the North American analog cellular phone standard operating at 800 MHz). The situation in Europe was far more complex; every country had its own standard (NMT-450 in Scandinavia, NETZ-C in Germany, TACS in the UK, R2000 in France). While cordless was nominally CT-0, different countries used different frequency plans, which led to a plethora of new "standards." These include, but are not limited to, CT-1, CT-1+, CT-2, CT-3, and DECT, as well as NMT-900, ETACS, NAMPS, GSM, TDMA, E-TDMA, CDMA, B-CDMA, DCS-1800, PHP, and JDC in cellular applications. Table 10.7 lists most of these acronyms and what they refer to.

Technologically, some proposed PCS standards are quite similar to cellular: DCS-1800 is a GSM (Europe's digital cellular standard at 1.8 GHz) standard, while CDMA is widely proposed for 800 MHz cellular and 1.8 GHz PCS networks. Some PCS standards have their roots in cordless technology.

Digital systems can offer better performance and/or higher capacity than analog systems. For cellular systems, the case for new digital standards is based on the need for more capacity—typical digital cellular systems provide 3 to 20 times improvement in the use of the frequency spectrum. It is less expensive to add a digital channel than an analog channel to a network, because one set of expensive radio hardware supports multiple channels.

Different trends in the United States and Europe emphasize that regulatory bodies either cannot or will not keep up with the industry, so multiple standards arise. In the United States, the FCC lacks the staffing and the inclination to set standards (caused by a smaller budget and deregulation). The lack of standards in the United States invited innovation among system designers. However, both TDMA and CDMA have been debated as to which is a better standard for digital cellular. Both TDMA and CDMA will be used in the United States. Only time will tell when and if standards are adopted.

Figure 10.16 shows a typical analog cellular antenna. It is a panel antenna suitable for high-traffic areas in the 820 to 960 MHz range.

Table 10.8 shows several characteristics for analog cellular telephones. The table gives frequency range, which form of multiple access is used, the duplex method, number of

Table 10.7   Deciphering the PCS Acronyms

AMPS	Advanced Mobile Phone Service
B-CDMA	Broadband CDMA
CT-X	Cordless Telephone "X" (0,1,1+,2,3)
CDMA (IS95)	Code-Division Multiple Access
CDPD	Cellular Digital Packet Data
DECT	Digital European Cordless Telephone
DCS-1800	Digital Communications System 1800 (GSM at 1.8 GHz)
E-TACS	Extended TACS
E-TDMA	Extended TDMA
FDMA	Frequency-Division Multiple Access
GSM	Groupe Speciale Mobile (now Global System for Mobile Communications)
IMTS	Improved Mobile Telephone Service (precursor to AMPS)
ISM	Industrial, Scientific, Medical Bands
IS54	Interim Standard 54 (dual-mode TDMA/AMPS)
IS95	Interim Standard 95 (dual-mode CDMA/AMPS)
JDC	Japanese Digital Cellular
N-AMPS	Narrowband AMPS (more analog channels)
NMT	Nordic Mobile Telephone (analog cellular in Scandinavia)
PCN	Personal Communications Network
PCS	Personal Communications Services
PHP	Personal Handy Phone (PCS in Japan)
PSTN	Public Switched Telephone Network
PTT	Post, Telephone, Telegraph (telephone company in Europe)
TACS	Total Access Communication System (UK's analog cellular)
TDMA	Time-Division Multiple Access

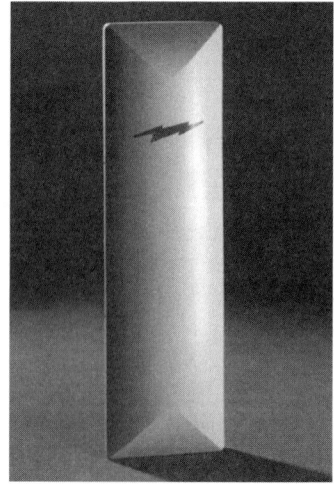

Figure 10.16   Analog cellular panel antenna (820–960 MHz). (Photo courtesy of Andrew Corporation.)

Table 10.8 Analog Cellular Telephones

Standard	AMPS	TACS	NMT
Mobile Frequency Range (MHz)	Rx: 869–894 Tx: 824–849	ETACS: Rx: 916–949 Tx: 871–904 NTACS: Rx: 860–870 Tx: 915–925	NMT–450: Rx: 463–468 Tx: 453–458 NMT–900: Rx: 935–960 Tx: 890–915
Multiple Access Method	FDMA	FDMA	FDMA
Duplex Method	FDD	FDD	FDD
Number of Channels	832	ETACS: 1000 NTACS: 400	NMT-450: 200 NMT-900: 1999
Channel Spacing	30 kHz	ETACS: 25 kHz NTACS:12.5 kHz	NMT-450: 25 kHz NMT-900:12.5 kHz
Modulation	FM	FM	FM

channels, channel spacing, and modulation type. Tables 10.9, 10.10, and 10.11 give data for digital cellular telephones, analog cordless telephones, and digital cordless telephones, respectively.

A simplified generic block diagram of the radio portion of a typical digital wireless personal communicator is shown in Figure 10.17. The T/R switch is used to select the transmit or receive function. The receiver is a typical heterodyne stage. The IF subsystem

Table 10.9 Digital Cellular Telephones

Standard	IS-54 North American Digital Cellular	IS-95 North American Digital Cellular	GSM Global System for Mobile Communications	PDC Personal Digital Cellular
Mobile Frequency Range (MHz)	Rx:869–894 Tx:824–849	Rx:869–894 Tx:824–849	Rx:935–960 Tx:890–915	Rx:810–826 Tx:940–956 Rx:1429–1453 Tx:1477–1501
Multiple Access Method	TDMA/FDM	CDMA/FDM	TDMA/FDM	TDMA/FDM
Duplex Method	FDD	FDD	FDD	FDD
Number of Channels	832 (3 users/channel)	20 (798 users/channel)	124 (8 users/channel)	1600 (3 users/channel)
Channel Spacing	30 kHz	1250 kHz	200 kHz	25 kHz
Modulation	$\pi/4$ DQPSK	BPSK/0QPSK	GMSK (0.3 Gaussian Filter)	$\pi/4$ DQPSK
Bit Rate	48.6 kb/s	1.2288 Mb/s	270.833 kb/s	42 kb/s

Table 10.10   Analog Cordless Telephones

Standard	CT-0 Cordless Telephone 0	JCT Japanese Cordless Telephone	CT-1/CT-1+ Cordless Telephone
Mobile Frequency Range (MHz)	2/48 (U.K.) 26/41 (France) 30/39 (Australia) 31/40 (The Netherlands, Spain) 46/49 (China, Taiwan, S. Korea, USA) 48/74 (China)	254/380	CT-1: 915/960 CT-1+: 887-932
Multiple Access Method	FDMA	FDMA	FDMA
Duplex Method	FDD	FDD	FDD
Number of Channels	10, 12, 15, or 20	89	CT-1: 40 CT-1+: 80
Channel Spacing	40 kHz	12.5 kHz	25 kHz
Modulation	FM	FM	FM

Table 10.11   Digital Cordless Telephones

Standard	CT-2/CT-2+ Cordless Telephone 2	DECT Digital European Cordless Telephone	PHP Personal Handy Phone	DCS-1800
Mobile Frequency Range (MHz)	CT-2: 864-868 CT-2+: 930/931-940/941	1880-1990	1895-1907	Rx: 1805-1880 Tx: 1710-1785
Multiple Access Method	TDMA/FDM	TDMA/FDM	TDMA/FDM	TDMA/FDM
Duplex Method	TDD	TDD	TDD	FDD
Number of Channels	40	10 (12 users/channel)	300 (4 users/channel)	750 (16 users/channel)
Channel Spacing	100 kHz	1.728 MHz	300 kHz	200 kHz
Modulation	GFSK (0.5 Gaussian Filter)	GFSK (0.5 Gaussian Filter)	$\pi/4$ DQPSK	GSMK (0.3 Gaussian Filter)
Bit Rate	72 kb/s	1.152 Mb/s	384 kb/s	270.833 kb/s

converts the IF signal to baseband and includes quadrature demodulation for the QPSK modulation format. The baseband processor does the final decoding into either data or reconstructed voice.

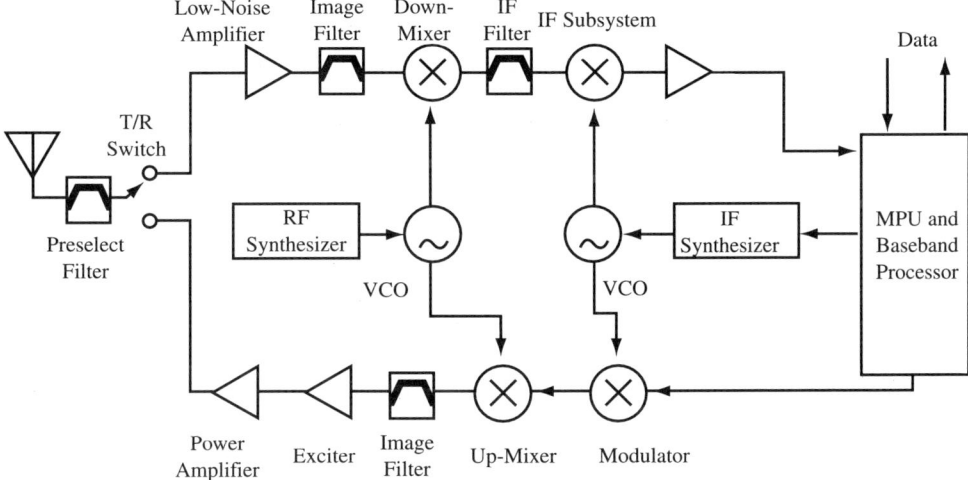

Figure 10.17   Simplified radio block diagram of PCS digital communicator.

On the transmit side, the coded signal from the baseband processor is applied to the modulator, which imparts the modulation to the same IF used in the receiver. The IF is converted up to the transmit frequency, amplified in the power amplifier chain, and routed to the antenna through the T/R switch. Some FDD systems still use a duplex filter in place of the T/R switch.

Mobile Satellite Services

The frequency spectrum has already been allocated, and several systems have been proposed, for low-Earth-orbit satellites (LEOs) and medium-Earth-orbit satellites (MEOs). These will be satellite-based competitors or supplements to the cellular systems and existing geostationary-Earth-orbit satellite systems (GEOs). GEO satellite systems have a drawback in that the distance is so great (36,000 km) that the propagation time delay produces overtalking and confusion for use in data and voice transmission.

Three classes of service can be identified for mobile satellite services. These include:

1. *Data transmission and messaging from very small, inexpensive satellites.* Sometimes these systems are called Little LEOs. These systems are the space equivalent of paging systems. Examples include Orbcomm, Starsys, and VITA. Orbcomm has already launched two experimental satellites at an altitude of 750 km.
2. *Voice and data communications from Big LEOs.* These include Iridium, Globalstar, Odyssey, and INMARSAT P. All of these providers expect to be in service by the year 2000. Iridium is a 66-satellite constellation in six orbit planes at an altitude of 747 km. Globalstar is a 48-satellite constellation in six orbit planes at an altitude of 1,390 km. Odyssey is the first system designed to employ orbiting satellites in medium-Earth-orbit (MEOs) or Intermediate Circular Orbits (ICOs). Odyssey will employ a 12-satellite system in three orbit planes at an altitude of 10,350 km. INMARSAT P is a 10-satellite system in two orbit planes at an MEO altitude of 10,350 km.

3. *Wideband data transmission.* This is an extension of the Global Information Initiative (GII). Systems proposed include Teledisc, Spaceway, and CyberStar. These systems will provide computer-to-computer links and video conferencing. Teledisc will provide wideband data from an LEO orbit of 700 km with an 840-satellite constellation in 21 orbit planes. Spaceway is a proposed 8-satellite constellation in one orbit plane in a GEO orbit of 35,000 km. CyberStar has proposed a similar system to Spaceway.

10.8   GLOBAL POSITIONING SYSTEM

Another application for the low-GHz band of frequencies is the Global Positioning System (GPS). The GPS carrier frequencies (1227.60 MHz and 1575.42 MHz) are in the L band. GPS was developed by the Department of Defense to simplify accurate navigation. This system uses a 24-satellite constellation to accurately determine the user's geographic position. Each satellite broadcasts a pseudorandom timing code that is based on a 10.23 MHz reference signal generated by an onboard atomic clock. Two services are available: the Standard Positioning Service (SPS) for civilian use, utilizing a single frequency C/A (course/acquisition) code, and the Precise Positioning Service (PPS) for military use, utilizing a dual frequency P-code (protected). The satellites are in orbit 10,900 miles above Earth with an orbital period of 12 hours.

GPS determines the distance to a satellite by measuring how long a radio signal takes to reach us from that satellite. Assumptions are made that the satellite and our receiver are generating the same code at the same time. Comparing how late the satellite's code is to our code determines how long it took for the satellite's signal to reach us. One's location is found through triangulation. Trignometry postulates that if three perfect measurements locate a point in three-dimensional space, then four imperfect measurements can eliminate any timing offset (as long as the offset is consistent). Because of a timing error by a receiver clock (they are not as accurate as the onboard satellite atomic clocks), an extra satellite range measurement is required, hence, four satellite measurements are taken.

GPS satellites not only transmit a pseudorandom code for timing purposes. They also transmit a "data message" about their exact orbital location and their system's health.

The receivers include a handheld unit that costs about $500 (pricing expected to drop). The diagram for such a unit is shown in Figure 10.18. The RF front end converts the antenna signal to an IF. The data is digitally processed to determine the user's position. Amplification is a mix of GaAs MESFETs, bipolar transistors, or silicon MMICs. Battery operated units employ very low current bipolar transistors. Mixer functions can be accomplished by using silicon MMICs or Schottky diodes. Silicon MMICs or PIN diodes are used for the automatic gain control (AGC). Bipolar transistors provide a low-phase noise device for the local oscillator.

Calculations of position are accurate to within 328 yards 99.99% of the time. At an accuracy of 95%, the range reduces to 109 yards. Military users have even better resolution of position. As developments improve, the accuracy will get even better.

The airline industry is expected to use this technology on long transoceanic flights. Shorter, more direct routes can then be taken, which will reduce the time of the flight and fuel costs and still provide for better safety.

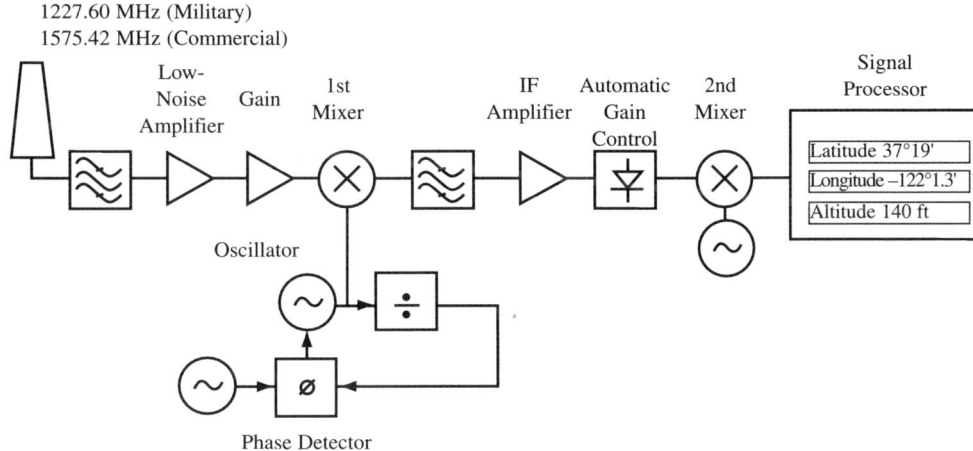

Figure 10.18   Receiver diagram for a typical Global Positioning System (GPS).

10.9   SUMMARY

1. The biggest growth area of microwave communications applications is in wireless technologies.
2. The usual form of radar is monostatic, utilizing one antenna for both transmit and receive functions.
3. Radar is divided into two main categories, pulse and continuous wave (CW). Pulsed radar can yield distance information, while CW radar is used for velocity information via the Doppler effect.
4. The moving target indicator (MTI) radar system can distinguish moving targets from stationary targets.
5. Typical microwave terrestrial links are in the C band. They offer similarities to a coaxial cable terminal. The most common antenna for a microwave terrestrial link is the hoghorn parabolic antenna.
6. Satellite communications utilizing geosynchronous satellites were first postulated by Arthur C. Clarke in 1945.
7. Geosynchronous or geostationary satellites are fixed in location 36,000 km above the earth's surface at the equator.
8. One of the largest users of satellites is INTELSAT, which provides international telecommunication services.
9. Domestic satellites are in the C and Ku bands. C-band satellites are used to provide television programming directly to the home and to cable TV operators.
10. The Ku band has been used by electronic news-gathering services. The Ku band is also being used with VSAT terminals. VSATs can provide two-way transfer of data and video.
11. DBS satellites provide direct-to-home (DTH) service of television programming. They employ higher-power transponders, using small receiver dishes.
12. Microcell technology is a new distribution feature of the PCS/PCN services.
13. Wireless standards are in a constant state of flux in both cordless and cellular technologies for both analog and digital formats.

14. Frequency bands have also been allocated for low- and medium-Earth-orbit satellites for mobile satellite services. These services will provide data transmission and messaging, voice and data communications, or wideband data transmission.
15. Another microwave communication application is the Global Positioning System (GPS), a 24-satellite constellation system.

Key Equations:

$$P_R = \frac{\lambda^2 P_T G^2 \sigma}{(4\pi)^3 r^4}$$ (Eq. 10.1)

$$R = ct/2$$ (Eq. 10.2)

$$V_r = f_d/3.9f_o$$ (Eq. 10.3)

PROBLEMS

1. A radar transmitter has a power of 25 kW and operates at a frequency of 12 GHz. Its signal reflects from a target 10 km away with a radar cross section of 8.0 m^2. The gain of the antenna is 25 dBi. Calculate the received power.

2. A radar transmitter has a power of 30 kW and operates at a frequency of 8.0 GHz. Its signal reflects from a target 20 km away with a radar cross section of 10 m^2. The gain of the antenna is 400. Calculate the received power.

3. A pulse sent to a target returns 8.0 μs later. How far away is the target in km?

4. The range to a target is 18 mi. How long will it take for a pulse sent to it to get back?

5. A 10 GHz radar receives an echo that is shifted up in frequency by 15 kHz. Calculate the target's velocity in mph.

6. Determine the Doppler shift in Hz when the velocity of the target is 75 mph and the transmitted frequency is 10.8 GHz.

7. Determine the azimuth and angle of elevation for a dish located in Houston, Texas, for a Satcom satellite located at 135° W longitude.

QUESTIONS

1. What are the two main categories of radar and how does each work?

2. Define *radar cross section.*

3. Why is the power of a radar system a function of r^4?

4. Describe how an MTI radar works.

5. What are the basic characteristics of a terrestrial communications link?

6. What is the most popular type of antenna for microwave links? Why?

7. Describe the features of the geosynchronous orbit.

8. Why are the uplink and downlink frequencies different in the C band?

9. What is a boresight pattern?

10. Describe orthogonal polarization and why it is used.

11. What is the difference between an LNA and an LNB?

12. Describe the solid-state devices used in a typical down-converter for a TVRO system.

13. Besides frequency, what are the differences in the Ku band and the C band?

14. Define *rain fade.* What can be done to compensate for it?

15. Why is circular polarization used in high power DBS?

16. What are VSATs and where are they being used?

17. What is the advantage of the digital architecture for the new high-power DBS systems?

18. Describe the PCS/PCN system and how it differs from a typical cellular cell.

19. Who will supply the link for the PCN?

20. Describe the wireless standards being used in Europe.

21. What are the advantages in digital over analog in wireless?

22. Describe three classes of service available for mobile satellite services.

23. What differences exist between geosynchronous and low-Earth-orbit satellites?

24. Describe the function of a GPS.

APPENDIX A
A PRIMER ON dBs

LOGARITHMS

Logarithms are used to simplify numerical computations involving exponents and powers. This is a valuable skill to learn in order to use decibels and dBm.

A logarithm is defined as "the exponent of the power to which the base must be raised in order to produce a given number."

Examples:

$$10^2 = 100 \qquad \text{or log } 100 = 2$$
$$10^3 = 1000 \qquad \text{or log } 1000 = 3$$
$$10^{4.25} = 17{,}782 \qquad \text{or log } 17{,}782 = 4.25$$
$$10^{-1} = 0.1 \qquad \text{or log } 0.1 = -1.0$$
$$10^{-2.5} = 0.003162 \qquad \text{or log } 0.003162 = -2.5$$

The whole-number part of the logarithm is called the *characteristic,* while the fractional part is called the *mantissa.*

Logarithms are commonly used in many scientific and technical calculations. For instance, amplifier gain, signal attenuation, and coupling factor are often expressed as the logarithm of a power ratio. Thus, it is to the technician's or technologist's benefit to be able to perform logarithmic computations.

The advent of the electronic calculator has made the calculation of a logarithm simple: Enter the number into the calculator, and press the LOG button. (Note: There are two log buttons on the calculator; one is labeled LOG, while the other is labeled LN. LOG refers to the common log, base 10, and LN refers to the natural log, base e, 2.718. Among other things, natural logs are used to do RC time constant problems, and they are the basis for the Neper.) Also note that some DAL (direct algebraic logic) calculators require you to press LOG first, then enter the number, and then press equal (=). The Sharp EL-506G is an example of such a calculator.

EXAMPLE A.1:

Find the log of 40.

1. Enter: 40
2. Press: LOG
3. Read display

   log 40 = 1.60206

   Note: The log of a negative number is undefined.

DETERMINING THE ANTILOG

If you are given the logarithm of a number and asked to find the number it represents, reverse the procedure used to find the log.

EXAMPLE A.2:

Find the antilog of 2.30103.

1. Enter: 2.30103
2. Press: INV or 2nd F
3. Press: LOG
4. Read display

   Antilog 2.30103 = 200

DECIBELS AND POWER RATIOS

The decibel is the common unit of measure for power amplification or attenuation. Many items of test equipment are calibrated in decibels (dB), yet many otherwise experienced personnel do not understand the term. Actually, the concept of the decibel is simple and the method of calculating dB is not difficult. The decibel is a logarithmic unit for expressing a power ratio. By using logs and dBs, the computations to find gain or loss are greatly simplified.

Gains, or amplification of power, are represented by *positive* dBs. Losses, or attenuation of power, are represented by *negative* dBs. Decibels can be added, subtracted, or combined algebraically to indicate the composite effect of more than one amplifier and/or attenuator.

The power ratio in decibels is as follows:

$$\text{Power Ratio}_{(dB)} = 10 \log (\text{Power out/Power in})$$

This formula may be found in various forms. Common examples are:

$$dB = 10 \log (P_o/P_{in}) \text{ and } dB = 10 \log (P_2/P_1)$$

EXAMPLE A.3:

An amplifier has an input of 5 watts, and the output is 10 watts. What is the gain in dB?

$$dB = 10 \log (10/5)$$
$$= 10 \log 2$$
$$= 3.01$$

EXAMPLE A.4:

A generator produces 100 mW of power, which goes through a 9 dB attenuator. What is the output power?

$$dB = 10 \log (P_o/P_{in})$$
$$P_o = [\text{antilog}(dB/10)](100 \text{ mW})$$
$$= [\text{antilog}(-9/10)](100 \text{ mW})$$
$$= 0.126(100 \text{ mW})$$
$$= 12.6 \text{ mW}$$

Note: The attenuation of 9 dB is –9 dB in the formula because it indicates a *loss* of power.

EXAMPLE A.5:

45 mW of power is measured on a wattmeter connected to a system with a known gain of 15 dB. How much power was applied to the circuit?

$$dB = 10 \log (P_o/P_{in})$$
$$P_{in} = P_o/\text{antilog}(dB/10)$$
$$= 45/\text{antilog}(15/10)$$
$$= 45/31.6$$
$$= 1.423 \text{ mW}$$

To find the overall gain of several stages, either multiply the power ratios together, or convert them to dB and add.

EXAMPLE A.6:

Assume that an amplifier circuit has four stages, each of which amplifies the input signal by two. What is the total gain of this circuit in dB?

$$dB = 10 \log (P_o/P_{in})$$
$$= 10 \log 2$$
$$= 3$$

Total Gain = 3 dB + 3 dB + 3 dB + 3 dB = 12 dB

Alternately: Total Gain = $2 \times 2 \times 2 \times 2 = 16$
$$dB = 10 \log (P_o/P_{in})$$
$$= 10 \log 16$$
$$= 12 \text{ dB}$$

Decibels do not represent actual power, but only the *ratio* of one power to another. To represent *actual power*, not a power ratio, dBm was developed. dBm represents the power level relative to one milliwatt (1 mW). Briefly, a dBm represents the power that would result if 1 mW were amplified (or attenuated) by that number of dB.

Thus, dBm indicates an arbitrary power level with a base of 1 mW. It may be found by the following:

$$dBm = 10 \log (P/1 \text{ mW})$$

EXAMPLE A.7:

If the power on a transmission line is measured to be 32 mW, what is the level in dBm?

$$dBm = 10 \log(32/1)$$
$$= 10 \,(1.50515)$$
$$= 15.05$$

EXAMPLE A.8:

If the power on a transmission line is measured as 0.005 mW, what is the level in dBm?

$$dBm = 10 \log(.005/1)$$
$$= 10 \,(-2.3)$$
$$= -23$$

EXAMPLE A.9:

The power at port 3 of a directional coupler is measured as –7.5 dBm. What power level does this represent?

$$dBm = 10 \log (P/1 \text{ mW})$$
$$P = [\text{antilog}(dBm/10)](1 \text{ mW})$$
$$= [\text{antilog} -0.75] \,(1 \text{ mW})$$
$$= 0.178 \text{ mW}$$

The advantage of dBm is that the effect of gains or losses in dB can be directly applied to dBm to obtain new power levels. For example, if a power at –4.7 dBm enters an amplifier with a gain of 15 dB, the output power is –4.7 + 15 = 10.3 dBm. Similarly, if 24 dBm of power is attenuated 6 dB, the new power level is 24 – 6 = 18 dBm. Problems involving dBms and dBs amount to nothing more than algebraic addition and subtraction.

One word of caution about dBm: *dBms do not add or subtract with each other.* Consider 10 dBm (which equals 10 mW) added to 10 dBm.

$$10 \text{ mW} + 10 \text{ mW} = 20 \text{ mW}$$
$$20 \text{ mW} = 13 \text{ dB}$$

However, $10 \text{ dBm} + 10 \text{ dBm} \neq 13 \text{ dB}$

Thus, you cannot add one value of dBm to another.

VOLTAGE

In some instances you will examine voltage ratios instead of power ratios. To work these in terms of their logarithmic values requires calculating the voltage ratio (V_{out}/V_{in}) using equal impedances. If the impedances are not equal, an adjustment to the formula must be made. We will assume equal impedances in our problems.

Because V^2 is proportional to power, the voltage ratio is multiplied by 2 (20 log, instead of 10 log). The new formula is:

$$dB = 20 \log (V_{out}/V_{in})$$

EXAMPLE A.10:

The input voltage to an amplifier is 2.5 mV, and the output voltage is 3,600 mV. What is the gain of the amplifier in dB?

$$dB = 20 \log (3600/2.5)$$
$$= 20 \log 1440$$
$$\approx 60$$

Note that there is a standard reference for actual voltages as there was for power. Here, dBmV represents voltage referenced to 1 mV. The formula is:

$$dBmV = 20 \log (V/1 \text{ mV})$$

EXAMPLE A.11:

What is the voltage level in dBmV when the voltage is 1.6 V?

$$dBmV = 20 \log (V/1 \text{ mV})$$
$$= 20 \log (1.6/.001)$$
$$= 20 \log 1600$$
$$= 64.1$$

Appendix B
Greek Alphabet

	Capital Letters	Small Letters
Alpha	A	α
Beta	B	β
Gamma	Γ	γ
Delta	Δ	δ
Epsilon	E	ε
Zeta	Z	ζ
Eta	H	η
Theta	Θ	θ
Iota	I	ι
Kappa	K	κ
Lambda	Λ	λ
Mu	M	μ
Nu	N	ν
Xi	Ξ	ξ
Omicron	O	o
Pi	Π	π
Rho	P	ρ
Sigma	Σ	σ
Tau	T	τ
Upsilon	Y	υ
Phi	Φ	φ
Chi	X	χ
Psi	Ψ	ψ
Omega	Ω	ω

Appendix C
Periodicals and Videos for Microwave

PERIODICALS

All offer free subscriptions.

1. *Microwaves & RF*
   Penton Publishing, Inc.
   1100 Superior Ave.
   Cleveland, OH 44114-2543
   (216) 696-7000
2. *Wireless Systems Design*
   Penton Publishing, Inc.
   1100 Superior Ave.
   Cleveland, OH 44114-2543
   (216) 696-7000
3. *Communications Technology*
   CT Publications Corp.
   1900 Grant St., Suite 720
   Denver, CO 80203
   (303) 839-1565
4. *Microwave & Wireless*
   J. F. White Publications, Inc.
   62 Cranberry Highway
   Orleans, MA 02653
   Fax (508) 240-3111

VIDEOS

Shelburne Films
54545 S.R. 681
Reedsville, OH 45772
(614) 378-6297

1. Installing Satellite Antennas
   VHS, 31 minutes
2. The World at 12 GHz
   VHS, 60 minutes
3. The Era of Direct Broadcast Satellites
   VHS, 56 minutes

BIBLIOGRAPHY

Blake, R., *Basic Electronic Communication*. St. Paul, Minn.: West Publishing Co., 1993.

Carr, J.J., *Elements of Microwave Electronics Technology*. Orlando, Fla.: Harcourt Brace Jovanovich, 1989.

Cheng, D.K., *Field and Wave Electromagnetics*. 2nd ed. Reading, Mass.: Addison-Wesley Publishing Co., 1989.

Cheung, W.S., and F.H. Levien, *Microwaves Made Simple*. Norwood, Mass.: Artech House, Inc., 1985.

Holz, G., "New Material Use Drives Low Cost and High Speed," *Advanced Packaging,* Summer 1993, pp. 42–43.

Ishii, T.K., *Microwave Engineering*. New York: Harcourt Brace Jovanovich, 1989.

Kennedy, G., and B. Davis, *Electronic Communication Systems*, 4th ed. New York: Glencoe, 1993.

Leff, B.J., "Making Sense of Wireless Standards and System Designs," *Microwaves & RF,* February 1994, pp. 113–118.

Monaco, F., *Introduction to Microwave Technology*. Columbus, Ohio: Merrill Publishing Co., 1989.

Rusch, R., "The Market and Proposed Systems for Satellite Communications," *Applied Microwave & Wireless,* Fall 1995, pp. 10–34.

Staff of Lab-Volt (Quebec) Ltd., *Microwave Technology,* Volume 1. Lab-Volt Ltd., 1988.

Veley, V.F., *Modern Microwave Technology*. Englewood Cliffs, N.J.: Prentice Hall, 1987.

GLOSSARY

Antenna Efficiency The ratio of the radiated power to the power delivered to an antenna. Typically greater than 90%.

Antenna Impedance A complex AC quantity that specifies the opposition to signal energy as measured at the feed point of an antenna.

Antenna Resistance see Antenna Impedance

Attenuation Any reduction in signal power or voltage. Attenuation is most commonly expressed in decibels (dB).

Backlobe Radiation A form of destructive radiation working against the main beam caused by feeding a parabolic reflector with an isotropic source.

Backward Wave Oscillator (BWO) A microwave oscillator capable of being tuned over a broad range of frequencies. Essentially a traveling wave tube with the attenuator removed.

Balanced Line A parallel-wire transmission line whose conductors experience the same electrical capacitance relative to ground or are equally isolated from ground.

Bandwidth The range of usable frequencies of a device between the 3 dB down (−3 dB half-power) points.

Beamwidth The angle measured between the −3 dB half-power points on the major lobe of an antenna's radiation pattern.

BNC (Bayonet Navy Connector) A type of bayonet connector commonly used in microwave work with coaxial cables to 4 GHz.

Broadside Antenna An array of simple dipoles fed in phase and having a radiation pattern whose maximum directivity is along the axis normal to the plane of the array.

Cassegrain Antenna A type of paraboloid in which the feed point is located at the vertex of the parabola and is directed against a secondary hyperbolic reflector.

Cavity Resonator A physical cavity formed in a solid material that, when excited, is capable of sustaining oscillations at microwave frequencies. Cavity resonators are functionally equivalent to LC resonant circuits, but they have a higher Q-value.

C-Band Satellites Satellites employing the 4 to 6 GHz range of frequencies. Used for TVRO and cable TV reception.

Characteristic Impedance (Z_O) The frequency-independent opposition offered to the propagation of TEM energy along a transmission line due to the distributed effects of inductance and capacitance. Also called the *surge impedance.*

Choke Flange A waveguide flange with an L-shaped channel whose total cross-sectional length is $\lambda/2$. Used to suppress wave reflections by acting like a short across the mating flange. Choke flanges mate with flat flanges only.

Circular Waveguide Waveguide having a circular cross section. Used for vertical runs.

Circulator A ferrite device having the property that only rotationally adjacent ports are coupled.

Coaxial Cable A type of transmission line consisting of an inner conductor surrounded by, but insulated from, an outer conductor.

Coupling Factor   A rating specifying the amount of energy coupled from the main arm to the auxiliary arm in a directional coupler. Typical values are 10, 20, or 30 dB.

Crystal Detector   Essentially a crystal diode (cat-whisker) square-law device used for sensing minute levels of microwave power. It forms an essential part of the slotted line and power meters.

Cutoff Frequency (f_c)   The lower limit of the frequency that can still propagate inside the waveguide. The dimension of "2a" (from the width of the rectangular waveguide) sets this frequency.

Decibel (dB)   A logarithmic unit used to express the ratio of two powers or voltages. (See Appendix A)

Dielectric Lens Antenna   A polystyrene or other dense dielectric material used to collimate spherical wave fronts at microwave. Stepped or zone configurations are often used to minimize weight and attenuation problems.

Dielectric Loss   Loss of energy along a transmission line due to the energy absorbed in the heating of the dielectric by the passage of a TEM wave.

Direct Broadcast Satellites (DBS)   Geosynchronous satellites that broadcast directly to a home TVRO system or other commercial venture.

Directional Coupler   A waveguide arrangement in which microwave energy moving along the main arm can be sampled through an auxiliary arm. Energy moving in the other direction is not coupled.

Directivity   A measure of the degree of isolation between the main and auxiliary arms of a directional coupler when the direction of energy flow is reversed. A typical factor is 30 dB.

Direct-To-Home (DTH)   Same as DBS, except only to the home TVRO.

Dish   see Parabolic Reflector

Dominant Mode   The mode for which the lowest possible frequency (f_c) can be propagated in a waveguide or other transmission line.

Doppler Radar   A type of radar using the Doppler effect to determine velocity information about the target.

Effective Radiated Power (ERP)   Represents the power input multiplied by the antenna gain (as a ratio) measured with respect to a half-wave dipole.

Elliptical Waveguide   One of the most commonly used waveguides, especially in microwave radio. It is flexible and comes in 400-foot rolls.

End-Fire Array   An antenna array consisting of a linear arrangement of simple dipoles fed 90° out of phase and having a radiation pattern of maximum directivity in the plane of the array.

Far-Field Distance   The distance from an aperture antenna that ensures that the induction (near field) is not affecting the receiving antenna.

Feed Antenna   One of several antenna types (usually a dipole or horn) that projects ("feeds") microwave radiation into the parabolic reflector.

Ferrite   An insulating material having magnetic properties that can be modified when placed in a magnetic field. Ferrites are used in microwave as isolators and circulators. Also used in the YIG oscillator/filter.

FET (Field-Effect Transistor)   The semiconductor equivalent of a triode vacuum tube. Operates by varying the conductivity of a semiconductor channel through changes in the electric field across the channel.

Folded Dipole   An antenna formed by connecting two simple dipole sections in parallel, thus increasing the radiation resistance from 73 to 300 Ω.

Frequency Domain   The representation of a complex waveform in such a way that the amplitude is a function of frequency rather than time. A spectrum analyzer uses this display type.

Frequency Meter   Used to measure the frequency of a source. Can be a cavity type, calibrated in frequency, that undergoes some power absorption at resonance.

Gain   The increase in power or voltage at the output of a device compared with its input or a reference. For an antenna it is compared to that of an isotropic antenna.

Gallium Arsenide (GaAs)   A microwave semiconductor compound having high ion mobility and thus high-frequency operating limits. Used to make the FET transistor and other devices.

Geostationary Orbit   The same as a geosynchronous orbit or the constellation of geosynchronous satellites.

Geosynchronous Orbit   An equatorial orbit for satellites 36,000 km above the earth's surface. The satellites rotate at the same rate as the earth, so they appear to be stationary.

Ground Wave   A vertically polarized propagated TEM wave that travels close to the ground. Ground waves are one of three identifiable modes of propagation from an antenna. The other two are the *sky wave* and the *space wave.*

Gunn Diode   A low-power microwave source using the transferred-electron device principle to initiate a domain excursion (current pulse) through the semiconductor at high-frequency rates. The Gunn diode also exhibits negative resistance.

Hertz Antenna   A half-wave dipole antenna.

Horn Antenna   A microwave antenna made by flaring a waveguide structure in either or both cross-sectional directions. The flaring ensures an impedance match to free-space impedance. Horn antennas are often used with parabolic reflector antennas.

Hybrid Tee   A multiple waveguide tee formed by the intersection of an E plane and H plane. Used as a discriminating port coupler. Sometimes called a *magic tee.*

IMPATT (IMPact Avalanche and Transit-Time) Diode   A plasma-mode device capable of producing a 180° phase shift between the domain and RF voltage and thereby exhibiting negative resistance. Used as a microwave source or amplifier.

Induction Field   The near-field EM field that surrounds an antenna but does not radiate. Its effect is local, but it is needed to establish the radiation field.

Insertion Loss   The amount of power lost in a device due to its presence in the path of energy flow (thus reflective loss).

Ionosphere   A heavily ionized layer of particles in the upper atmosphere capable of reflecting and refracting EM propagation.

Iris   A partition located within the E or H field of a waveguide that causes a discontinuity and effectively changes the impedance at that point in the guide.

Isolator   A ferrite device that uses the property of the precessional frequency effect to absorb microwave energy passing in one direction but not in the other.

Isotropic Source   A hypothetical radiator of TEM waves that radiates uniformly in all directions.

j-Operator   A rectangular complex number used in conjunction with a Smith chart to plot impedances.

Klystron   A thermionic device that utilizes velocity modulation and is used as a microwave source or as an amplifier.

Ku-Band Satellites   Satellites employing the 12 to 14 GHz range of frequencies. DBS and VSAT systems employ this range of frequencies.

Magnetron   A high-power microwave oscillator using the principle of sustained interaction and energy exchange between the electrons circulating within the device and the attendant RF field. Used extensively in radar transmitters in which high peak power is required.

Marconi Antenna   A quarter-wave, vertical antenna utilizing the reflection properties of a ground plane.

MESFET (Metal Semiconductor Field-Effect Transistor)   An FET that uses a Schottky barrier gate to extend the frequency range of operation.

Microstrip   A parallel-wire transmission line that can be fabricated as part of a printed circuit board. Very common with low-power microwave circuits.

Microwave Integrated Circuit (MIC)   Gallium arsenide and MESFET technology combined to produce integrated devices to function at microwave frequencies.

Microwaves   A range of frequencies generally identified as those from 1 to 100 GHz.

Mode   The manner in which the E and H fields arrange themselves in a given waveguide operation at a given frequency.

Monolithic Microwave Integrated Circuit (MMIC)   MIC circuits in which all components, both active and passive, are fabricated directly within the substrate material.

Near Field   The region within the influence of the induction field of an antenna.

Network Analyzer   A test instrument that measures the transfer and/or impedance functions through sine wave testing.

Noise   A signal caused either internally or externally that competes with the effect of the propagated signal. Current pulses generated by heat within a device are a major source of noise.

Normalized Value   The value of Z, R, or X obtained by dividing the parameter by the characteristic impedance (Z_O) for use on the Smith chart.

Parabolic Reflector   A type of microwave antenna formed as a surface of revolution from a parabola. This antenna forms a narrow beam and is highly directional. It has a high gain.

Parallel-Lead Transmission Line   Also called twin-lead. An inexpensive transmission line often used with VHF television receivers. It has higher dielectric and radiation losses than coaxial cable.

Parametric Amplifier (PARAMP)   A low-noise, low-power narrowband microwave amplifier utilizing a varactor diode to achieve gain.

Parasitic Array   An arrangement of passive antenna elements with the induction field of the active element such that a significant modification of the radiation pattern is achieved. The Yagi-Uda is an example of a parasitic array.

Personal Communications Network (PCN)   The network through which PCS transmits. The link could be done via microwave, local carrier exchange, or cable TV networks.

Personal Communications Services (PCS)   A wireless technology employing microcells. With PCS a person can communicate to anyone anywhere via a small handheld unit.

Phased Array   An antenna consisting of many elemental antennas, each of which is fed with a ferrite phase shifter so that the beam can be electronically steered and the antenna can remain stationary.

Planar Transistor   A type of diffused microwave transistor in which the emitter, base, and collector regions are all brought out to the same plane surface.

Plane Wave   Any finite, usually small, local area of a spherical wave front. At any appreciable distance from the source, the wave front may be considered as flat.

Polarization   The orientation of the electric field of an antenna relative to the earth's surface and antenna structure.

Post   A screw inserted into the wall of a waveguide. The depth of insertion can be varied, thus providing a variable reactance.

Power Density   A measure of the distribution of radiated power over a given area.

Precessional Frequency Effect   Related to the tilt of a spinning electron causing a dampening effect (energy loss). The effect is activated by the direction a microwave signal travels through the device.

Pyramidal Horn Antenna   A type of microwave antenna made by flaring a waveguide in both directions.

Q-value   A figure of merit for a resonant cavity or conventional resonant circuit.

Radar   An acronym for *r*adio *d*etection *a*nd *r*anging, and a means of gathering information about a target.

Radiation Loss   Loss of microwave energy due to simple radiation from a conductor carrying the signal. It is most prevalent in parallel-wire lines and least in coaxial lines.

Radiation Pattern   A polar diagram of field strength measurements made at a fixed distance from an antenna in a given plane.

Radio Horizon   The point beyond the true (geometrical) horizon at which the reception of line-of-sight wave is just discernible. It is about 1/3 farther than the optical horizon.

Radome   A fiberglass structure that encloses an antenna assembly to protect it from the elements.

Rectangular Waveguide   A waveguide having a rectangular cross section and used for propagating EM waves by reflection.

Reflection Coefficient   A complex polar number, $\Gamma = \rho \angle \theta$, associated with a transmission line and a load in the rectangular form $Z_L = R \pm jX$ where ρ is the ratio of reflected voltage to incident voltage. Note that $0 < \rho < 1$ and $0 < \theta < 180°$. In terms of line and load impedances, Γ is usually given by $(Z_L - Z_0)/(Z_L + Z_0)$.

Reflex Klystron   A type of klystron that utilizes a reentrant cavity and is used as a source oscillator.

Return Loss   The ratio of incident power to reflected power. For a perfect match, $RL = \infty$.

Ridged Waveguide   A waveguide formed with longitudinal ridges such that the cutoff wavelength may be closer to the free-space wavelength with the result that Z_o becomes very large. Used for impedance matching.

Rotary Joint   A mechanical union of two sections of circular waveguide designed to rotate relative to one another.

Skin Effect   The tendency of electrons to confine themselves to the outer surface of a conductor due to induced voltages in the center of the conductor. This reduces cross-sectional area and increases the resistance of the conductor.

Sky Wave   One of three identifiable propagation modes from a terrestrial antenna. Sky waves propagate into the ionosphere where they are reflected and refracted back to Earth.

Slotted Line   A microwave laboratory instrument used for measuring the distance between nodes of a standing wave pattern.

Smith Chart   Named for P. H. Smith, American engineer. A transmission line admittance or impedance calculator.

Space Wave   One of three identifiable propagation modes from a terrestrial antenna. Usually travels line-of-sight and therefore is affected by the horizon line.

S-Parameters   Ratios of various properties from a two-port microwave device. They in turn describe the reflection and transmission properties of the device. S-parameters are usually in the form of output data from a network analyzer.

Spectrum Analyzer A swept superheterodyne receiver capable of resolving and displaying the sinusoidal components of a complex wave.

Square-Law Detector A type of crystal detector in which the output current is proportional to the square of the input voltage.

Stripline A type of parallel-wire transmission line formed as a multilayer printed circuit board. Appears as a flattened coaxial cable.

Stub A short length of transmission line, usually shorted at one end and attached at an appropriate distance from the load for the purpose of matching a complex load to the transmission line. Actual values are found using the Smith chart or via computer software.

Surface Acoustical Wave (SAW) Device A microwave device using the piezoelectric properties of quartz and other materials to produce a narrowband resonant filter.

SWR The standing wave ratio of the maximum to minimum voltage on a transmission line. Also called the voltage standing wave ratio (VSWR).

TEM Wave A transverse electromagnetic (TEM) wave propagated from a two-wire transmission line into free space. It exists both on the line and in free space.

Time-Domain The representation of a waveform in such a way that its amplitude is displayed as a function of time. The common oscilloscope presents such information.

Transmission Line An arrangement of two or more conductors, having a precise geometry, used to convey microwave energy from source to load with a minimum amount of loss.

Traveling Wave Tube (TWT) A microwave thermionic device used primarily as a wideband amplifier. Its heart is a slow-wave structure in which interaction takes place between an electron beam and an RF signal. Used on board satellites as the high-power amplifier (HPA).

Tunnel Diode A microwave diode that exhibits negative resistance, thus allowing it to be used as an oscillating device.

TVRO (Television Receive Only) The typical home TV receiving system that includes an antenna, a low-noise amplifier, and a video receiver/demodulator or integrated receiver demodulator.

Unbalanced Line A coaxial cable whose conductors do not experience the same capacitance relative to ground.

Varactor A microwave device utilizing the properties of a reverse-biased PN junction to form a voltage-controlled capacitor.

Very Small Aperture Terminals (VSAT) A Ku-band hardware system where two-way data and video are sent via small dishes.

VSWR see SWR

Waveguide A generally hollow metallic structure through which microwave energy propagates by reflection rather than conduction.

Wavelength The distance traveled by a point on a periodic TEM wave in the time required to complete one cycle.

Wireless Technology A new form of low-power microcell technology utilizing Personal Communications Services/Personal Communications Networks just above and below 1 GHz.

X-Band A range of microwave frequencies from 8.2 to 12.4 GHz.

YIG Oscillator/Filter A common microwave component that uses a yttrium iron garnet ferrite material in a tunable bandpass filter or in oscillators. The operating frequency is controlled by an external electromagnetic field.

SOLUTIONS TO CHAPTER PROBLEMS

CHAPTER 1

1. $v = 2.75 \times 10^{10}$ cm/s
2. $v = 2.0 \times 10^8$ m/s
3. $\in_r = 1.52$
4. $T = 10$ ns, 1 ns, and .1 ns (100 ps)
5. $f = 50$ Hz, 5.0 kHz, 200 MHz, and 83.3 GHz
6. $\lambda = 6.0$ km, 333.3 m, 12m, 12.5 cm, and 9.09 mm
7. $f = 2.4$ THz
8. $\lambda = 2.41$ cm
9. $\lambda = .95$ in
10. $t = 187$ μs
11. $d = 9.67$ mi
12. $P_D = 12.28$ nW/m^2
13. $r = 50,000$ km
14. $R = 34.3$ mi
15. $R = 45.6$ mi
16. $R = 70.33$ km

CHAPTER 2

1. $Z_O = 864$ Ω
2. $C = 181$ pF/m
3. $v = 1.33 \times 10^8$ m/s
4. $\Gamma = 0.111$
5. $\Gamma = -0.61$
6. $\Gamma = 0.31$
7. $\Gamma = 0.31 \angle -115°$
8. $\rho^2 = 0.096$
9. $RL = 33.3$; $RL_{(dB)} = 15.2$ dB
10. $RL = 10.4$; $RL_{(dB)} = 10.2$ dB
11. $RL_{(dB)} = 10.2$ dB
12. VSWR = 1.5
13. VSWR = 1.9
14. $\rho = 0.2$
15. $Z_{max} = 142.5$ Ω; $Z_{min} = 39.5$ Ω
16. $Z'_O = 70.7$ Ω

CHAPTER 3

1. a. 1.5–j0.5
   b. 0.67+j1.33
   c. 0.3–j0.8
   d. 1.0+j0.75
   e. 0.7–j1.5

2. see Smith chart

3. VSWR = 1.77
   $\Gamma = 0.28 \angle -33.7°$
   $\rho^2 = 0.077$

4. VSWR = 4.75
   $\Gamma = 0.65 \angle 67°$
   $\rho^2 = 0.42$

5. $Z_{max} = 280$ Ω
   $Z_{min} = 8.9$ Ω

6. $Y_{LN} = 0.63$–j0.48
   $Y_L = 6.3$–j4.8 mS

7. relative $\lambda = 0.333$, λ toward generator

8. $Z = 21+j50.5\ \Omega$

9. $\phi = -36.4°$

10. $Z_{LN} = 1.53–j0.68$

11. $d = 0.035\ \lambda$, $l = 0.149\ \lambda$

12. $d = 0.351\ \lambda$, $l = 0.170\ \lambda$

13. $d = 0.5$ cm, $l = 2.12$ cm

14. $d = 5.79$ cm, $l = 2.8$ cm

CHAPTER 4

1. $Z_O = 283\ \Omega$

2. $Z_O = 206\ \Omega$

3. $Z_O = 137.7\ \Omega$

4. $Z_O = 101.8\ \Omega$

5. $\lambda_c = 1.5$ in

6. $\lambda_c = 38.1$ mm

7. $f_c = 7.87$ GHz

8. $f_c = 8.33$ GHz

9. $\theta = 74.8°$

10. $\theta = 22.7°$

11. $\lambda_g = 37.95$ mm

12. $\lambda_c = 6.52$ cm

CHAPTER 5

1. $CF_{(dB)} = 30$ dB

2. $CF_{(dB)} = 20$ dB

3. Power out = 16 mW

4. Power out = 48 mW

5. Power out = 475.524 μW

6. Power in = 1,500 mW

7. $D_{(dB)} = 58$ dB

CHAPTER 6

No problems

CHAPTER 7

No problems

CHAPTER 8

1. $L = 0.157$ m

2. $Eff = 0.91$ or 91%

3. $P_x = 910$ W

4. $D = 2.113$

5. $G = 2$

6. $ERP = 2,000$ W

7. $L = 0.71$ m

8. $G = 19,542$

9. $G_{dB} = 42.9$ dBi

10. $G = 10,417$

11. $BW = 1.68°$

12. $G = 9.99$

13. $G = 35$

14. $\phi_v = 36.1°$

15. $\phi_h = 28.3°$

CHAPTER 9

1. $P_n = 3.45 \times 10^{-16}$ W

2. $P_n = 6.9 \times 10^{-16}$ W

3. $P_n = 3.45 \times 10^{-16}$ W

4. $SNR_{(dB)} = 16.99$ dB

5. $(S + N)/N = 200$

6. $(S + N)/N_{(dB)} = 23$ dB

7. $NF = 2.0$

8. $NF_{(dB)} = 3.0$ dB

9. $S/N_i = 450$

10. $T_e = 290°$K

11. $T_e \approx 290°$K

12. $NF_T = 1.707$

13. $T_e = 205°$K

14. $f = 14.5$ GHz

15. 100 mW = 20 dBm

16. 40 dBm = 10 W

17. 50 W = 16.99 dBW

18. 50 W = 46.99 dBm

19. $IL = 0.97$ dB

20. $IL = 2$ dB

CHAPTER 10

1. $P_R = 63$ pW

2. $P_R = 21.3$ pW

3. $R = 1,932$ km

4. $t = 0.12\ \mu$s

5. $V_r = 384.6$ mph

6. $f_d = 3,159$ Hz

7. angle of elevation = 35°, azimuth = 59° W of south

Index